Biochemistry of Halogenated Organic Compounds

BIOCHEMISTRY OF THE ELEMENTS

Series Editor: Earl Frieden
Florida State University
Tallahasse, Florida

Biochemistry of Halogenated Organic Compounds

Kenneth L. Kirk

National Institutes of Health
Bethesda, Maryland

PLENUM PRESS • NEW YORK AND LONDON

Library of Congress Cataloging-in-Publication Data

Kirk, Kenneth L.
    Biochemistry of halogenated organic compounds / Kenneth L. Kirk.
        p.   cm. -- (Biochemistry of the elements ; v. 9B)
    Includes bibliographical references and index.
    ISBN 0-306-43657-4
    1. Organohalogen compounds. 2. Biochemistry.   I. Title.
II. Series.
QP801.H34K57 1991
574.19'2--dc20   91-10708
                                                                CIP

ISBN 0-306-43657-4

© 1991 Plenum Press, New York
A Division of Plenum Publishing Corporation
233 Spring Street, New York, N.Y. 10013

Printed in the United States of America

To my wife, Susan; and children, David and Rebecca

Preface

Biochemistry of Halogenated Organic Compounds has been written as a general reference source for researchers in several related areas, including organic chemists, medicinal chemists, pharmacologists, toxicologists, and medical researchers. The development of halogenated compounds as medicinal agents and pharmacological tools and the fascinating biochemical processes that have been discovered and studied using these analogues have generated extremely active areas of research and an enormous volume of literature. Thus, halogenated organic compounds pervade every aspect of biochemistry, a fact made apparent by the numerous reviews and monographs available on individual topics—halogenated nucleosides, halogenated carbohydrates, and so forth.

Given the quantity of material already written on these topics, some of which material is quite current, it might be asked whether a one-volume review of these subjects is useful, or possible. Having now completed this work, I feel the answer to both questions is an emphatic yes. There are fascinating stories to be related in each area, and, where appropriate, I have attempted to develop these topics from a historical perspective. For example, the discovery of the anticancer activity of fluorouracil, the unraveling of the several mechanisms of its action, and the development of a host of later generations of anticancer and antiviral agents based on the parent fluoro-, iodo-, bromo-, and trifluoromethylpyrimidines were, and are, contributions of major magnitude to medical science. So, too, the work of Sir Rudolf Peters and of Kun and co-workers, in defining the mechanisms of toxicity of fluoroacetate and fluorocitrate, has had a profound impact on biochemistry and medicinal chemistry. In these and many other examples, sufficient scientific and historical backgrounds are given to allow the reader—even if unacquainted with the subject—to understand and appreciate fully the material being presented. Recent progress in each area is then reviewed to illustrate directions of current research and perhaps to provoke new ideas for research among those who are active in these areas. As a unified presentation, written by a single author, I believe that this work achieves a unique balance between detailed accounts (as available in specific reviews) and a general overview (presented in biochemical textbooks).

Because of the amount of source material, and the breadth of the topics, the preparation of *Biochemistry of Halogenated Organic Compounds* presented many tactical problems. As with the preparation of Volume 9A, examples had to be chosen from many possibilities to illustrate concepts. This brings into play the natural bias of the author as to what is most relevant. For example, since much of my own research has involved the synthesis and biological evaluation of organic compounds containing fluorine, this element may have received more than its fair share of exposure. The chapters are divided according to classes of compounds. Historical considerations prompted me to treat carboxylic acids first, primarily because of the use of iodoacetate, iodoacetamide, and bromoacetate in the early years of biochemistry and the importance of fluoroacetate, fluorocitrate, and the concept of "lethal synthesis" in biochemistry. In Chapter 2, halogenated aldehydes and ketones are reviewed. The important topic of halogenated steroids, retinoids, and other isoprenoids are covered in Chapter 3, while halogenated analogues of the arachidonic acid cascade are reviewed in Chapter 4. In Chapter 5, the fascinating area of halogenated nucleosides and nucleotides is addressed—from fluorouracil to such newly discovered effective antiviral agents as FIAC and FEAU. The link between Chapter 5 and Chapter 6, biochemistry of halogenated carbohydrates, clearly is the presence of a pentose. The discussion of the biochemistry of halogenated carbohydrates begins with the degradation of glycogen to hexoses and follows a logical sequence through glucose metabolism to trioses, just upstream from many of the carboxylic acids discussed in Chapter 1. The use of halogenated carbohydrates in the study of glycoproteins is also reviewed in this chapter. The final two chapters cover halogenated amino acids and amines, respectively, areas that have profited immensely from the exploitation of halogenated analogues.

As in Volume 9A, the amount of material to be covered precludes a thorough review of the subjects. I have attempted to provide the reader with sufficient material to appreciate the scientific and historical importance of research in each area.

Acknowledgments

I wish to thank friends and colleagues who have given me unwavering encouragement and stimulation during preparation of the two volumes that make up *Biochemistry of the Halogens.* I especially thank my family for their love, encouragement, and remarkable patience throughout this project—without their support, it would not have been possible to sustain the many months of intense effort that this project required.

Contents

Biochemistry of Halogenated Carboxylic Acids

1.1 Introduction

Mono-, di-, and tricarboxylic acids are key intermediates in several metabolic pathways. The ability of halogenated analogues of many of these acids to mimic their behavior, or to inhibit critical enzymes required for the processing of these acids, has made such analogues important research tools for the study of a broad spectrum of biological processes. For example, in the 1930s, extensive use of iodoacetic acid as an inhibitor was an important strategy in the elucidation of metabolic pathways. In another example having major impact on biochemistry, the studies on the mechanism of fluoroacetate toxicity carried out by Sir Rudolf Peters in the 1950s spurred intense interest in the development of fluorinated compounds as antimetabolites and mechanistic probes. Over the past few decades, an enormous number of halogenated carboxylic acids have been synthesized, and virtually every aspect of energy metabolism has been studied using such analogues. In this review, topics have been arranged progressively from simple to more complex carboxylic acids, with one exception. Because of the intimate biochemical and historical relationship between fluoroacetate and fluorocitrate, the biochemistry of these compounds will be considered together.

1.2 Biochemistry of Halogenated Monocarboxylic Acids Containing No Additional Functionality (I)

1.2.1 Bromoacetic Acid and Iodoacetic Acid as Metabolic Inhibitors

1.2.1.1 Historical Background

In 1874, Steinmauer described several toxic manifestations of bromoacetic acid injected into frogs, including respiratory and cardiac

1

depression and weakness and paralysis of muscles. The toxic properties of bromoacetate were confirmed in 1930 by Lundsgaard, who observed that iodoacetate was even more potent. Lundsgaard also made the critical observation that iodoacetate is capable of inhibiting the formation of lactate in muscles under a variety of conditions. Research in the 1930s confirmed and extended these initial observations, and iodoacetate along with fluoride became among the most important tools used in the elucidation of the glycolytic pathway. As the sequence of biochemical events in glycolysis became known, the site of action of iodoacetate was also determined. Thus, the coupling between the reduction of pyruvate to lactate and the oxidation of glyceraldehyde 3-phosphate to diphosphoglyceric acid was found to be very sensitive to iodoacetate. Lactate dehydrogenase was shown early to be insensitive to iodoacetate, and subsequent research over several years confirmed that glyceraldehyde-3-phosphate dehydrogenase (GPD) is quite sensitive to iodoacetate. In the oxidation of glyceraldehyde 3-phosphate to phosphoglyceric acid, the NADH required for the reduction of pyruvate to lactate is produced. Inhibition of GPD thus blocks lactate formation during anaerobic glycolysis [for an entertaining and thorough review of these historical events in the early years of biochemistry, the review by Webb (1966) is recommended].

1.2.1.2 Mechanism of Iodoacetate Inhibition of Glyceraldehyde-3-phosphate Dehydrogenase

The rapid reaction of iodoacetate with nucleophiles, in particular, with SH groups, was recognized early as the likely mechanistic basis for enzyme inhibition. GPD consists of four identical subunits, each containing a highly reactive sulfhydryl group (Cys-149) (Conway and Koshland, 1968; Harris and Perham, 1968, and references therein). Enzymatic oxidation of glyceraldehyde 3-phosphate requires an initial addition of an enzyme-bound thiol group to the aldehyde carbonyl group followed by hydride abstraction (Fig. 1-1). The high reactivity of this essential SH (Cys-149) has been associated with ion-pair formation with a neighboring histidine imidazole group (Polgár, 1975). Carboxymethylation of this essential SH group results in irreversible inactivation of the enzyme, and the reactivity of this group renders GPD particularly sensitive to iodoacetate and iodoacetamide (Foxall et al., 1984; MacQuarrie and Bernhard, 1971; Webb, 1966, and references therein).

As noted, early development of SH alkylating agents involved primarily the use of bromoacetate. The relative reactivities of haloacetic acids and amides toward sulfhydryl groups have been determined in several

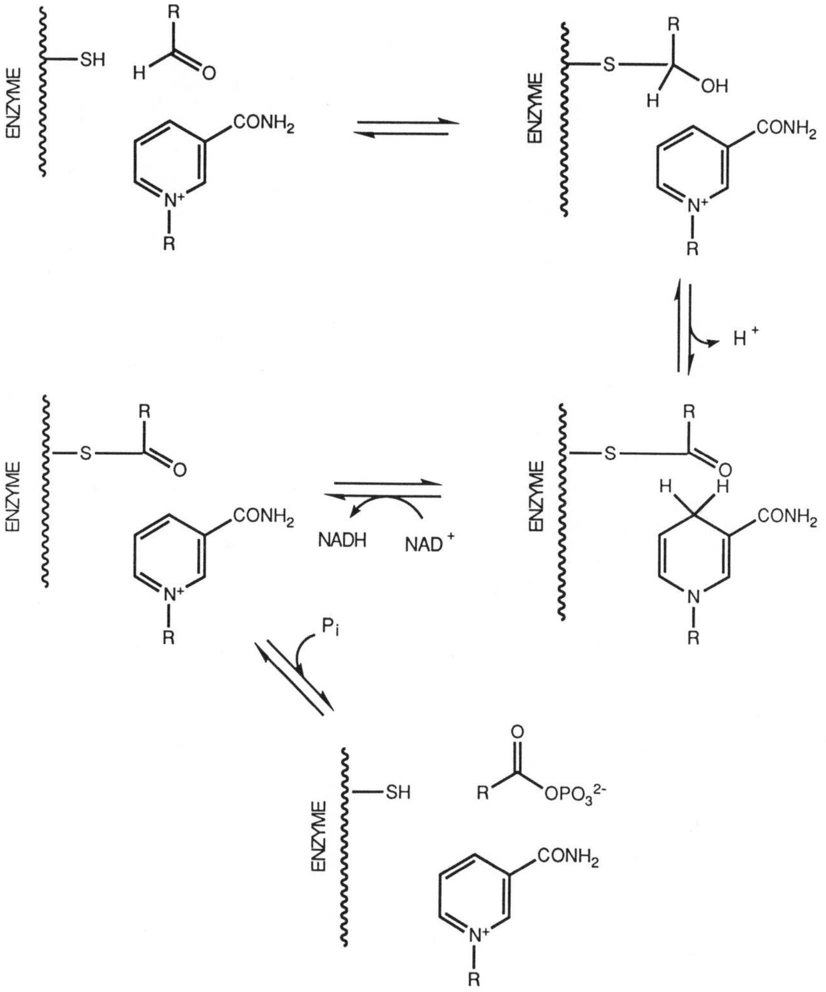

Figure 1-1. The mechanism of oxidation of glyceraldehyde 3-phosphate by glyceraldehyde-3-phosphate dehydrogenase.

systems and follow the expected order for halogen substitution of $I > Br \gg Cl \gg F$ [reviewed by Webb (1966)].

1.2.1.3 Applications as Reagents for Carboxymethylation

As nonspecific carboxymethylating reagents, iodoacetate, iodo-acetamide, and bromoacetate have been used as tools to study the properties of a vast number of proteins and peptides. In one major application, protection of cysteine residues by carboxymethylation before protein degradation is used routinely in procedures for amino acid sequence determinations (Crestfield *et al.*, 1963). Haloacids also react with the imidazole group of histidine residues in proteins (Heinrikson *et al.*, 1965). Finally, while GPD is very sensitive to iodoacetate, the ability of this reagent to affect other enzymes involved in glucose metabolism was recognized early, and the effects of iodoacetate and iodoacetamide on most aspects of mammalian metabolism were the subjects of early extensive investigations (Webb, 1966).

In recent work, Shapiro *et al.* (1988) have identified His-13 and His-114 as sites of alkylation of human angiogenin by [^{14}C]bromoacetate. Creatine kinase was shown in early work to be alkylated by iodoacetate and iodoacetamide, and modification of this enzyme by these reagents has been studied extensively. Wang *et al.* (1988) have presented a recent detailed kinetic analysis of iodoacetamide inactivation of creatine kinase and studied the effects of formation of enzyme–substrate complexes on this inactivation. Iodoacetate and iodoacetamide have been used to probe the functional role of the active-site sulfhydryl residue of aconitase (Kennedy *et al.*, 1988). Chemical modification of cytoplasmic malate dehydrogenase with iodoacetate was used to identify cooperativity between two subunits of this dimeric enzyme (Zimmerle *et al.*, 1987). These and many other applications serve to emphasize the past and present importance of these compounds as reagents for chemical modification of biological molecules.

1.2.1.4 Physiological Responses to Iodoacetic Acid

The toxic symptoms of iodoacetate poisoning noted in early work reflect the several physiological consequences of disruption of metabolic functions [reviewed by Webb (1966)]. Included in the biological sequelae of iodoacetate treatment is an enhancement of potassium efflux from cells, and iodoacetate and iodoacetamide have been used extensively as research tools to study the cellular effects of this increase (Lew and Ferreira, 1978). Blockade of metabolism by iodoacetate results in a depletion of cellular ATP. This in turn inhibits active transport within the cell. Potassium loss is accelerated both by an increase in intracellular calcium caused by a turning off of the calcium pump and by a blockade of potassium reuptake. A recent study has shown that iodoacetate can also inhibit calcium-dependent efflux of potassium by irreversible alkylation of a membrane-bound protein (Plishker, 1985).

1.2.1.5 Haloacetylated Molecules as Affinity Labels and as Medicinal Agents

The susceptibility of the α-haloacyl functionality to nucleophilic attack has also been exploited by attaching this group through an acyl bond to other organic moieties which serve to deliver the reactive halogen to specific biological sites. Wilchek and Givol (1977) have reviewed the use of haloacetyl reagents for affinity labeling of proteins. Examples of the development of medicinal agents based on this strategy, presented in later chapters, include such analogues as haloalkyl steroids, nucleosides, and carbohydrates.

1.2.2 Dichloroacetic Acid

1.2.2.1 Metabolic Effects of Dichloroacetic Acid

Dichloroacetate (DCA) has striking and diverse effects on mammalian metabolism. Initial *in vivo* effects reported in the 1960s included an increase in respiratory quotient and a lowering of blood glucose in diabetic rats. Subsequent *in vitro* studies demonstrated that dichloroacetate stimulates oxidation of glucose, leading to a lowering of blood glucose levels, as well as levels of three-carbon glucose precursors, lactate, pyruvate, and alanine (Crabb and Harris, 1979; Whitehouse *et al.*, 1974; and references therein). In part because of potential medical applications of DCA as a hypoglycemic, hypolactatemic, and hypolipidemic agent (see Section 1.2.2.6), over the past several years there has been much research on the biochemistry and pharmacology of DCA, and this will be summarized below [reviewed by Crabb *et al.* (1981)].

1.2.2.2 Biochemical Basis for Stimulation of Glucose Metabolism by Dichloroacetic Acid

The stimulation of carbohydrate oxidation at the expense of lipid fuels was found to be caused by inhibition by DCA of pyruvate dehydrogenase kinase (PDH kinase), the enzyme that, through phosphorylation, deactivates the pyruvate dehydrogenase (PDH) complex (Whitehouse *et al.*, 1974). Biological sequelae of this inhibition derive from the fact that mammalian systems are unable to synthesize glucose from acetyl-CoA. Thus, the irreversible oxidative decarboxylation of pyruvate to acetyl-CoA, catalyzed by PDH, commits a two-carbon unit to oxidation to carbon dioxide by the Krebs cycle or to incorporation into lipids (Stryer, 1988). DCA activates heart mitochondrial PDH half-maximally at a concentration of 4 μM. The inhibition of PDH kinase activity is uncompetitive in the absence of ADP but noncompetitive for ATP in the presence of ADP. Several other halogenated acids, including mono- and trichloroacetate, 2-chloropropionate, 3-chloropropionate, and 2,2'-dichloropropionate, activate pyruvate dehydrogenase by the same mechanism (reviewed by Crabb *et al.*, 1981).

1.2.2.3 Effects of Dichloroacetate-Derived Glyoxylate and Oxalate

DCA has metabolic effects in the liver that apparently are unrelated to PDH. Transamination of leucine and oxidation of α-ketoisocaproate are promoted by DCA, but chloropropionate has no similar action on branched-chain amino acid catabolism. Enhanced leucine metabolism has been shown to be caused in part by enzymatic dechlorination of dichloroacetate to glyoxylate. The latter serves as an amino group acceptor to increase the transamination of leucine, the rate-limiting step in leucine catabolism. Glyoxylate also is oxidized to oxalate, an inhibitor of several enzymes, including lactate dehydrogenase and pyruvate carboxylase, key enzymes in gluconeogenesis in the liver. In isolated hepatocytes, oxalate inhibits gluconeogenesis from lactate under conditions of low bicarbonate levels, probably by inhibition of pyruvate carboxylase (the metabolic interactions of DCA, oxalate, and glyoxylate with leucine catabolic pathways and glucose synthesis are shown in Fig. 1-2). The high bicarbonate levels of blood make this mechanism of action appear less likely *in vivo*. Nonetheless, metabolism of DCA to oxalate complicates interpretation of the effects of the former on liver metabolism. Oxalate administered to suckling rats had no effect on blood glucose levels, so a role of oxalate in the hypoglycemic activity of DCA is not likely (Crabb and Harris, 1979; Crabb *et al.*, 1981). Whereas stimulation of leucine oxidation in liver appears to be mediated primarily by an enhanced rate of transamination, DCA also stimulates branched-chain α-ketoacid dehydrogenase by inhibition of the associated kinase (see also Section 1.2.3). Although this behavior parallels that of PDH toward DCA, the sensitivity to DCA is an order of magnitude less (Paxton and Harris, 1984, and references therein).

1.2.2.4 Biochemical Basis for Dichloroacetate-Induced Lowering of Blood Glucose

Plasma glucose levels are regulated, in part, by the alanine and Cori cycles. Activation of PDH in peripheral tissues by DCA disrupts these cycles by reducing the availability of lactate, pyruvate, and alanine for gluconeogenesis in the liver. Blood glucose lowering is seen when a con-

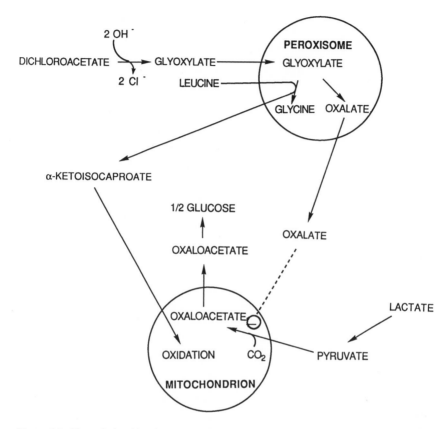

Figure 1-2. The relationships between DCA, oxalate, glyoxylate, and leucine catabolism. (Adapted, with permission, from Crabb *et al.*, 1981).

tinuous supply of these three-carbon precursors from muscle is needed for gluconeogenesis, for example, in diabetes and fasting (Crabb and Harris, 1979).

1.2.2.5 Dichloroacetic Acid and Lipid Metabolism

DCA also can affect lipid metabolism. Whereas 2-chloropropionate increases lipogenesis and ketogenesis in hepatocytes by an exclusive action on PDH kinase, the effects of DCA depend on experimental conditions and the extent to which metabolism to oxalate occurs (Demaugre *et al.*, 1983). Recently, an indirect action of DCA on lipid metabolism has been demonstrated. DCA, in the absence of insulin, enhances the activity of malic enzyme, a key lipogenic enzyme that catalyzes the oxidation of malate to pyruvate. Induction of increased malic enzyme activity was associated with altered mRNA profiles, alterations similar to those produced previously by high glucose concentrations. A metabolite of glucose had been implicated as the regulatory signal, and the effects of DCA suggest that a product of pyruvate metabolism is this signal (Mariash and Schwartz, 1986).

1.2.2.6 Therapeutic Potential of Dichloroacetic Acid

The metabolic effects described above and other metabolic effects of DCA reflect in part the critical role played by pyruvate in metabolism of several classes of biomolecules—carbohydrates, amino acids, lipids, and ketone bodies are examples. The review by Crabb et al. (1981) is recommended for a more thorough discussion of the sometimes quite complex issues that have arisen in the studies of dichloroacetate. This review also contains a discussion of the therapeutic potential of dichloroacetate for management of diabetes mellitus, hyperlipidemia, and hyperlactatemia. While toxicity of DCA has complicated human trials, results have been promising in certain areas. For example, the treatment of hyperlactatemia, a common and usually fatal disorder of acid–base metabolism caused by such conditions as hypotension and sepsis and characterized by abnormally high serum levels of lactic acid, has been encouraging. DCA prevents or reverses hyperlactatemia in animal models and has increased survival time in recent clinical trials, with no evidence of toxicity (Stacpoole et al., 1988). Vary et al. (1988) have reported a recent comparison of the effects of DCA on metabolism of normal and septic animals.

1.2.3 2-Chloroisocaproic Acid—A Stimulator of Branched-Chain Amino Acid Catabolism

The branched-chain α-ketoacid decarboxylase complex (BCKDH) catalyzes the oxidative decarboxylation of α-ketoacids formed from leucine, isoleucine, and valine. Similar to PDH, this enzyme complex is regulated by phosphorylation of a serine residue on the α subunit of the E1 component. As noted above, BCKDH activity is stimulated by DCA but BCKDH is less sensitive than PDH. On the other hand, 2-chloroisocaproic acid, a substrate analogue, is a very potent inhibitor of the complex-associated kinase responsible for regulatory phosphorylation and is a very potent activator of the complex in intact tissue (Harris et al., 1982). The important relationship between branched-chain amino acid catabolism and regulation of such processes as protein synthesis and insulin release, as well as the altered plasma and tissue levels of the branched-chain amino acids in certain pathological states, has prompted much interest in mechanisms for regulation of BCKDH [reviewed by Harris *et al.* (1986)]. 2-Chloro-isocaproic acid has been exploited effectively as a tool for studying effects of BCKDH stimulation in several recent investigations related to these issues (for example, Paxton *et al.*, 1988; Han *et al.*, 1987; Toshima *et al.*, 1985).

1.3 Biochemistry of Halogenated Monocarboxylic Acids (II): Fluoroacetic Acid, Fluorocitric Acid, and "Lethal Synthesis"

1.3.1 Isolation and Identification of Fluoroacetic Acid as the Toxic Principle of *Dichapetalum cymosum*

The biochemical studies of fluoroacetic acid in the 1950s marked the beginning of rapid progress in applications of fluorinated compounds in biochemistry and medicine. Fluoroacetic acid (also known as compound 1080) was first synthesized in the 19th century and was studied in Germany as a potential insecticide. However, the impetus given to biochemical studies of this toxic compound came from the demonstration by Marius in the 1940s that biosynthesis of fluoroacetic acid by the South African plant *Dichapetalum cymosum* (gifblaar) was the source of toxicity of this plant. In fact, mortality in grazing lifestock is a serious problem in South Africa and other areas of the Southern Hemisphere where this plant and other members of this genus are indigenous and can contain up to 2000 ppm fluoroacetate [reviewed by Peters (1972) and Goldman (1969)].

In a series of classical biochemical experiments, Sir Rudolf Peters demonstrated a mechanism for fluoroacetate poisoning involving a process of "lethal synthesis" (Peters, 1972, and references therein). Thus, fluoroacetate itself had no effect on acetate metabolism in guinea pig kidney particles but caused a marked increase in citrate levels. In 1953, the isolation by Peters and co-workers of fluorocitrate from kidney homogenates treated with fluoroacetate and the demonstration that this compound was a potent inhibitor of aconitase, the enzyme that catalyzes the dehydration of citrate to aconitate, confirmed the proposed "lethal synthesis" of an inhibitor of the citric acid cycle (Fig. 1-3) [reviewed by Peters (1972)].

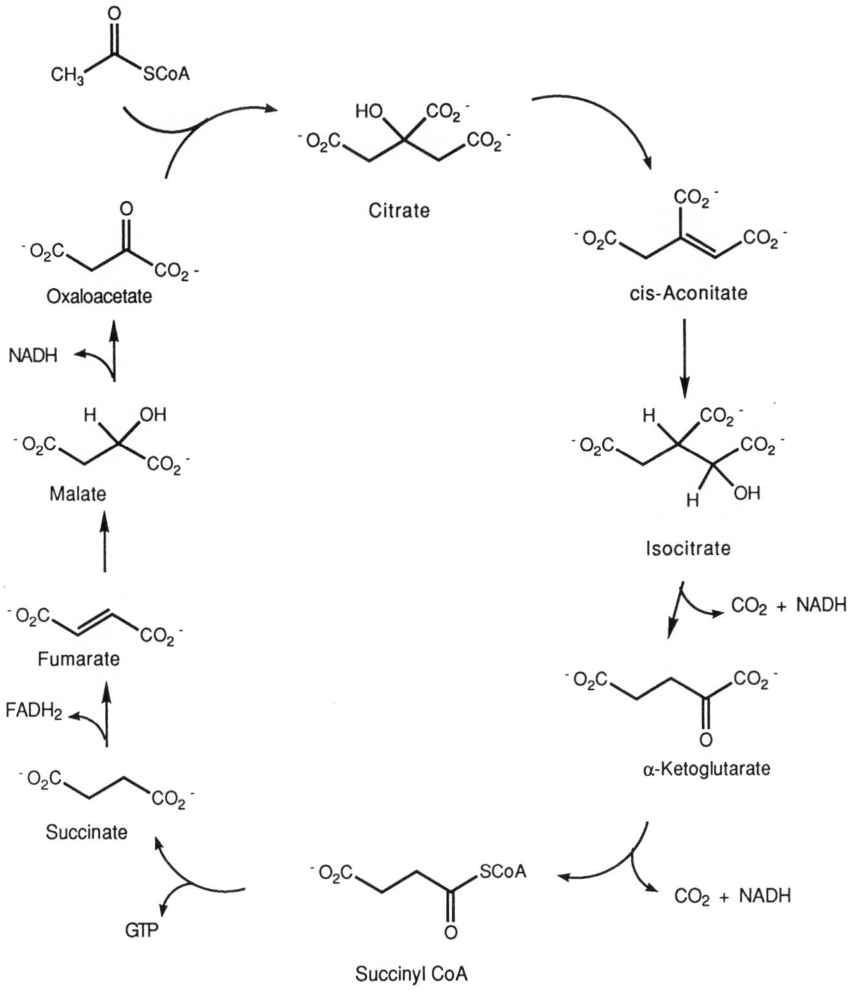

Figure 1-3. The citric acid cycle.

The concept of letal synthesis and the recognition of the potent antimetabolic effects elicited by the presence of the $C-F$ bond were two important outcomes of this work, spurring interest in the design of other analogues incorporating these concepts. That aconitase inhibition by fluorocitrate apparently is not the basis for the acute toxicity of fluoroacetate (see Section 1.3.3) does not diminish the importance of the impact this work had on future research.

1.3.2 Biosynthesis and Biological Properties of Fluorocitric Acid

Fluoroacetyl-CoA serves as a substrate for citrate synthase with a V_{max} of 0.3% of that for acetyl-CoA (Kun, 1976). The stereochemical aspects of the citrate synthase-catalyzed condensation of fluoroacetyl-CoA with oxaloacetate are interesting quite apart from the biochemical significance of the product, in that one of four possible stereoisomers is formed [reviewed by Walsh (1983)]. A series of synthetic and enzymatic studies culminated in the identification of the toxic isomer as $(-)$-$(2R,3R)$-erythro-fluorocitrate (Carrell et al., 1970, and references therein). The formation of a single stereoisomer requires that citrate synthase differentiate between the pro-R and pro-S protons on carbon-2 of fluoroacetyl-CoA. Using $(2R)$-2-deutero- and $(2S)$-2-deuterofluoroacetate, Keck et al. (1980) established that the pro-R proton is removed. With net inversion at C-2, the resulting carbanion attacks the si face of oxaloacetate, producing $(2R,3R)$-erythro-fluorocitrate (Fig. 1-4). This result is consistent with the stereochemical behavior previously determined for [^{2}H,^{3}H]acetyl-CoA (Luthy et al., 1969).

Fluorocitrate produces an initial competitive inhibition of aconitase, with K_i values of 64 μM for the mitochondrial isozyme and 18 μM for the cytoplasmic isozyme (Eanes et al., 1972). Mechanistic interpretations of this inhibition have been complicated by the realization that $(2R,3R)$-fluorocitrate has fluorine substituted on the arm of citrate (pro-S) that is not acted on by the enzyme. The reasonable assumption that fluorocitrate with fluorine substituted on the pro-R arm might be even more toxic was shown to be false by the demonstration that citrate synthase produced non-inhibitory isomers $(2R,3S$ and $2S,3S)$ from acetyl-CoA and racemic fluorooxaloacetate (Keck et al., 1980; Fanshier et al., 1962). The formation

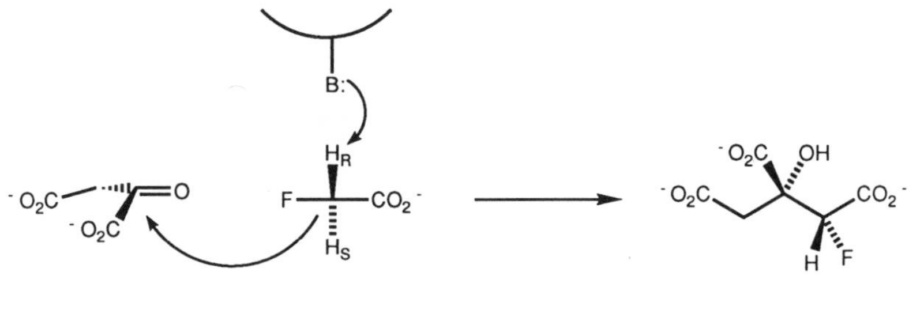

2R,3R-Erythro-fluorocitrate

Figure 1-4. Stereochemistry involved in the biosynthesis of $(2R,3R)$-erthyro-fluorocitrate (Keck et al., 1980).

of an enzyme–ferrous–substrate ternary complex in the active site has been proposed. Analysis of crystallographic data and model building revealed that only the inhibitory 2*R*,3*R* isomer of fluorocitrate could form a comparable ternary structure. Based on this, the model for inhibition and inactivation shown in Fig. 1-5 was proposed (Carrell *et al.*, 1970). However, no evidence for alkylation of aconitase by fluorocitrate has been found (Eanes *et al.*, 1972). Glusker and co-workers have published an analysis of possible intermolecular interactions of fluorocitrate and of the $C-F$ bond in general (Murray-Rust *et al.*, 1983).

1.3.3 An Alternative Mechanism for Fluorocitrate Toxicity

Both the competitive inhibition (K_i values reported by several groups ranging from 60 to 200 μM) and time-dependent inactivation of aconitase are reversed readily by addition of citrate, a fact difficult to reconcile with the irreversible toxicity of fluorocitrate. Furthermore, an estimated cerebral concentration of the active isomer of fluorocitrate of only 10 nM results from the lethal dose of 0.2–0.5 μg racemate injected into the brain of an adult rat (Kun, 1976).

In exploring the mechanism of toxicity further, Eanes *et al.* (1972) showed that preincubation of intact isolated liver mitochondria with low concentrations of fluorocitrate irreversibly inhibited the formation of isocitrate from externally added citrate or *cis*-aconitate. Disruption of the

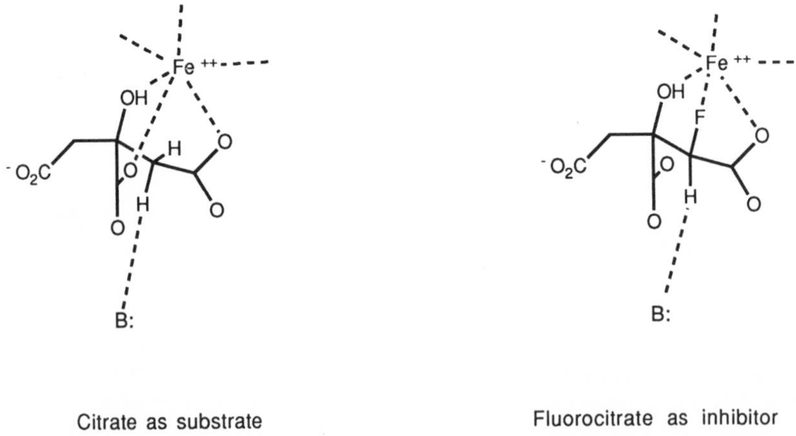

Citrate as substrate   Fluorocitrate as inhibitor

Figure 1-5. Models showing citrate as a substrate and fluorocitrate as an inhibitor of aconitase (Carrell *et al.*, 1970).

mitochondrial membrane completely restored activity, suggesting that this inhibition was a result of a blockade of citrate transport. More evidence was provided by the use of $(-)$-*erythro*-fluoromalate, an analogue which strongly stimulates the membrane-bound tricarboxylate carrier mechanism, but which itself is unable to penetrate the inner mitochondrial membrane (Skilleter et al., 1972). Preincubation with $2 \mu M$ fluorocitrate completely inhibited the action of fluoromalate, again suggesting an action on the tricarboxylic acid carrier (Eanes et al., 1972). The concentrations used in these studies were well below levels that inhibit aconitase.

Subsequent studies gave additional evidence supporting the proposal that inhibition of citrate mitochondrial transport results from preincubation of mitochondria with nanomolar concentrations of fluorocitrate. Preincubation was found to be necessary to observe inhibition, and this was shown to reflect the irreversible binding of fluorocitrate to mitochondrial proteins (Kirsten et al., 1978). Identification of a mitochondrial enzyme responsible for the formation of a citrate–glutathione thioester and the demonstration that this enzyme is irreversibly inhibited by fluorocitrate suggested that this is a molecular site of toxicity of fluorocitrate. Two proteins were found to form thioesters with fluorocitrate—citryl-S-glutathione synthase and hydrolase. The proposal was made that the synthase may be a constituent of the mitochondrial tricarboxylate transport system. The concentrations of fluorocitrate that inhibit the citryl-S-glutathione enzymes, and which block citrate transport, are of the same order of magnitude as that found in mitochondria of fluorocitrate-poisoned rats (Kun et al., 1977). These data strongly suggest that the acute toxic effects of fluorocitrate are related to inhibition of mitochondrial citrate transport. Furthermore, the synthesis of acetylcholine in nerve cells is known to depend on mitochondrial citrate as a source of acetyl-CoA. Inhibition of this process by fluorocitrate would be consistent with its apparent action as a neurotoxin (Kirsten et al., 1978).

1.3.4 Physiological Consequences of Toxic Mechanisms of Fluoroacetic and Fluorocitric Acids

As Kun (1976) has stressed, the extrapolation of results from *in vitro* enzymology to *in vivo* cellular processes is very difficult, in particular where distributive, branching, and multienzymatic processes are interacting. In the intact organism, the inhibition by fluorocitrate of two critical biochemical steps would be expected to alter many biological processes dependent on the proper functioning of these steps. Thus, the classic symptom of fluoroacetate poisoning, tissue citrate accumulation, has been attributed to both toxic mechanisms—inhibition of aconitase and inhibition of citrate transport in mitochondria. Other toxic effects include heart ATP accumulation, hyperglycemia, and hypocalcemia (Bosakowski and Levin, 1986, and references therein). Bosakowski and Levin (1986) have determined that elevated citrate levels can be used as a measure of fluoroacetate and fluorocitrate toxicity, at least in the rat and dog.

Fluoroacetate and fluorocitrate have been used as biochemical tools to study certain aspects of metabolism. For example, in brain slices fluoroacetate has been shown to have selective action in glia as a consequence of selective uptake and utilization of acetate (and fluoroacetate) for the Krebs cycle in this compartment (Szerb and Issekutz, 1987, and references therein). The excitatory neurotransmitter glutamate is deactivated by conversion to glutamine in glial cells, and fluoroacetate has been used to block this deactivation. Szerb and Issekutz (1987) found that, in rat hippocampal slices, not only is glutamine synthesis in glial cells inhibited by fluoroacetate, but so also is uptake of synaptically released glutamate. They suggest that lack of this uptake may contribute to convulsions characteristic of *in vivo* fluoroacetate toxicity. Similarly, fluorocitrate has been used as a selective toxic agent to abolish glial cell metabolism for an *in vivo* study of brain tissue (Paulsen *et al.*, 1987).

Approaches to pro-insecticides are based on *in vivo* conversion of substrates to fluorocitrate (Chapter 2) (for example, Prestwich *et al.*, 1984). Mechanisms evolved by certain species to detoxify fluoroacetate (discussed in Chapter 9 of Vol. 9A of this series) are related to glutathione availability.

1.4 Fluorinated Carboxylic Acids as Mechanistic Probes

1.4.1 Additional Biochemistry of Fluoroacetic Acid

1.4.1.1 Acetyl-CoA Synthase

Implicit in the involvement of fluoroacetate in the process of lethal synthesis of fluorocitrate is the substrate activity of fluoroacetate with acetyl-CoA synthase. Patel and Walt (1987) verified this and obtained a V_{max} of 10% of that for acetate.

1.4.1.2 Fluoroacetic Acid as a Prochiral Stereochemical Probe

Research on the mechanism of fluorocitrate toxicity demonstrated clearly the value of fluorine as a stereochemical probe. In a series of investigations on the stereochemistry of enzymatic reactions responsible for the synthesis and metabolism of biologically important carboxylic acids, Kun, Walsh, Nowak, and others have taken advantage of chirality introduced by the substitution of a $C-F$ bond for a $C-H$ bond [reviewed by Walsh (1983)]. An earlier technique to determine the stereochemical fate of the achiral methyl group or prochiral methylene group during enzymatic processing was based on the introduction of chirality using isotopes of hydrogen (Cornforth et al., 1969; Luthy et al., 1969). Disadvantages to this approach include the presence and uncertain magnitude of isotope effects and difficulties often associated with the determination of the stereochemical outcome of the reactions. On the other hand, the fluoromethyl group is prochiral, and stereospecific replacement of a second $C-H$ group, for example, through enzymatic carboxylation or condensation, will result in a chiral configuration. A similar replacement of the prochiral methylene group with the chiral fluoromethylene group has been used in stereochemical analyses.

Several examples of the study of the stereochemical course of enzymatic transformations with fluorinated analogues will be discussed in later sections. An additional example using fluoroacetate is provided by research involving malate synthase. The malate synthase-catalyzed condensation of fluoroacetyl-CoA ($V_{max} = \sim 5\%$) with glyoxylate produces an approximately equal mixture of two diastereomeric fluoromalates. This indicates that, unlike citrate synthase, malate synthase does not discriminate between the *pro-R* and *pro-S* protons of fluoroacetyl-CoA in the deprotonation step. Attack on the *si* face of glyoxylate results in the *R*-configuration for the hydroxyl group (Fig. 1-6) (Marletta et al., 1981). A similar achiral synthesis of fluoromalate was reported by Keck et al. (1980). This represents a fundamental difference between the mechanisms of malate synthase and citrate synthase.

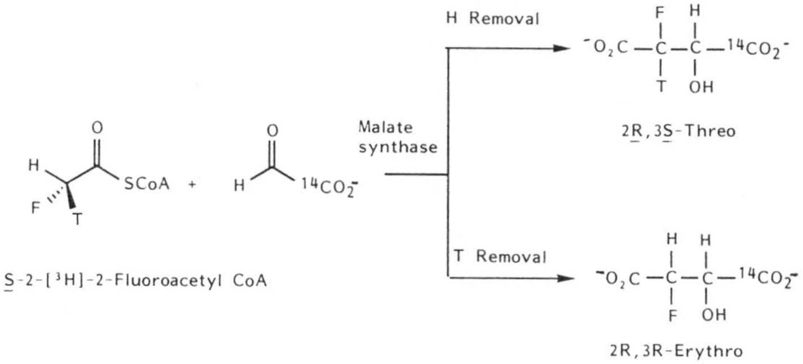

Figure 1-6. Demonstration of the achiral nature of the malate synthase-catalyzed condensation of 2-fluoroacetyl-CoA and glyoxylate (Marletta et al., 1981).

1.4.2 3-Fluoropropionic Acid as a Mechanistic Probe

In all biotin-dependent carboxylations, a transfer of CO_2 from carboxybiotin to substrate is required (Fig. 1-7). A concerted mechanism, not requiring base at the active site, and a carbanion mechanism have both been given serious consideration for this process. 3-Fluoropropionyl-CoA has been used as a probe for the presence of a C-2 carbanion in the carboxylations of propionyl-CoA catalyzed by propionyl-CoA carboxylase and transcarboxylase (Stubbe *et al.*, 1980; Stubbe and Abeles, 1977). Both

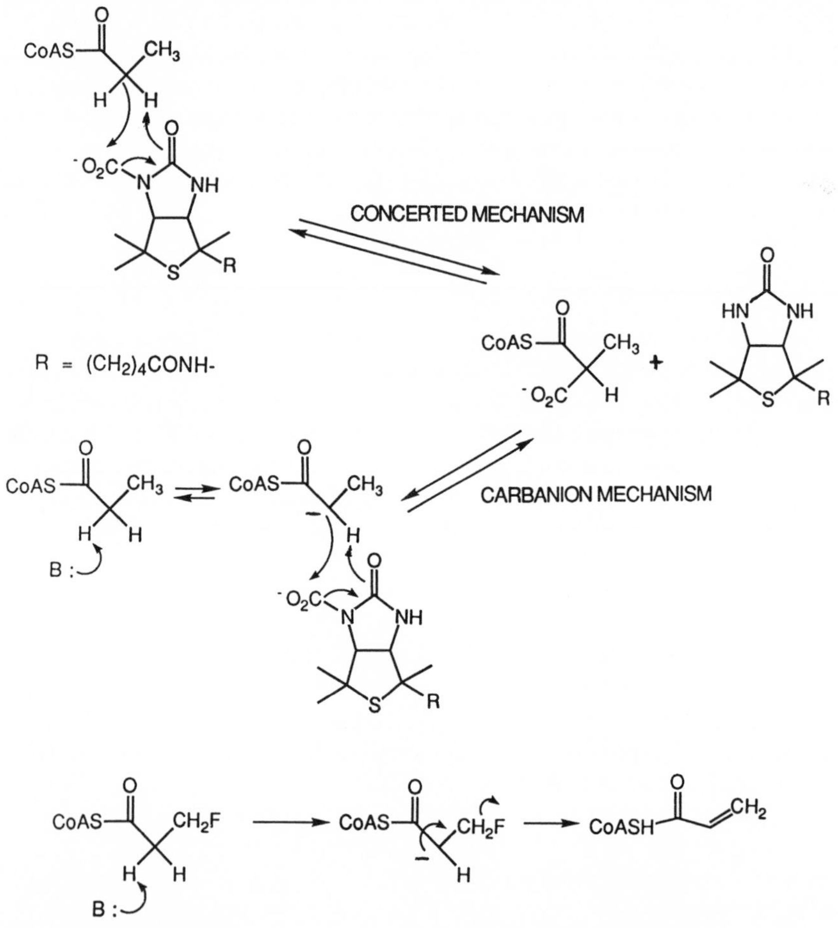

Figure 1-7. Concerted and carbanion mechanisms suggested for biotin-catalyzed carboxylation reactions. The elimination of HF from the substrate 3-fluoropropionyl-CoA is evidence that an intermediate carbanion is formed (Stubbe *et al.*, 1980).

enzymes catalyzed the elimination of HF from the fluorinated substrate, with no evidence for the formation of fluoromethylmalonyl-CoA, suggesting a carbanionic intermediate (Fig. 1-7). Although HF elimination could also accompany decarboxylation of enzyme-produced fluoromethylmalonyl-CoA, the relatively slow rate of decarboxylation of the closely related acetylaminoethanethiol derivative of fluoromethylmalonate was cited as evidence against this alternative explanation.

1.4.3 3-Fluoropyruvic Acid as a Mechanistic Probe

3-Fluoropyruvate (F-Pyr) has been the subject of extensive investigations as a substrate analogue for pyruvate, a reflection of the critical role of pyruvate as the link between carbohydrate metabolism, the Krebs cycle, and several biosynthetic processes. The prochiral methyl group of F-Pyr has provided a wealth of information on the stereochemistry of several enzymatic reactions involving pyruvate.

1.4.3.1 Stereochemistry of Enolization Catalyzed by Pyruvate Kinase

A group of phosphoenolpyruvate (PEP)-utilizing enzymes transfer phosphate from PEP to an acceptor molecule with concomitant addition of a group (e.g., H^+ or CO_2) to the 3-position of PEP. An important member of this group, pyruvate kinase, catalyzes the synthesis of ATP from PEP and ADP. In the reverse reaction, enolization of fluoropyruvate catalyzed by pyruvate kinase, there is no discrimination between the two prochiral protons (Goldstein et al., 1978).

1.4.3.2 Stereochemistry of Carboxylation of Pyruvate

In contrast to the above result, complete stereospecificity was observed in the biotin-dependent transcarboxylase-catalyzed synthesis of 3-fluoro-oxaloacetate from F-Pyr (Fig. 1-8). Malate dehydrogenase reduction of the product formed exclusively (2R,3R)-erythro-fluoromalate, demonstrating that carboxylation of F-Pyr involves complete discrimination between the two prochiral protons of F-Pyr. Retention of configuration, as previously observed in the carboxylation of chiral [3-^{1}H,^{2}H,^{3}H]pyruvate, would require stereospecific removal of the pro-S hydrogen (Fig. 1-8) (Goldstein et al., 1978).

In a similar experiment, fluorooxaloacetate formed from enzymatic carboxylation of fluoropyruvate by transcarboxylase from Propionibacterium shermanii, using malonyl-CoA as the cosubstrate, was trapped

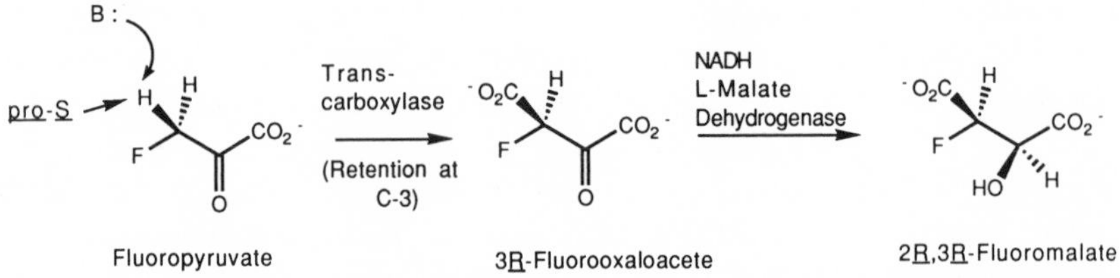

Fluoropyruvate   3R-Fluorooxaloacete   2R,3R-Fluoromalate

Figure 1-8. Fluoropyruvate as a probe for the stereochemistry of the carboxylation of pyruvate. Retention of configuration during carboxylation and formation of (2R,3R)-fluoromalate requires that the *pro-S* proton be removed (Goldstein *et al.*, 1978).

in situ by reduction with malate dehydrogenase. In this case, ^{19}F-NMR analysis confirmed that the fluoromalate formed had exclusively the 2R,3R configuration, in agreement with the previous results. When the reaction was done in D_2O, enzyme-dependent incorporation of deuterium was specific for only one of the enantiotopic protons of the prochiral methyl protons, showing that the reaction proceeds stereospecifically (Hoving *et al.*, 1985).

While many precedents suggested that carboxylation of pyruvate should occur with retention of configuration at the methyl group, coupled enzymatic processes produced direct proof of this. Enzyme I serves an important role in PEP-dependent sugar transport in bacteria by catalyzing the transfer of a phosphoryl group from PEP to the phosphocarrier protein in the bacterial cell wall. In this process, a proton is transferred to the C-3 atom of enolpyruvate to give pyruvate. Duffy and Nowak (1984) had studied the effect of stereochemistry of several PEP analogues, including (Z)- and (E)-3-fluorophosphoenolpyruvate (F-PEP), on reactions catalyzed by several PEP-utilizing enzymes and had found the Z-isomer to have more activity. Hoving *et al.* (1985) used ^{19}F-NMR spectroscopy to show that enzyme I-catalyzed equilibration of (Z)-F-PEP and F-Pyr in D_2O results in incorporation of deuterium in only one of the two enantiotopic positions of the fluoromethyl group. This was assumed to be the *pro-R* position based on previous results with 2-oxobutyrate (Hoving *et al.*, 1983), a result of protonation at the 2-*re*,3-*si* face of the enolate. The so-derived chiral fluoropyruvate was used as a substrate for transcarboxylase in a reaction coupled with malate dehydrogenase. NMR identification of C-3-deuterated (2R,3R)-fluoromalate as the product demonstrates that carboxylation occurs with retention of configuration. These stereochemical interrelationships are shown in Fig. 1-9.

Using similar strategy, Hoving *et al.* (1985) showed that pyruvate kinase transfers a proton to the 2-*si* face of (Z)-F-PEP, producing as the

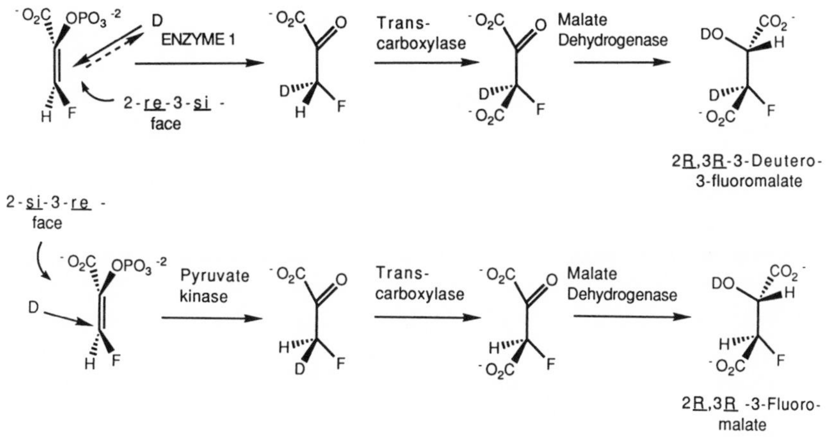

Figure 1-9. Demonstration that enzyme I and pyruvate kinase have opposite stereochemical selection in protonation of (Z)-F-PEP (Hoving *et al.*, 1983, 1985).

final product protonated (2R,3R)-fluoromalate (Fig. 1-9). This result is more in keeping with the usual mode of addition to phosphoenolpyruvate (Kuo and Rose, 1982).

1.4.3.3 Stereochemistry of PEP Carboxykinase-Catalyzed Carboxylation

Fluorine-19 NMR spectroscopy was used by Hwang and Nowak (1986) to study the stereochemistry of carboxylation catalyzed by PEP carboxykinase from two different sources. The mammalian enzyme converts oxaloacetate to PEP as an important early step in gluconeogenesis whereas the enzyme isolated from certain anaerobic bacteria catalyzes the formation of oxaloacetate from pyruvate and CO_2. These fundamentally different roles suggested that the mechanisms might have evolved differently. In fact, fluorooxaloacetate produced from (Z)-F-PEP by each enzyme gave (2R,3R)-3-fluoromalate after malate dehydrogenase reduction, indicative of carboxylation on the *si* side of the enzyme-bound substrate (Fig. 1-10).

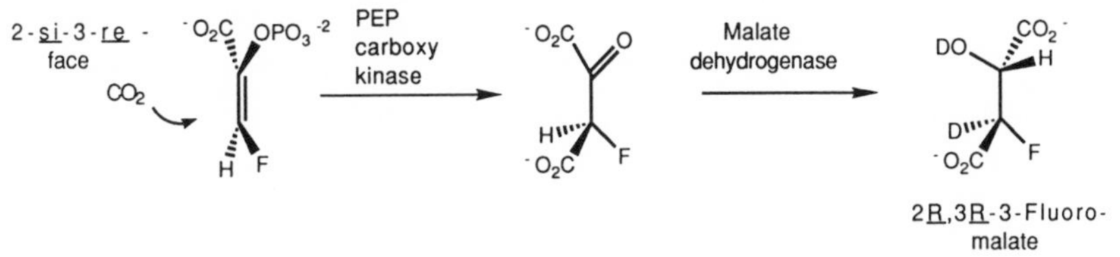

Figure 1-10. Use of (Z)-F-PEP to determine the stereochemistry of PEP carboxykinase-catalyzed carboxylation reactions (Hwang and Nowak, 1986).

1.4.3.4 Fluoropyruvic Acid as a Substrate Analogue for Pyruvate Dehydrogenase

F-Pyr was used as a substrate analogue to study the mechanism of the bacterial enzyme complex which catalyzes the oxidative decarboxylation of pyruvate to CO_2 and acetyl-CoA. A parallel reaction path with F-Pyr would result in the formation of fluoroacetyl-CoA, imparting toxicity to F-Pyr by a process of lethal synthesis. In fact, F-Pyr is a substrate for the pyruvate dehydrogenase component of the complex (V_{max} is 10% of that for pyruvate decarboxylation), but, in a thiamine pyrophosphate (TPP)-dependent process (Fig. 1-11), acetate, CO_2, and fluoride are formed (Leung and Frey, 1978). A mechanistic detail was addressed by taking advantage of the requisite formation of acetyl-TPP in the fluoropyruvate reaction. At issue was whether tautomerization of hydroxyethyl-TPP to

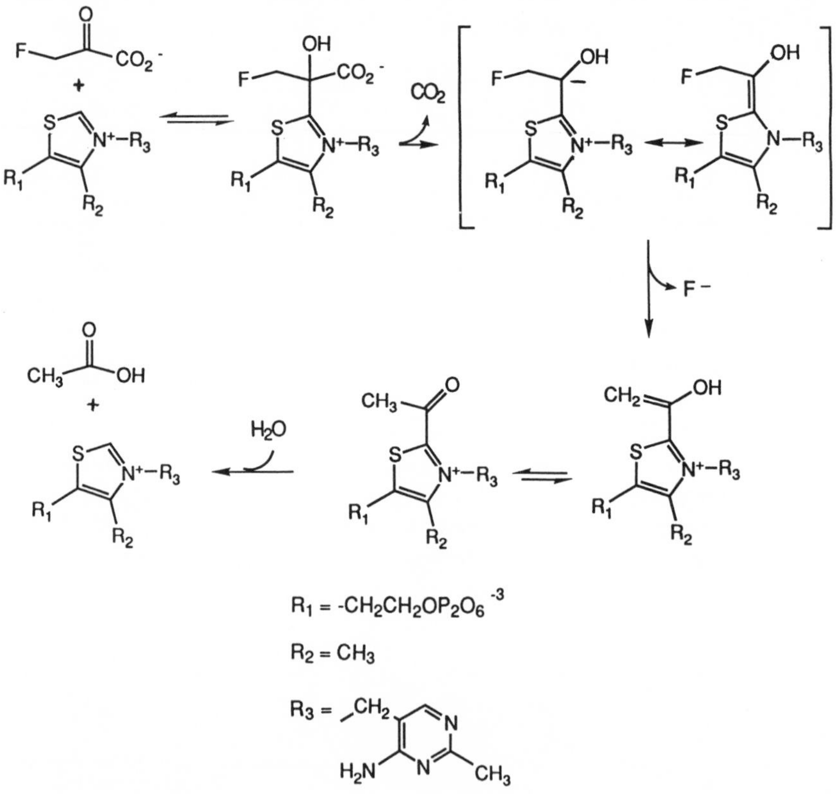

$$R_1 = -CH_2CH_2OP_2O_6{}^{-3}$$

$$R_2 = CH_3$$

$$R_3 = \text{(CH}_2\text{-pyrimidine)}$$

Figure 1-11. Fluoropyruvate as a substrate for pyruvate dehydrogenase (Leung and Frey, 1978).

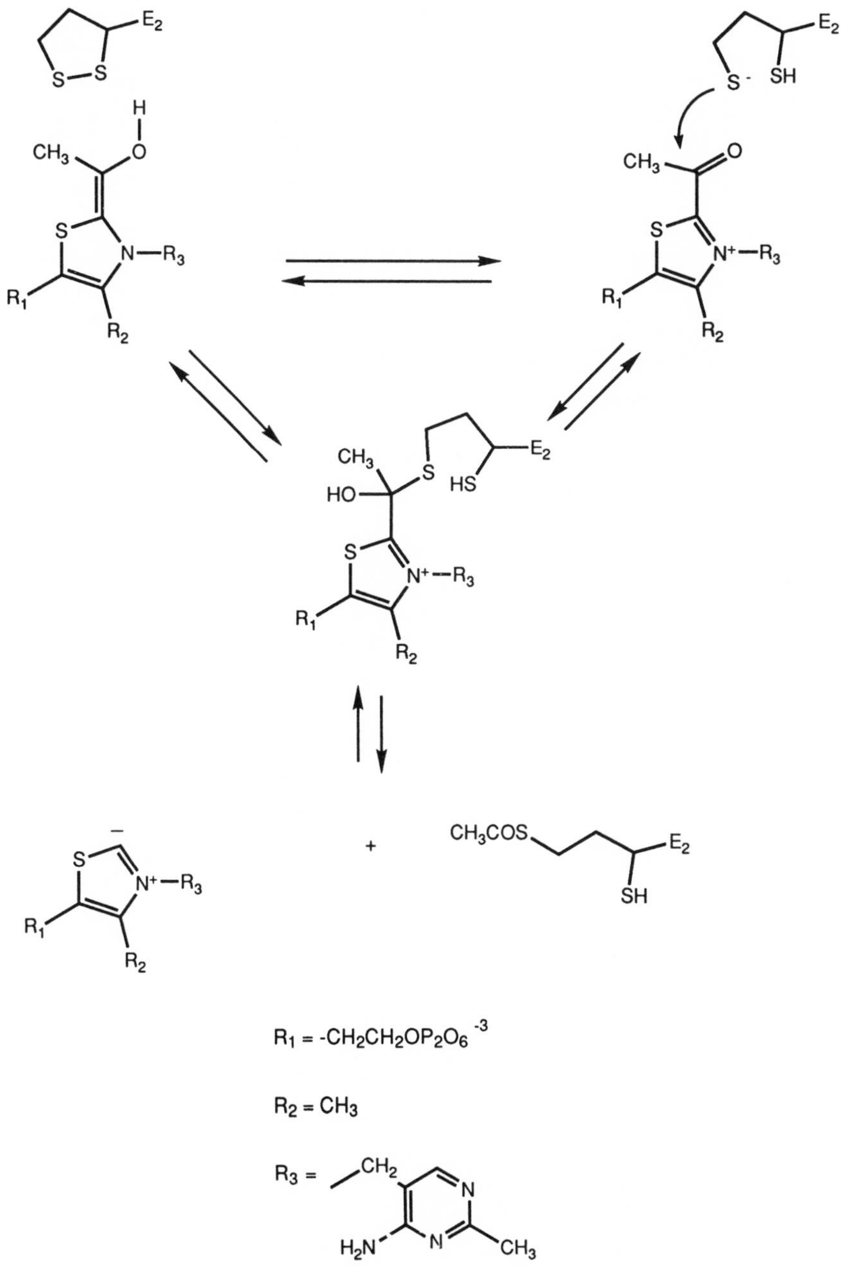

Figure 1-12. Demonstration, using fluoropyruvate to generate acetyl-TPP, that acetyl-TPP is an intermediate in the acylation of dihydrolipoamide (Flournoy and Frey, 1986).

acetyl-TPP is required for dihydrolipoyl transacetylase-catalyzed acylation of dihydrolipoamide (Fig. 1-12). Acetyl-TPP was generated from fluoro-pyruvate by the pyruvate dehydrogenase system in the presence of dihydrolipoamide. Efficient formation of S-acetyllipoamide under a variety of conditions supports the existence of acetyl-TPP as an intermediate during decarboxylation of the natural substrate (Flournoy and Frey, 1986).

1.5 Bromopyruvic Acid as an Affinity Labeling Reagent

3-Bromopyruvic acid (Br-Pyr), similar to α-haloacetic acids, possesses a reactive halogen that can be displaced by enzyme-bound nucleophiles. With many enzymes, in particular those that use pyruvate as a substrate, Br-Pyr has an avid affinity for the active site and selectively modifies nucleophilic residues in or near this site. Thus, Br-Pyr has often been found to fulfill many of the criteria that indicate that a reagent is functioning as a true affinity label—including specific substrate protection of the enzyme against inactivation by the reagent (Wold, 1977). Examples of the use of Br-Pyr as an affinity label are given below.

1.5.1 2-Keto-3-deoxy-6-phosphogluconic Aldolase

2-Keto-3-deoxy-6-phosphogluconic (KDP-gluconic) aldolase is a bacterial enzyme that catalyzes the reversible condensation of pyruvate and D-glyceraldehyde 3-phosphate to form KDP-gluconate, as well as an exchange between the methyl protons of pyruvate and water (Fig. 1-13).

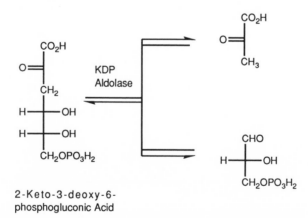

2-Keto-3-deoxy-6-
phosphogluconic Acid

Figure 1-13. KDP-gluconic-aldolase-catalyzed aldol reaction.

Aldolases have attracted particular attention because these enzymes can provide information on biochemical ·mechanisms available for the formation and breaking of carbon–carbon bonds. The extensive studies of KDP aldolase by Meloche and co-workers illustrate effectively the use of Br-Pyr as an affinity label [Meloche and Glusker, 1973, and references therein; reviewed by Hartman (1977)]. Pyruvate-sensitive inactivation of KDP-gluconic aldolase by Br-Pyr is accompanied by esterification of a protein-bound carboxyl group and alkylation of cysteine, with a total incorporation of one molecule of Br-Pyr for each molecule of enzyme (Meloche, 1967, 1970). An observed dependence of the ratio of cysteine and carboxyl labeling on ionic strength led Meloche to conclude that two different active-site conformers were being labeled. Under conditions in which enzyme-carboxyl esterification predominates (low ionic strength), the *pro-R* hydrogen of $(3R,S)$-[3-^{3}H$_2$]bromopyruvate is selectively exchanged, showing that Br-Pyr is functioning as a substrate as well as an inactivator. The kinetics of exchange and concurrent inactivation taken with the stereospecificity of exchange led to the proposal of single base-mediated proton activation in aldolase catalysis (Meloche and Glusker, 1973). According to this mechanism, subsequent to Schiff base formation between pyruvate and a lysyl amino group on the enzyme, a suitably positioned glutamyl residue operates in both proton transfer steps required to effect the condensation reaction (Fig. 1-14). Br-Pyr, as a substrate analogue, can suffer proton exchange but also is subject to nucleophilic attack by the active-site glutamate (Fig. 1-14). Subsequent research demonstrated that three different pyruvate aldolases catalyze protonation of the *re* face of the enzyme-bound bromopyruvyl eneamine, showing a consistency in the evolution of this process. The stereochemical integrity of protonation–deprotonation catalyzed by KDP-gluconic aldolase was confirmed by the findings that 2-keto-(3*S*)-3-bromobutyrate, but not the 3*R* isomer, is deprotonated by the enzyme and that only the 3*S* enantiomer alkylates the enzyme (Monti *et al.*, 1979).

1.5.2 The Pyruvate Dehydrogenase Multienzyme Complex

Halopyruvates have been studied as active-site-directed inhibitors of PDH for several years. Maldonado *et al.* (1972) reported that Br-Pyr inactivates the complex from *E. coli* in a TPP-dependent process, possibly by alkylation of lipoic acid residues. Bisswanger (1981) found that several analogues, including bromo-, fluoro-, and hydroxypyruvate, function as potent competitive inhibitors. Lowe and Perham (1984) have shown that Br-Pyr is an effective irreversible inhibitor of the intact complex, and also

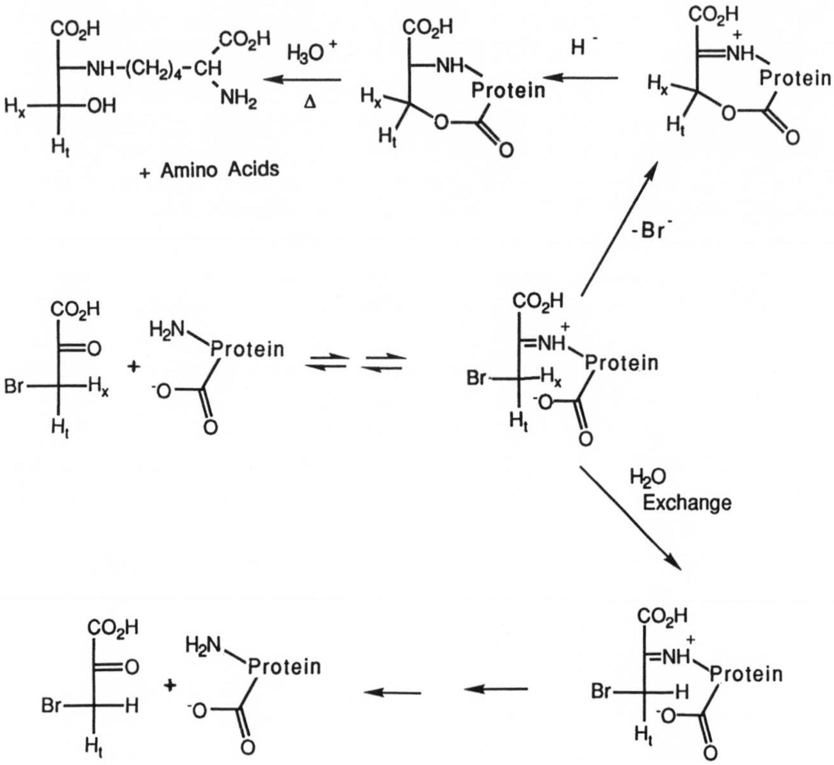

Figure 1-14. Br-Pyr as an active-site-directed inhibitor of KDP-gluconic aldolase. See text for details (Meloche and Glusker, 1973).

of the E1 component. In the intact complex, decarboxylation of Br-Pyr apparently gives S-(bromoacetyl)dihydrolipoic acid residues, and additional Br-Pyr then reacts with the generated free SH groups (Fig. 1-15). The reaction of the isolated E1 component with Br-Pyr was assumed to be similar to its interaction with fluoropyruvate (Fig. 1-11).

1.5.3 Aspartate Aminotransferase

Okamoto and Morino (1973) reported the irreversible inactivation of the PLP-dependent enzyme L-aspartate:2-oxoglutarate aminotransaminase by Br-Pyr. The inactivation was dependent on the presence of either L-cysteine or L-aspartate, and the cysteine associated with the active site was alkylated. Based on these and other data, Okamoto and Morino con-

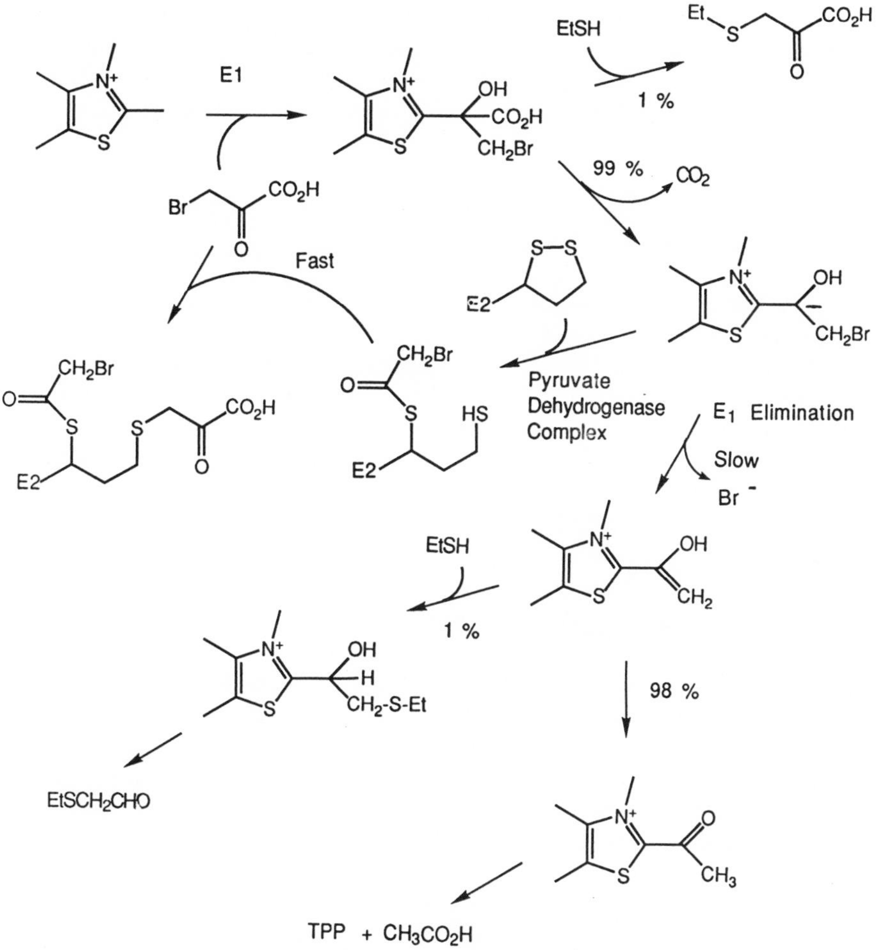

Figure 1-15. Br-Pyr as an irreversible inhibitor of PDH (see text for details). (Adapted with permission from P. N. Lowe and R. N. Perham, Bromopyruvate as an active-site-directed inhibitor of the pyruvate dehydrogenase multienzyme complex from *Escherichia coli*, *Biochemistry* 23:91–97. Copyright 1984, American Chemical Society.)

cluded that Br-Pyr was serving as an affinity label. However, Birchmeier and Christen (1974) showed that, during catalysis of transamination, aspartate aminotransaminase displays enhanced reactivity toward a range of sulfhydryl reagents not structurally related to the substrate. Several lines of evidence from their work led them to conclude that inactivation by Br-Pyr was occurring by a syncatalytic mechanism. Thus, Br-Pyr and an amino acid substrate form covalent enzyme–substrate intermediates, one of which contains a cysteinyl residue that acquires greatly enhanced reactivity relative to the nonworking enzyme. This reactive cysteinyl residue then is alkylated by Br-Pyr functioning as a nonselective sulfhydryl reagent.

1.5.4 Bromopyruvic Acid as an Affinity Label—Additional Studies and Summary

Br-Pyr has been used as an affinity label for several other enzymes. A partial list includes pyruvate; phosphate dikinase (Yoshida and Wood, 1978), inactivation of which involves modification of essential histidyl residues; and glutamate apodecarboxylate (Fonda, 1976), inactivation of which involves cysteinyl alkylation. In each of these cases, initial non-covalent binding of Br-Pyr to a cationic group at or near the active site is followed by alkylation. The pyruvate binding site of pyruvate carboxylase has been labeled with Br-Pyr (Hudson *et al.*, 1975). Br-Pyr serves as substrate for a transhydrogenation reaction with lactate catalyzed by L-lactate dehydrogenase from bakers' yeast and also functions as an affinity label in inactivating the enzyme. Pyruvate, as well as F- and Cl-Pyr, also takes part in the L-lactate dehydrogenase-catalyzed transhydrogenation reaction in a thermodynamically controlled process (Urban *et al.*, 1983).

These selected examples are illustrative of the knowledge that can be derived from the use of affinity labeling of enzyme active sites. Br-Pyr has been particularly useful, a reflection in part of the central role that pyruvate plays in metabolism. Many other examples could be given, and references to these studies can be found in the publications cited above (e.g., Yoshida and Wood, 1978; Hartman, 1977).

1.5.5 Bromopyruvic Acid as a Cross-Linking Reagent

The bifunctional nature of bromopyruvate has been exploited in a recent procedure for the chemical cross-linking of nucleic acids and proteins. Such procedures have proven effective in providing information on the spatial arrangements of nucleic acids and proteins in biological systems. Treatment of single-stranded polynucleotides with a mixture of hydrazine and bisulfite results in selective conversion of cytosine residues to the hydrazino analogue. As demonstrated with the *E. coli* 30S ribosomal subunit, Br-Pyr functions as a sulfhydryl reagent and as a reactive ketone in reactions that lead to cross-linking of the modified cytosine residue with protein sulfhydryl groups (Fig. 1-16) (Nitta *et al.*, 1984).

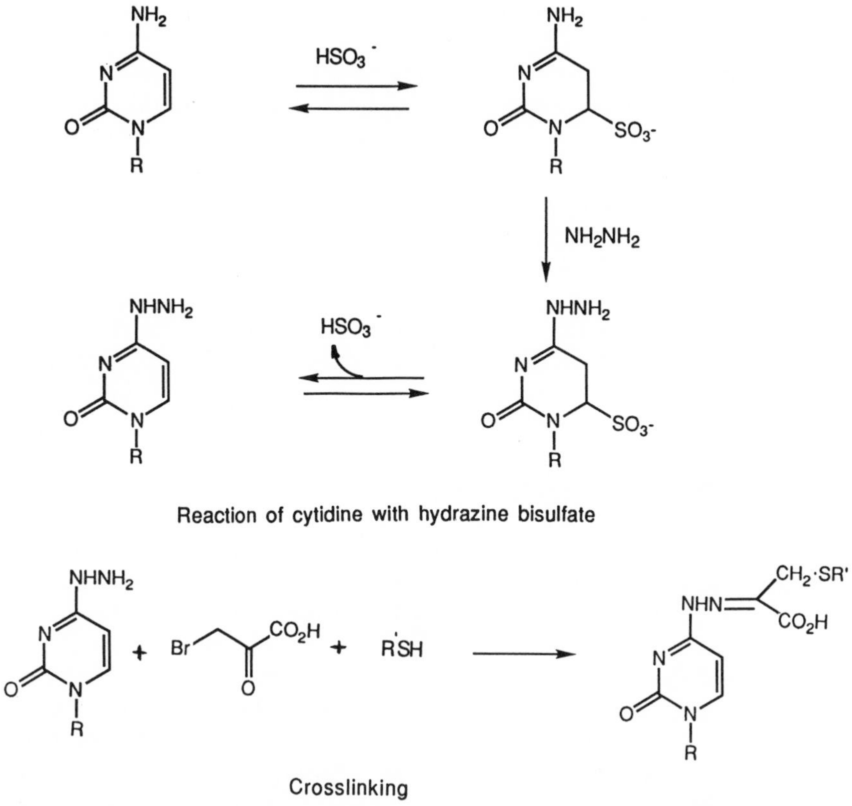

Reaction of cytidine with hydrazine bisulfate

Crosslinking

Figure 1-16. Br-Pyr functions as a reactive ketone and as a sulfhydryl reagent in cross-linking single-stranded polynucleotides (Nitta *et al.*, 1984).

1.5.6 Phosphoenol-3-bromopyruvic Acid

(*Z*)-Phosphoenol-3-bromopyruvate (Br-PEP) is one of several PEP analogues synthesized to study the chemical specificity of pyruvate kinase. Both Br-PEP and (*Z*)-F-PEP are substrates for yeast pyruvate kinase, and the reaction that produces Br-Pyr and ATP from Br-PEP and ADP is accompanied by irreversible inhibition of the enzyme (Blumberg and Stubbe, 1975). Br-PEP also is a potent competitive inhibitor ($K_i = 30 \, \mu M$) of phosphoenolpyruvate carboxylase ($K_m = 300 \, \mu M$) and is slowly carboxylated by the enzyme. The product, bromooxaloacetate, alkylates an enzyme-bound nucleophile, resulting in inactivation (O'Leary and Diaz, 1982).

1.6 ω-Fluoromonocarboxylates

A series of ω-fluorocarboxylic acids have been found in plant species that biosynthesize fluoroacetate (see Chapter 7 in Vol. 9A of this series for a proposed mechanism for the incorporation of fluorine into such plant metabolites). This suggests that fluoroacetyl-CoA can be used in the plant fatty acid synthase complex. The toxicity of carboxylic acids substituted at the terminal position with fluorine is determined by the number of carbons separating fluorine from the carboxyl group. In the series $FCH_2(CH_2)_nCO_2^-$, if n is even, the product of metabolism is toxic fluoroacetate; if n is odd, 3-fluoropyruvate is formed. The latter product is subject to both enzymatic and nonenzymatic defluorination (Goldman, 1969).

1.7 Biochemistry of Halogenated Dicarboxylic Acids

In his pioneering work on the use of fluorocarboxylic acids, including fluorocitrate, as mechanistic probes, Kun prepared and studied substrate and inhibitory properties of a series of fluorinated analogues of biochemically significant dicarboxylic acids. The pattern that emerged from this work showed that the presence of one or two fluorine atoms did not interfere with binding to the enzyme. Some of the analogues, however, were considerably poorer substrates than the natural relative and thus constituted effective reversible inhibitors of the enzymes in question. Analogues included β-fluorooxaloacetate, β,β'-difluorooxaloacetate, β-fluoroglutarate, β,β'-difluoromalate, $(-)$-*erythro*-fluoromalate, and L-$(+)$-β-fluorolactate. A concise review of these studies has been presented by Kun (1976). A particularly useful outcome of these experiments was the recognition that fluoromalate was not oxidized appreciably by mitochondrial enzymes. Thus, as discussed in Section 1.3.3, studies of fluorocitrate inhibition of the tricarboxylic acid transport system that involves stimulation of transport by fluoromalate were simplified by the metabolic stability of this analogue (Kirsten *et al.*, 1978; Kun *et al.*, 1977).

The use of malate dehydrogenase-catalyzed reduction of fluorooxaloacetate to fluoromalate to determine the stereochemical course of several enzymatic conversions has been discussed in previous sections. An additional example of this strategy is found in the study of fumarase. Whereas β-halomalates are produced by fumarase-catalyzed hydration of chloro-, bromo-, and iodofumarate, hydration of monofluorofumarate occurs by addition of hydroxide almost exclusively to the carbon bearing the fluorine substituent. Elimination of HF produces oxaloacetate

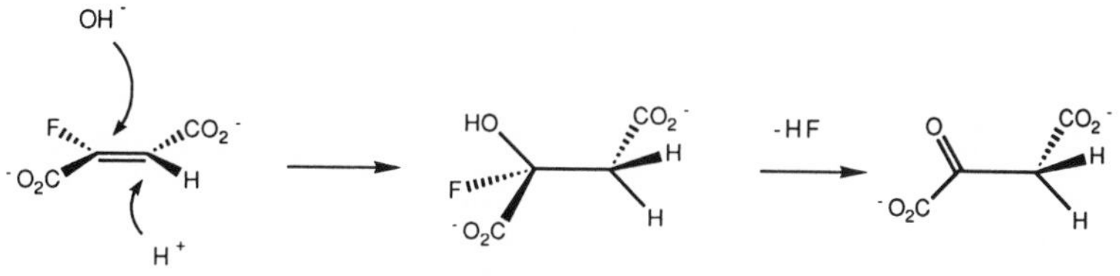

Figure 1-17. Formation of oxaloacetate following fumarase-catalyzed hydration of β-fluorofumarate [Teipel *et al.*, 1986; see Walsh (1983) for a discussion of the mechanism of HF loss].

(Fig. 1-17) (Walsh, 1983; Teipel *et al.*, 1968). A similar sequence with difluorofumarate will produce 3-fluorooxaloacetate, which will be chiral at C-3 if the substrate interacts with the enzyme in a stereospecific manner. The isolation of *threo*-fluoromalate by malate dehydrogenase trapping of the product confirms the stereospecific nature of hydration and defines the direction of addition (Fig. 1-18) [Marletta *et al.*, 1982; reviewed by Walsh (1983)].

β-Methylaspartase from *Clostridium tetanomorphum* catalyzes the reversible elimination of ammonia from methylaspartate or aspartate to give β-methylfumarate and fumarate, respectively (Fig. 1-19). In the reverse steps, stereospecific addition occurs by N-attack at the *si* face of C-2 and protonation from the *re* face of C-3. In an attempt to use this enzyme for the synthesis of halogenated analogues of aspartic acid, fluoro-, chloro-, bromo-, and iodofumaric acids were examined as substrates. Chlorofumarate

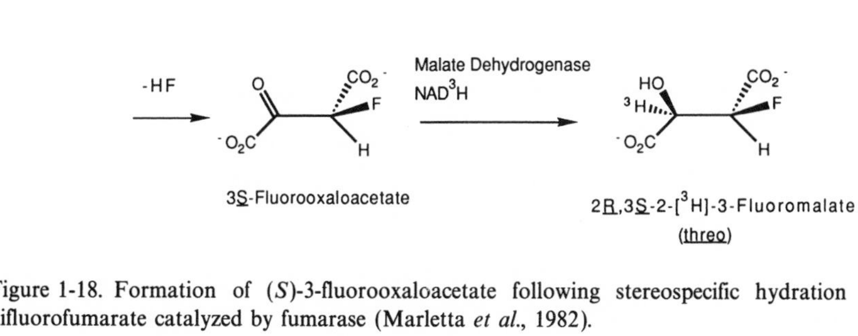

Figure 1-18. Formation of (S)-3-fluorooxaloacetate following stereospecific hydration of difluorofumarate catalyzed by fumarase (Marletta *et al.*, 1982).

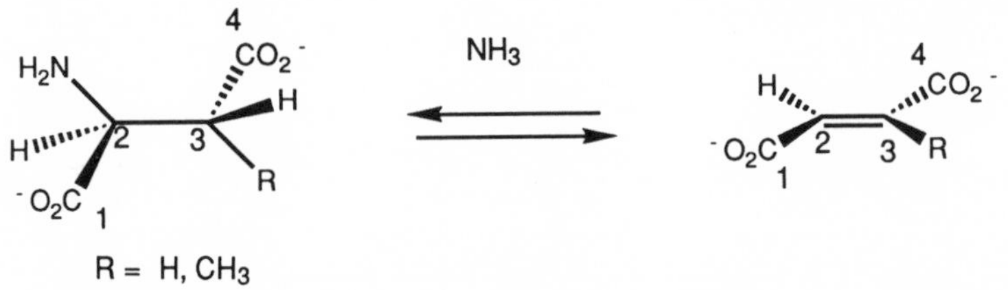

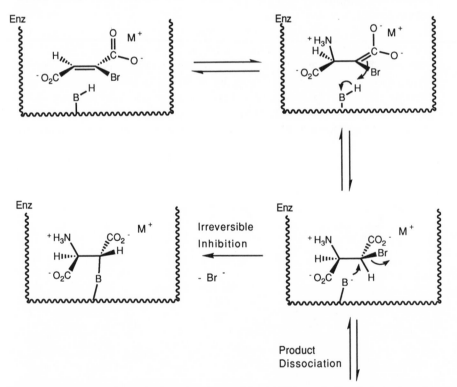

Figure 1-19. β-Methylaspartase-catalyzed reversible elimination of ammonia from β-methylaspartate or aspartate to produce methylfumarate and fumarate, respectively.

was converted smoothly into chloroaspartate, fluoroaspartate, was formed in low yield from fluorofumarate, and iodofumarate was not a substrate. Bromoaspartate, formed initially from bromofumarate, proved to be an irreversible inhibitor of the enzyme, apparently a result of alkylation of an active-site nucleophile during the inactivation process (Fig. 1-20) (Akhtar *et al.*, 1986).

Figure 1-20. Proposed mechanism for irreversible inhibition of β-methylaspartase by bromofumarate (Akhtar *et al.*, 1986).

1.8 Additional Biochemical Studies with Halogenated Tricarboxylic Acids

1.8.1 2-Fluorocitric Acid

Both *erythro* isomers of 2-fluorocitrate are substrates for ATP citrate lyase ("citrate cleavage enzyme") with V_{max} rates of 3% and 0.1% for the (+)- and (−)-isomers, respectively [Marletta *et al.*, 1981; reviewed by Walsh (1983)]. As noted in Section 1.3.2, enzymatic cleavage of (2S,3S)-*erythro*-fluorocitrate to acetyl-CoA and (3S)-fluorooxaloacetate constituted direct proof of the configuration at the 2 position of the toxic 2R,3R isomer of fluorocitrate. Cleavage of the (−)-isomer produces fluoroacetyl-CoA and oxaloacetate. ATP citrate lyase also catalyzes the hydrolysis of ATP to provide a thermodynamic driving force for the cleavage reaction. Thus, in the initial phosphorylation step of the enzymatic process, ATP is converted to ADP and a phosphoenzyme is formed. With citrate as substrate, ADP and oxaloacetate are formed in stoichiometric amounts (Fig. 1-21). However, with both fluorinated substrates, increased ATPase activity relative to cleavage is observed. The slower reaction relative to citrate has been attributed to possible diffusion of fluorocitryl-CoA from the active site prior to bond cleavage, uncoupling most of the ATPase activity from the retro-aldol reaction (Rokita *et al.*, 1982; Walsh, 1983).

The behavior of 2-fluorocitrate with aconitase, a second citrate-processing enzyme, was discussed thoroughly in Section 1.3.3.

1.8.2 3-Deoxy-3-fluorocitric Acid (3-Fluorocitrate)

Fluorine in 3-fluorocitrate serves as a replacement for the hydroxyl group. Rokita *et al.* (1982) have also examined the behavior of 3-fluorocitrate with the four citrate-processing enzymes, three of which carry out retro-aldol reactions. This research has been reviewed by Walsh (1983) and will be summarized briefly here.

1.8.2.1 Citrate Synthase

While the retro-aldol reaction is not available to 3-fluorocitrate, it was bound effectively by citrate synthase ($K_i = 550 \, \mu M$) and served as a competitive inhibitor of citrate ($K_m = 250 \, \mu M$).

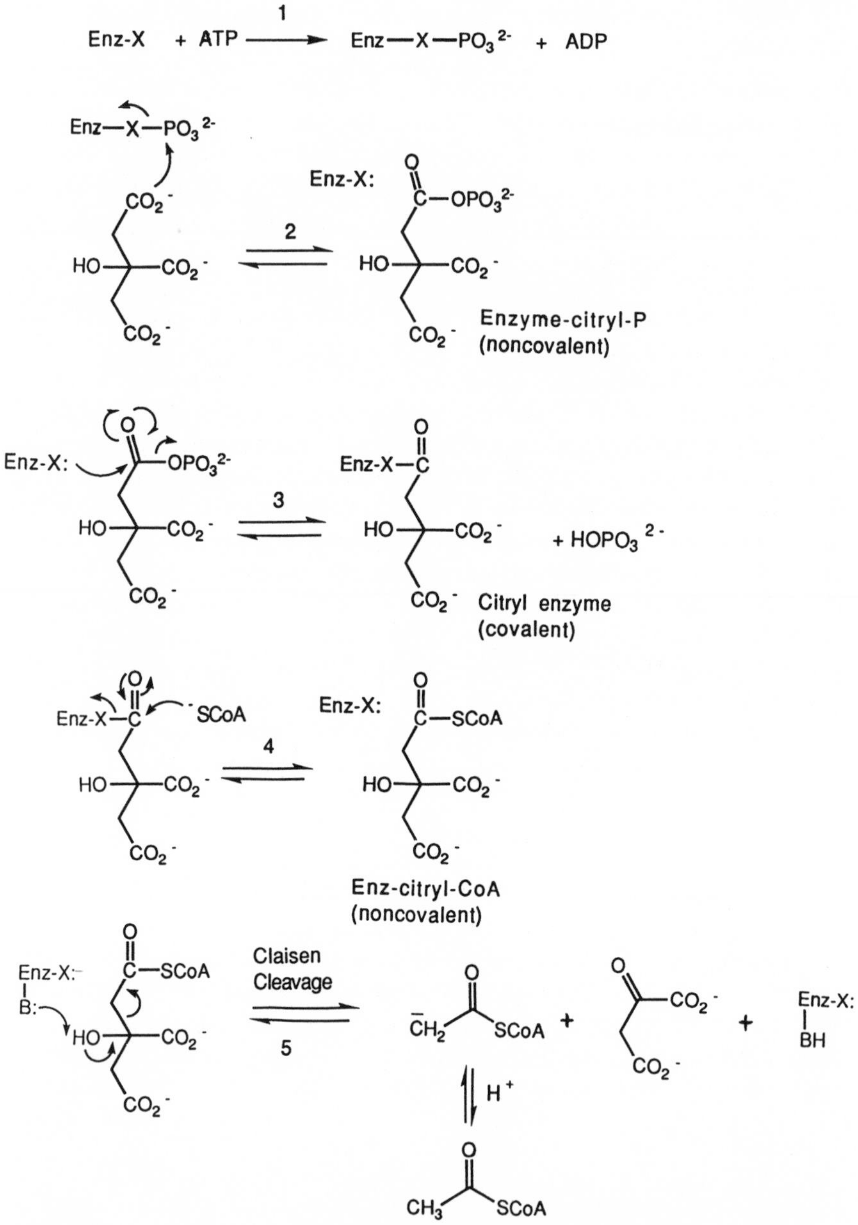

Figure 1-21. The mechanism for ATP-citrate lyase-catalyzed cleavage of citrate. Diffusion of 2-fluorocitryl-CoA (following step 4) before bond breaking uncouples ATP hydrolysis from the retro-aldol reaction (Rokita *et al.*, 1982; Walsh, 1983).

1.8.2.2 ATP-Citrate Lyase

In the presence of 3-fluorocitrate and CoA, ATP-citrate lyase exhibited ATPase activity, with a V_{max} of 8.9% of that for citrate. The requirement for CoA in order to produce sustained enzyme activity suggested that 3-fluorocitryl-CoA is formed and diffuses from the active site (see Fig. 1-21). This was confirmed by HPLC analysis (Rokita et al., 1982).

1.8.2.3 Bacterial Citrate Lyase

In the resting state, bacterial citrate lyase is in the form of an acetyl-S-enzyme. An acyl exchange reaction through an acetyl-citryl mixed anhydride produces a citryl-S-enzyme species. This is followed by a retro-aldol reaction, forming oxaloacetate and regenerating the acetyl-S-enzyme. The absence of the retro-aldol path allows 3-fluorocitrate to function as a suicide substrate for this enzyme by two mechanisms. First, the acetyl-3-fluorocitryl mixed anhydride diffuses from the enzyme leaving inactive desacetyl enzyme. Acyl exchange of the mixed anhydride to yield the non-cleavable 3-fluorocitryl enzyme also results in irreversible inactivation (Fig. 1-22) (Rokita et al., 1982).

1.8.2.4 Aconitase

In a slow process relative to citrate processing ($V_{max} = 0.1\%$), aconitase catalyzes the elimination of HF from 3-fluorocitrate (Rokita et al., 1982). The product, aconitate, is rehydrated by aconitase to citrate and isocitrate.

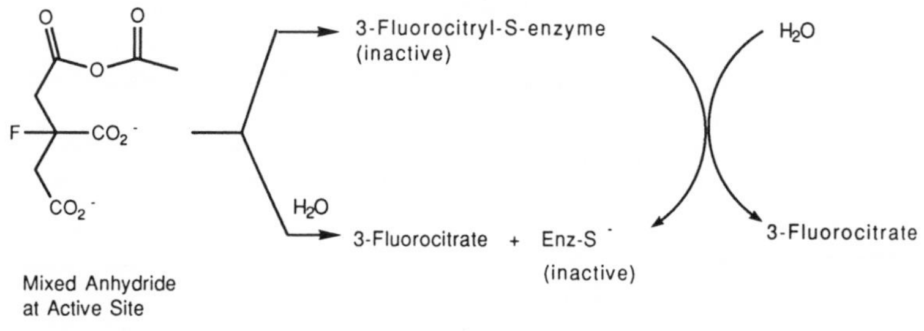

Figure 1-22. 3-Fluorocitrate functioning as a suicide substrate for bacterial citrate lyase. Formation of acetyl-3-fluorocitryl mixed anhydride followed by acyl exchange gives a dead-end 3-fluorocitryl-S-enzyme that cannot undergo the retro-aldol reaction. Diffusion of the mixed anhydride from the active site followed by hydrolysis also produces inactive (desacetyl) enzyme (Rokita et al., 1982).

1.8.3 Chlorocitric Acid

In a program designed to identify compounds capable of acting in the periphery as an appetite suppressant, $(-)$-*threo*-chlorocitrate was discovered to exhibit anorectic activity (Sullivan *et al.*, 1981). A long-lasting, fully reversible suppression of food intake in both normal and obese rats was produced, and there was no evidence of tolerance or effects on the central nervous system. This activity has been attributed to delayed gastric acid secretion (Triscari and Sullivan, 1981). High-dose chronic treatment produced dose-related collapse/recovery or lethality, which may reflect a chlorocitrate-induced hyperlactatemia (Bosakowski and Levin, 1987).

1.9 Biochemistry of Halogenated Carboxylic Acids—Summary

Topics stressed in this chapter have been related primarily to halogenated acids that are analogues of metabolic intermediates (acetate, citrate, and pyruvate, for example). An attempt has been made to provide an up-to-date review of research in each area, and, for that reason, certain historically significant developments have been given quite brief treatment (the identification of the toxic isomer of fluorocitrate, for example). The organization of topics in this volume must be arbitrary to a degree because of overlapping subjects. Thus, bromopyruvic acid has been considered in this chapter, but, as a halogenated ketone, it also is relevant to Chapter 2. Similarly, the biochemistry of halogenated analogues of products of the arachidonic acid cascade is considered separately in Chapter 4. Halogenated simple carboxylic acids (e.g., trifluoroacetic acid) are products of the metabolism of halogenated precursors having lower oxidation states such as halogenated anesthetics and hydrocarbons. This topic was covered in Chapters 8 and 9 of Vol. 9A of this series. Finally, as will be the case in the following chapters, omissions, both accidental and elected due to space limitations, are inevitable.

References

Akhtar, M., Cohen, M. A., and Gani, D., 1986. Enzymic synthesis of 3-haloaspartic acids using β-methylaspartase: Inhibition by 3-bromoaspartate, *J. Chem. Soc., Chem. Commun.* 1986:1290–1291.

Birchmeier, W., and Christen, P., 1974. The reaction of cytoplasmic aspartate amino-transferase with bromopyruvate. Syncatalytic modification simulates affinity labeling, *J. Biol. Chem.* 249:6311–6315.

Bisswanger, H., 1981. Substrate specificity of the pyruvate dehydrogenase complex from *Escherichia coli*, *J. Biol. Chem.* 256:815–822.

Blumberg, K., and Stubbe, J., 1975. Chemical specificity of pyruvate kinase from yeast, *Biochim. Biophys. Acta* 384:120–126.

Bosakowski, T., and Levin, A. A., 1986. Serum citrate as a peripheral indicator of fluoroacetate and fluorocitrate toxicity in rats and dogs, *Toxicol. Appl. Pharmacol.* 85:428–436.

Bosakowski, T., and Levin, A. A., 1987. Comparative acute toxicity of chlorocitrate and fluorocitrate in dogs, *Toxicol. Appl. Pharmacol.* 89:97–104.

Carrell, H. L., Glusker, J. P., Villafranca, J. J., Mildvan, A. S., Dummel, R. J., and Kun, E., 1970. Fluorocitrate inhibition of aconitase: Relative configuration of the inhibitory isomer by X-ray crystallography, *Science* 170:1412–1414.

Conway, A., and Koshland, D. E., Jr., 1968. Negative cooperativity in enzyme action. The binding of diphosphopyridine nucleotide to glyceraldehyde 3-phosphate dehydrogenase, *Biochemistry* 7:4011–4023.

Cornforth, J. W., Redmond, J. W., Eggerer, H., Buckel, W., and Gutschow, C., 1969. Asymmetric methyl groups, and the mechanism of malate synthase, *Nature* 221:1212–1213.

Crabb, D. W., and Harris, R. A., 1979. Mechanism responsible for the hypoglycemic actions of dichloroacetate and 2-chloropropionate, *Arch. Biochem. Biophys.* 198:145–152.

Crabb, D. W., Yount, E. A., and Harris, R. A., 1981. The metabolic effects of dichloroacetate, *Metabolism* 30:1024–1039.

Crestfield, A. M., Moore, S., and Stein, W. H., 1963. The preparation and enzymatic hydrolysis of reduced and S-carboxymethylated proteins, *J. Biol. Chem.* 238:622–627.

Demaugre, F., Buc, H. A., Cepanec, C., Moncion, A., and Leroux, J.-P., 1983. Comparison of the effects of 2-chloropropionate and dichloroacetate on ketogenesis and lipogenesis in isolated rat hepatocytes, *Biochem. Pharmacol.* 32:1881–1885.

Duffy, T. H., and Nowak, T., 1984. Stereoselectivity of interaction of phosphoenolpyruvate analogues with various phosphoenolpyruvate-utilizing enzymes, *Biochemistry* 23:661–670.

Eanes, R. Z., Skilleter, D. N., and Kun, E., 1972. Inactivation of the tricarboxylate carrier of liver mitochondria by (−)-erythrofluorocitrate, *Biochem. Biophys. Res. Commun.* 46:1618–1622.

Fanshier, D. W., Gottwald, L. K., and Kun, E., 1962. Enzymatic synthesis of monofluorocitrate from β-fluoro-oxaloacetate, *J. Biol. Chem.* 237:3588–3596.

Filler, R., and Naqvi, S. M., 1982. Fluorine in biomedicinal chemistry. An overview of recent advances and selected topics, in *Biomedicinal Aspects of Fluorine Chemistry* (R. Filler and Y. Kobayashi, eds.), Kodansha Ltd., Tokyo; Elsevier Biomedical Press, Amsterdam, pp. 1–32.

Flournoy, D. S., and Frey, P. A., 1986. Pyruvate dehydrogenase and 3-fluoropyruvate: Chemical competence of 2-acetylthiamin pyrophosphate as an acetyl group donor to diyhydrolipoamide, *Biochemistry* 25:6036–6043.

Fonda, M. L., 1976. Bromopyruvate inactivation of glutamate apodecarboxylase. Kinetics and specificity, *J. Biol. Chem.* 251:229–235.

Foxall, D. L., Brindle, K. M., Campbell, R. D., and Simpson, I. J., 1984. The inhibition of erythrocyte glyceraldehyde-3-phosphate dehydrogenase. *In situ* PMR studies, *Biochim. Biophys. Acta.* 804:209–215.

Goldman, P., 1969. The carbon–fluorine bond in compounds of biological interest, *Science* 164:1123–1133.

Goldstein, J. A., Cheung, Y.-F., Marletta, M. A., and Walsh, C., 1978. Fluorinated substrate analogues as stereochemical probes of enzymatic reaction mechanisms, *Biochemistry* 17:5567–5575.

Han, A. C., Goodwin, G. W., Paxton, R., and Harris, R. A., 1987. Activation of branched-chain α-ketoacid dehydrogenase in isolated hepatocytes by branched-chain α-ketoacids, *Arch. Biochem. Biophys.* 258:85–94.

Harris, J. I., and Perham, R. N., 1968. Glyceraldehyde 3-phosphate dehydrogenase from pig muscle, *Nature* 219:1025–1028.

Harris, R. A., Paxton, R., and DePaoli-Roach, A. A., 1982. Inhibition of branched chain α-ketoacid dehydrogenase kinase activity by α-chloroisocaproate, *J. Biol. Chem.* 257:13915–13918.

Harris, R. A., Paxton, R., Powell, S. M., Goodwin, G. W., Kuntz, M. J., and Han, A. C., 1986. Regulation of branched-chain α-ketoacid dehydrogenase complex by covalent modification, *Adv. Enzyme Regul.* 25:219–237.

Hartman, F. C., 1977. Haloketones as affinity labeling reagents, *Methods Enzymol.* 46:130–153.

Heinrikson, R. I., Stein, W. H., Crestfield, A. M., and Moore, S., 1965. The reactivities of the histidine residues at the active site of ribonuclease toward halo acids of different structures, *J. Biol. Chem.* 240:2921–2934.

Hoving, H., Nowak, T., and Robillard, G. T., 1983. *Escherichia coli* phosphoenolpyruvate-dependent phospho-transcarboxylase system: Stereospecificity of proton transfer in the phosphorylation of enzyme I from (Z)-phosphoenolbutyrate, *Biochemistry* 22:2832–2838.

Hoving, H., Crysell, B., and Leadlay, P. F., 1985. Fluorine NMR studies on stereochemical aspects of reactions catalyzed by transcarboxylase, pyruvate kinase, and enzyme I, *Biochemistry* 24:6163–6169.

Hudson, P. J., Keech, D. B., and Wallace, J. C., 1975. Pyruvate carboxylase: Affinity labelling of the pyruvate binding site, *Biochem. Biophys. Res. Commun.* 65:213–219.

Hwang, S. H., and Nowak, T., 1986. Stereochemistry of phosphoenolpyruvate carboxylation catalyzed by phosphoenolpyruvate carboxykinase, *Biochemistry* 25:5590–5595.

Keck, R., Haas, H., and Retey, J., 1980. Synthesis of stereospecifically deuterated fluoroacetic acid and its behavior in enzymic aldol-type condensations, *FEBS Lett.* 114:287–290.

Kennedy, M. C., Spoto, G., Emptage, M. H., and Beinert, H., 1988. The active site sulfhydryl of aconitase is not required for catalytic activity, *J. Biol. Chem.* 263:8190–8193.

Kirsten, E., Sharma, M. L., and Kun, E., 1978. Molecular toxicology of (−)-*erythro*-fluorocitrate: Selective inhibition of citrate transport in mitochondria and the binding of fluorocitrate to mitochondrial proteins, *Mol. Pharmacol.* 14:172–184.

Kun, E., 1976. Fluorocarboxylic acids as enzymatic and metabolic probes, in *Biochemistry Involving Carbon–Fluorine Bonds* (R. Filler, ed.), ACS Symposium Series, No. 28, American Chemical Society, Washington, D.C., pp. 1–22.

Kun, E., Kirsten, E., and Sharma, M. L., 1977. Enzymatic formation of glutathione-citryl thioester by a mitochondrial system and its inhibition by (−)-*erythro*-fluorocitrate, *Proc. Natl. Acad. Sci. USA* 74:4942–4946.

Kuo, D. J., and Rose, I. A., 1982. Utilization of enolpyruvate by the carboxybiotin form of transcarboxylase: Evidence for a nonconcerted mechanism, *J. Am. Chem. Soc.* 104:3235–3236.

Leung, S. L., and Frey, P. A., 1978. Fluoropyruvate: An unusual substrate for *Escherichia coli* pyruvate dehydrogenase, *Biochem. Biophys. Res. Commun.* 81:274–279.

Lew, V. L., and Ferreira, H. G., 1978. Calcium transport and the properties of a calcium-activated potassium channel in the red cell membrane, *Curr. Top. Membr. Transport* 10:217–277.

Lowe, P. N., and Perham, R. N., 1984. Bromopyruvate as an active-site-directed inhibitor of the pyruvate dehydrogenase multienzyme complex from *Escherichia coli*, *Biochemistry* 23:91–97.

Luthy, J., Retey, J., and Arigoni, D., 1969. Preparation and detection of chiral methyl groups, *Nature* 221:1213–1215.

MacQuarrie, R. A., and Bernhard, S. A., 1971. Mechanism of alkylation of rabbit muscle glyceraldehyde 3-phosphate dehydrogenase, *Biochemistry* 10:2456–2466.

Maldonado, M. E., Oh, K.-J., and Frey, P. A., 1972. Studies on *Escherichia coli* pyruvate dehydrogenase complex. I. Effect of bromopyruvate on the catalytic activities of the complex, *J. Biol. Chem.* 247:2711–2716.

Mariash, C. N., and Schwartz, H. L., 1986. Effect of dichloroacetate on rat hepatic messenger RNA activity profiles, *Metabolism* 35:452–456.

Marletta, M. A., Srere, P. A., and Walsh, C., 1981. Stereochemical outcome of processing of fluorinated substances by ATP citrate lyase and malate synthase, *Biochemistry* 20:3719–3723.

Marletta, M. A., Cheung, Y.-F., and Walsh, C., 1982. Stereochemical studies on the hydration of monofluorofumarate and 2,3-difluorofumarate by fumarase, *Biochemistry* 21:2637–2644.

Meloche, H. P., 1967. Bromopyruvate inactivation of 2-keto-3-deoxy-6-phosphogluconic aldolase. I. Kinetic evidence for active site specificity, *Biochemistry* 6:2273–2280.

Meloche, H. P., 1970. Reaction of the substrate analog bromopyruvate with two active-site conformers of 2-keto-3-deoxy-6 phosphogluconic aldolase, *Biochemistry* 9:5050–5055.

Meloche, H. P., and Glusker, J. P., 1973. Aldolase catalysis: Single base-mediated proton activation, *Science* 181:350–352.

Monti, C. T., Waterbor, J. W., and Meloche, H. P., 1979. Interaction of the chiral pyruvate analog, 2-keto-3-bromobutyrate with pyruvate lyases. 2-Keto-3-deoxygluconate-6-phosphate aldolase of *Pseudomonas putida*, *J. Biol. Chem.* 254:5862–5865.

Murray-Rust, P., Stallings, W. C., Monti, C. T., Preston, R. K., and Glusker, J. P., 1983. Intermolecular interactions of the C−F bond: The crystallographic environment of fluorinated carboxylic acids and related structures, *J. Am. Chem. Soc.* 105:3206–3214.

Nitta, N., Kuga, O., Yui, S., Tsugawa, A., Negishi, K., and Hayatsu, H., 1984. A new reaction useful for chemical cross-linking between nucleic acids and proteins, *FEBS Lett.* 166:194–198.

Okamoto, M., and Morino, Y., 1973. Affinity labeling of aspartate aminotransferase isozymes by bromopyruvate, *J. Biol. Chem.* 248:82–90.

O'Leary, M. H., and Diaz, E., 1982. Phosphoenol-3-bromopyruvate. A mechanism-based inhibitor of phosphoenolpyruvate carboxylase from maize, *J. Biol. Chem.* 257:14603–14605.

Patel, S. S., and Walt, D. R., 1987. Substrate specificity of acetyl coenzyme A synthase, *J. Biol. Chem.* 262:7132–7134.

Paulsen, R. E., Contestabile, A., Villani, L., and Fonnum, F., 1987. An *in vivo* model for studying function of brain tissue temporarily devoid of glial cell metabolism: The use of fluorocitrate, *J. Neurochem.* 48:1377–1385.

Paxton, R., and Harris, R. A., 1984. Clofibric acid, phenylpyruvate, and dichloroacetate inhibition of branched-chain α-ketoacid dehydrogenase *in vitro* and in perfused rat heart, *Arch. Biochem. Biophys.* 231:58–66.

Paxton, R., Harris, R. A., Sener, A., and Malaisse, W. J., 1988. Branched chain α-ketoacid dehydrogenase and pyruvate dehydrogenase activity in isolated rat pancreatic islets, *Horm. Metabol. Res.* 20:317–322.

Peters, R., 1972. Some metabolic aspects of fluoroacetate especially related to fluorocitrate, in *Ciba Foundation Symposium: Carbon–Fluorine Compounds: Chemistry, Biochemistry, and Biological Activities*, Elsevier, Amsterdam, 1972, pp. 55–76.

Plishker, G. A., 1985. Iodoacetic acid inhibition of calcium-dependent potassium efflux in red blood cells, *Am. J. Physiol.* 248:C419–C424.

Polgár, L., 1975. Ion-pair formation as a source of enhanced reactivity of the essential thiol group of D-glyceraldehyde-3-phosphate dehydrogenase, *Eur. J. Biochem.* 51:63–71.

Prestwich, G. D., Yamaoka, R., Phirwa, S., and DePalma, A., 1984. Isolation of 2-fluorocitrate produced by *in vivo* dealkylation of 20-fluorostigmasterol in an insect, *J. Biol. Chem.* 259:11022–11026.

Rokita, S. E., Srere, P. A., and Walsh, C. T., 1982. 3-Fluoro-3-deoxycitrate: A probe for mechanistic study of citrate-utilizing enzymes, *Biochemistry* 21:3765–3774.

Shapiro, R., Strydom, D. J., Weremowicz, S., and Vallee, B. L., 1988. Sites of modification of human angiogenin by bromoacetate at pH 5.5, *Biochem. Biophys. Res. Commun.* 156:530–536.

Skilleter, D. N., Dummel, R. J., and Kun, E., 1972. Specific enzyme inhibitors. XIV. Effects of enzymically synthesized (−)-*erythro*-fluoromalic acid on malate dehydrogenase and on anion carriers of liver mitochondria, *Mol. Pharmacol.* 8:139–148.

Stacpoole, P. W., Lorenz, A. C., Thomas, R. G., and Harman, E. M., 1988. Dichloroacetate in the treatment of lactic acidosis, *Ann. Internal Med.* 108:58–63.

Stryer, L., 1988. *Biochemistry*, 3rd ed., W. H. Freeman and Company, New York, pp. 379–382.

Stubbe, J., and Abeles, R. H., 1977. Biotin carboxylations—concerted or nonconcerted? That is the question! *J. Biol. Chem.* 252:8338–8340.

Stubbe, J., Fish, S., and Abeles, R. H., 1980. Are carboxylations involving biotin concerted or nonconcerted? *J. Biol. Chem.* 255:236–242.

Sullivan, A. C., Dairman, W., and Triscari, J., 1981. (−)-*threo*-Chlorocitric acid: A novel anorectic agent, *Pharmacol. Biochem. Behav.* 15:303–310.

Szerb, J. C., and Issekutz, B., 1987. Increase in the stimulation-induced overflow of glutamate by fluoroacetate, a selective inhibitor of the glial tricarboxylic acid cycle, *Brain Res.* 410:116–120.

Teipel, J. W., Hass, G. M., and Hill, R. L., 1968. The substrate specificity of fumarase, *J. Biol. Chem.* 243:5684–5694.

Toshima, K., Kuroda, Y., Yokota, I., Naito, E., Ito, M., Watanabe, T., Takeda, E., and Miyao, M., 1985. Activation of branched-chain α-ketoacid dehydrogenase by α-chloroisocaproate in normal and enzyme deficient fibroblast, *Clin. Chim. Acta* 147:103–108.

Triscari, J., and Sullivan, A. C., 1981. Studies on the mechanism of a novel anorectic agent, (−)-*threo*-chlorocitric acid, *Pharmacol. Biochem. Behav.* 15:311–318.

Urban, P., Alliel, P. M., and Lederer, F., 1983. On the transhydrogenase activity of baker's yeast flavocytochrome b_2, *Eur. J. Biochem.* 134:275–281.

Vary, T. C., Siegel, J. H., Tall, B. D., and Morris, J. G., 1988. Metabolic effects of partial reversal of pyruvate dehydrogenase activity by dichloroacetate in sepsis, *Circ. Shock* 24:3–18.

Walsh, C., 1983. Fluorinated substrate analogues: Routes of metabolism and selective toxicity, *Adv. Enzymol.* 55:197–289.

Wang, Z.-X., Preiss, B., and Tsou, C.-L., 1988. Kinetics of inactivation of creatine kinase during modification of its thiol groups, *Biochemistry* 27:5095–5100.

Webb, J. L., 1966. *Enzyme and Metabolic Inhibitors*, Vol. III, Academic Press, New York, pp. 1–283.

Welch, J. T., 1987. Advances in the preparation of biologically active organofluorine compounds, *Tetrahedron* 43:3123–3197.

Whitehouse, S., Cooper, R. H., and Randle, P. J., 1974. Mechanism of activation of pyruvate dehydrogenase by dichloroacetate and other halogenated carboxylic acids, *Biochem. J.* 141:761–774.

Wilchek, M., and Givol, D., 1977. Haloacetyl derivatives, *Methods Enzymol.* 46:153–157.

Wold, F., 1977. Affinity labeling—an overview, *Methods Enzymol.* 46:3–14.

Yoshida, H., and Wood, H. G., 1978. Crystalline pyruvate, phosphate dikinase from *Bacteroides symbiosus.* Modification of essential histidyl residues by bromopyruvate inactivation, *J. Biol. Chem.* 253:7650–7655.

Zimmerle, C. T., Tung, P. P., and Alter, G. M., 1987. Ligand-induced symmetry between active sites of cytoplasmic malate dehydrogenase: A chemical modification study, *Biochemistry* 26:8535–8541.

Biochemistry of Halogenated Aldehydes and Ketones

2

2.1 Introduction

The susceptibility of a halogen situated at the α position of an aldehyde or ketone to nucleophilic displacement plays a central role in much of the biochemical behavior of these compounds. Examples to be reviewed in this chapter include the reaction of α-chloro- and α-bromo-aldehydes with nucleic acid components and the use of α-halocarbonyl compounds as affinity labels. The presence of halogen also affects the chemistry of the neighboring carbonyl group, making the carbonyl carbon more electrophilic, an effect especially pronounced with α-fluorine substitution. This has been used to advantage in the design of several transition state analogue enzyme inhibitors, based on the ready formation of tetrahedral intermediates from such halogen-substituted analogues.

2.2 Formation of Etheno-Substituted Nucleotides from Halogenated Analogues of Aliphatic Aldehydes and Ketones

2.2.1 Introduction

In 1971, the facile reaction of chloroacetaldehyde (CAA) and bromoacetaldehyde (BAA) with adenine and cytosine moieties was reported (Kochetkov *et al.*, 1971). The recognition that products of this reaction, exemplified by $1,N^6$-ethenoadenosine (ε-adenosine) and $3,N^4$-ethenocytidine (ε-cytidine) (Fig. 2-1), formed from adenosine and cytidine, respectively, are highly fluorescent has had far-reaching consequences (Barrio *et al.*, 1972a). For example, in aqueous solution at pH 7.0, ε-adenosine fluoresces at 415 mm with a quantum yield of 0.56 (Secrist *et al.*, 1972a). The parent bases play critical roles in such processes as enzyme–coenzyme interactions (FAD), nucleic acid–protein interactions,

ε-Adenosine   ε-Cytidine

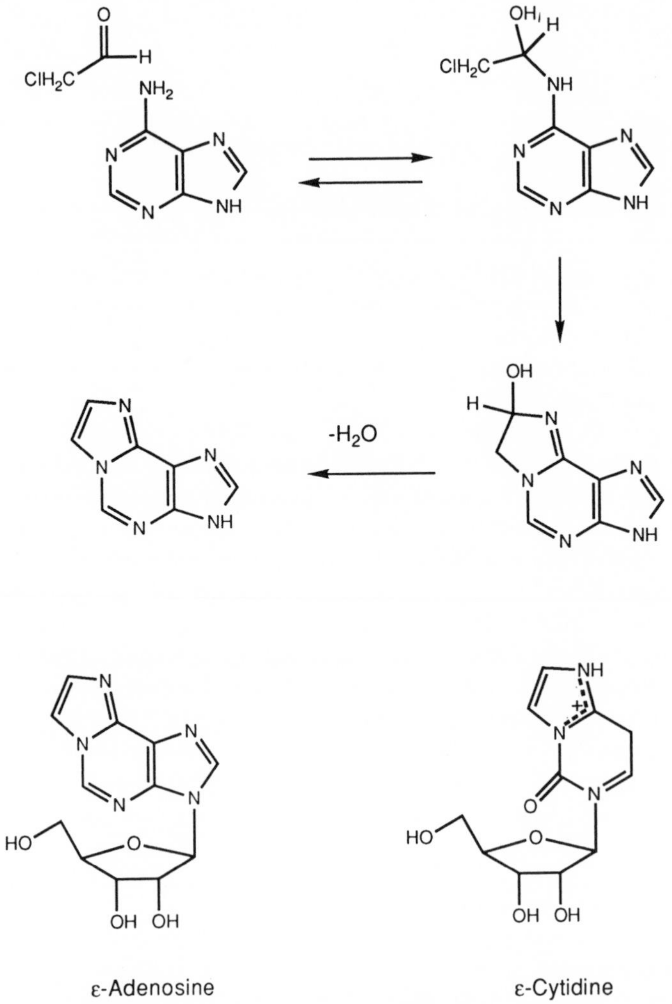

Figure 2-1. Mechanism for the formation of etheno adducts between chloroacetaldehyde and base components of nucleic acids. Structures of ε-adenosine and ε-cytidine (hydrochloride).

and energy production (ATP) and transduction (cAMP) and as components of polynucleotides. The fact that the fluorescent etheno adducts retain biological activity in many of these processes has rendered them very useful in a host of biochemical investigations. A survey of the biochemistry of these etheno adducts is beyond the scope of this book—Leonard (1984) has provided a particularly thorough review of research in this area. In this section, examples will be presented of formation of etheno adducts of the several classes of biomolecules containing purines and pyrimidines. The relevance of etheno adducts—formed from products of metabolism of such xenobiotics as vinyl chloride—to carcinogenicity and mutagenicity was discussed in Chapter 9 of Vol. 9A in this series.

2.2.2 Mechanism of Etheno-Adduct Formation

The reaction of CAA and BAA with adenine and cytosine moieties occurs readily in weakly acidic aqueous solution, whereas guanine, which also possesses an exocyclic amine, is much less reactive. Initial reversible formation of a carbinolamine by reaction of the carbonyl group with the exocyclic nitrogen is followed by cyclization to an isolable cyclic carbinolamine and dehydration to the etheno adduct. Reaction of higher homologues of CAA and BAA (α-bromovaleraldehyde and α-bromo- and α-chloropropionaldehyde, for example) leads to 7-substituted products in the ε-adenine series and to 3-substituted products in the ε-cytosine series. α-Haloketones such as phenacyl bromide and ring-substituted phenacyl bromides, bromoacetone, and chloroacetone form 8-substituted products in the ε-adenine series and 2-substituted products in the ε-cytosine series. Formation of these substituted etheno adducts is consistent with reaction of the carbonyl group, rather than the α-halo substituent, with the more nucleophilic exocyclic nitrogen (Fig. 2-1) (Leonard, 1984).

2.2.3 Formation of Etheno Adducts of Nucleotides and Dinucleotides

Early demonstration of the potential use of ε-adenine-containing molecules came from studies of simple adenine nucleotides. 3'- and 5'-ε-AMP, ε-ADP, ε-ATP, and 3',5'-ε-cAMP are prepared easily from the parent nucleotide by reaction with CAA. In a series of enzyme studies, it was shown that these highly fluorescent ε-adenine nucleotides functioned effectively in various roles as phosphoryl, pyrophosphoryl, and adenyl donors and as allosteric effectors (Secrist et al., 1972a; Leonard, 1984). In addition, cyclic-3',5'-ε-AMP was shown to be a substrate for phosphodiesterase and to act similarly to cyclic-3',5'-AMP in activating protein kinase, but with 10-fold lower potency (Secrist et al., 1972b). In a similar study, the 8-phenyletheno analogue of cyclic-3',5'-AMP, prepared by using phenacyl bromide as the derivatizing agent, was found to be significantly better than the parent as a substrate for protein kinase, a result attributed to increased lipophilicity of the analogue (Meyer et al., 1973). An etheno analogue of nicotinamide adenine dinucleotide (NAD^+), prepared by reaction of NAD$^+$ with chloroacetaldehyde, showed activity as a substitute for NAD$^+$ in dehydrogenase-catalyzed reactions (Barrio et al., 1972b).

These and other early results which demonstrated the ability of ε-nucleotides to interact with biological systems opened the way for the use of these analogues in a host of studies. The review by Leonard (1984) is recommended.

2.2.4 Reaction of Haloacetaldehydes with Polynucleotides

2.2.4.1 Reaction of Haloacetaldehydes with Oligo- and Polynucleotides

Oligonucleotides containing ε-adenine and ε-cytosine have been made by stepwise enzymatic incorporation of the modified nucleotides. An example of direct derivatization of oligonucleotides is seen in the reaction of poly(riboadenylic acid) [poly(rA)] with CAA. Modified ε-poly(rA), produced by this reaction, is an effective inhibitor of avian myeloblastosis virus DNA polymerase. Potential use of this polymer as an inhibitor of oncogenic viral polymerases was suggested (Chirikjian and Papas, 1974). Kuśmierek and Singer (1982) studied the effect of CAA modification of poly(ribocytidylic acid) [poly(rC)] and poly(deoxycytidylic acid) [poly(dC)] on transcription fidelity and found that the presence of ε-cytosine as well as the cycle carbinolamine precursor [ε-cytosine(H$_2$O)] led to misincorporations.

2.2.4.2 Formation of Etheno Analogues of RNA

A major factor in the use of CAA and BAA in the study of nucleic acid structure and function is the observation that reaction is limited to accessible bases. Thus, adenine and cytosine residues that are involved in normal B-DNA helical base pairing, or otherwise in tertiary structure stabilization through base pairing or stacking, are unreactive. For example, this selectivity of reaction has been used to locate six accessible cytosines and five accessible adenines in formylmethionine transfer RNA (tRNA^{fMet}) from E. coli (Schulman and Pelka, 1976). Loss of modified cytosine residues during chromatography of oligonucleotides produced by ribonuclease digestion of modified tRNA^{fMet} was attributed to degradation of ε-cytosine. However, in a similar study with tRNA^{Phe}, Krzyzosiak et al. (1981) demonstrated that this apparent complication results from unexpected stability of the intermediate cyclic carbinolamine formed from reaction of CAA and cytosine residues. A maturation step (heating in water) to complete formation of ε-adducts avoided this problem and increased the value of CAA as a probe for accessible RNA bases. In a modification, tRNA^{Val} from *Drosophila melanogaster* was denatured prior to treatment with CAA. The resulting loss of secondary structure in the ε-adduct-containing RNA led to more efficient ribonuclease cleavage and facilitated sequence determination (Addison *et al.*, 1982).

2.2.4.3 Reaction of Haloacetaldehydes with DNA

The expectation that participation of adenine in Watson–Crick base pairing in DNA would render these residues unreactive toward CAA was confirmed by Kimura *et al.* (1977). The selectivity of chloro- and bromoacetaldehyde for reaction with unpaired DNA adenine and cytosine residues has been used very effectively to probe DNA structure and function. Applications have included determinations of the accessibility of adenine and cytosine in various structural forms of DNA and in single-stranded DNA complexed with protein and investigations of the potential role of exposed bases as regulatory sites of DNA and as targets for carcinogens.

Kohwi-Shigematsu *et al.* (1978) found that DNA complexed with DNA helix-destabilizing protein—a protein having high affinity for single-stranded DNA—reacts readily with CAA. These and other data indicated that DNA bases are left uncovered in this DNA–protein complex.

In addition to B-DNA (the normal right-handed Watson–Crick double helix), DNA has been shown recently to exist in a variety of unusual structures, both *in vitro* and *in vivo*. These include left-handed DNA (Z-DNA), which exists in a double-helical conformation containing antiparallel chains held together by Watson–Crick base pairing. The requirement that nucleosides in Z-DNA must alternate between *syn* and *anti* configurations destabilizes this conformation relative to all *anti* B-DNA. Since the *syn* configuration is more favorable for pyrimidine than for purine bases, Z-DNA is favored by sequences having alternating pyrimidine and purine bases, with alternating d(CG) being most favorable [reviewed by Rich *et al.* (1984)]. Additional unusual secondary structures of DNA include such conformations as supercoiled structures, cruciform structures, and bent and slipped structures [reviewed by Wells (1988)].

Investigation of the relevance of these structures to *in vivo* DNA function has been an area that has attracted recent intense interest. Indeed, the location of sequences that can adopt alternate secondary structures in biologically critical regions of natural DNA has suggested that the transformation to and from such structures may constitute a biological signal [for a discussion of these points, see McLean and Wells (1988) and Wells (1988)]. Many of these investigations have included the use of BAA or CAA to probe for segments of these structures that are available for interactions with other molecules. For example, a segment containing accessible adenine residues, as shown by ε-adduct formation, may also be expected to be available for interaction with a regulatory protein. In one example of this approach, Kohwi-Shigematsu et al. (1983) found that transcriptionally active adult chicken β^A globin gene contains effectively unpaired DNA bases as determined by their ability to react with BAA. Reaction occurred at specific sites within the 5' flanking sequence, thought to be a regulatory region. These "unpaired" bases were not detected if the gene was in a transcriptionally inactive state. Similarly, Bode et al. (1986) have investigated induction-dependent conformational changes in human interferon β genes in mouse host cells in vivo. Reaction of BAA at specific sites within the regulatory region of the gene was one of several induction-dependent changes of chromatin structure detected.

Many other studies on the relationship between DNA secondary structure and DNA function as well as the factors which influence conformational changes have been reported recently. BAA and CAA are two of several reagents often used to map sites of reactivity. Diethyl oxalate and nucleases also are used extensively in these studies. Examples include the research of Wells and co-workers (Collier et al., 1988, and references therein), Kohwi-Shigematsu et al. (1988), Vogt et al. (1988), and Furlong et al. (1989). It is evident that the facile reaction of adenine and cytosine residues with BAA and CAA has had and will continue to have many applications in DNA research.

2.3 Halogenated Aldehydes and Ketones as Enzyme Inhibitors

2.3.1 Introduction

Formation of etheno adducts by reaction of BAA and CAA with adenine and cytosine is one of many examples of the biochemical exploitation of the reactivity of bifunctional α-halocarbonyl compounds. Because of its relevance to carboxylic acid metabolism, the extensive use of bromopyruvic acid as an affinity label was discussed in Chapter 1. Other examples of affinity labeling and enzyme inhibition based on halogenated aldehydes and ketones will be examined in this section. A major development in this area is the successful design of several transition state analogue inhibitors, some with therapeutic potential.

2.3.2 Irreversible Enzyme Inhibition with Halomethyl Ketones

The proteases, enzymes that catalyze the hydrolytic cleavage of peptide bonds, are classified according to the functional groups responsible for carrying out substrate bond scission. Thus, serine, thiol, and acid proteases have reactive serine hydroxyl, cysteine thiol, and aspartate or glutamate carboxyl groups, respectively. As part of a program to identify specific amino acid residues at the active center of enzymes, Schoellman and Shaw (1963) synthesized the chymotrypsin substrate analogue 1-N-tosylamido-2-phenyl chloromethyl ketone [Tos-PheCH$_2$Cl (TPCK); Fig. 2-2]. Stoichiometric inactivation of chymotrypsin, a serine protease, by TPCK, accompanied by loss of one histidine residue, provided the first direct evidence for the presence of histidine at the active site. Tos-LysCH$_2$Cl similarly was shown to be a specific reagent for lysine (Shaw, 1970, and references therein). These developments were important in emphasizing the potential of using the selectivity of enzyme binding sites in the design of reagents for selective modification of enzymes. Shaw (1970) provides an account of the background and early developments in this important area of enzyme biochemistry. Powers (1977) has listed several halomethyl ketone inhibitors of proteolytic enzymes along with the enzymes studied with each. The labeling of proteins by this method has been used extensively for such purposes as characterization of active-site functional groups and introduction of spectroscopic and radioactive labels. It has been particularly effective in the study of protease function. In this approach, the structural variability available through alteration in amino acid residues in the reagent permits optimization of reactivity with the enzyme active site.

Figure 2-3 depicts the generally accepted mechanism of chymotrypsin-catalyzed peptide bond cleavage. Inhibition of chymotrypsin by halomethyl ketones has been shown to involve formation of a reversible enzyme-inhibitor complex, addition of Ser-195 to the carbonyl group to form a hemiketal, and irreversible alkylation of His-57 [reviewed by Powers

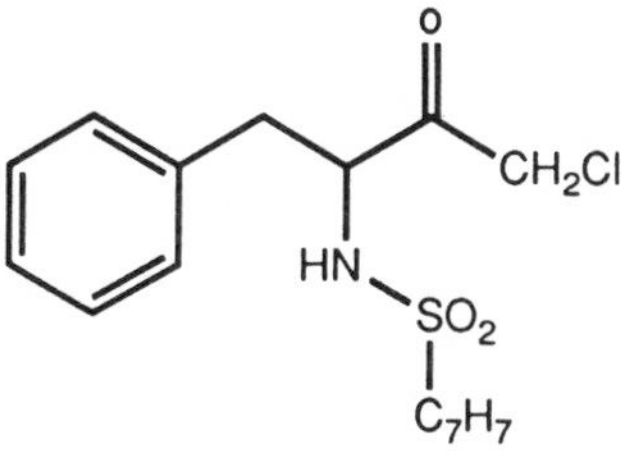

Figure 2-2. Tos-PheCH$_2$Cl.

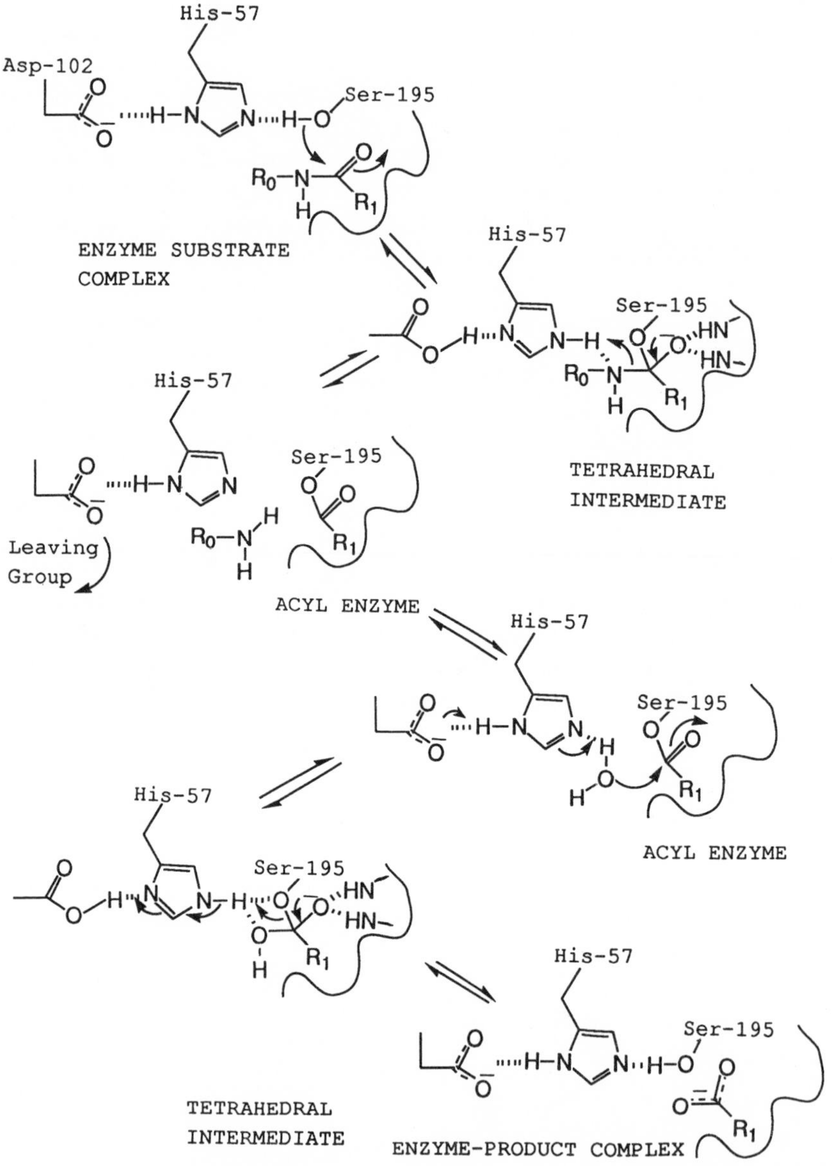

Figure 2-3. General mechanism for chymotrypsin-catalyzed hydrolyses.

(1977)]. In order to gain further information on the precise timing of these events leading to inactivation of serine proteases, McMurray and Dyckes (1987) studied the inhibition of trypsin by a series of peptide analogues (Lys-Ala-LysCH$_2$X, where X = H, CH$_2$CO$_2$CH$_3$, COCH$_3$, OCOCH$_3$, and F). A linear relationship between the free energy of binding and the electron-withdrawing power of X indicates that reversible hemiketal formation represents the initial step in inactivation and that this is followed by alkylation of the enzyme (Fig. 2-4).

Halomethyl ketone substrate analogues continue to be used in numerous studies on proteolytic enzymes. For example, destruction of

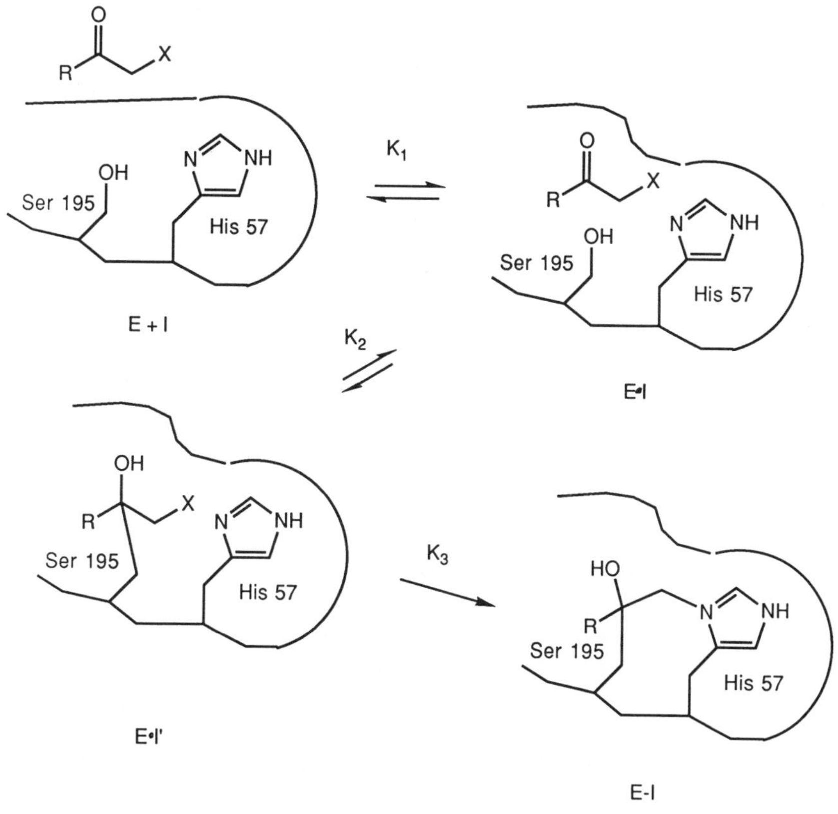

Figure 2-4. Initial step in inactivation of serine proteases by halomethyl ketones involves reversible hemiketal formation. This is followed by alkylation of the enzyme. (Adapted, with permission, from J. S. McMurray and D. F. Dyckes, Evidence for hemiketals as intermediates in the inactivation of serine proteinases with halomethyl ketones, *Biochemistry* 25:2298–2301. Copyright 1987 American Chemical Society).

the antiviral activity of rabbit and mouse fibroblast interferons by chloromethyl analogues of phenylalanine, including Tos-PheCH$_2$Cl, was accompanied by modification of a specific histidine residue. Tos-LysCH$_2$Cl had no effect. The possibility was explored that interferons may be proteases similar to chymotrypsin (McCray and Weil, 1982). [^{125}I]Iodotyrosylated L-Ala-L-Lys-L-Arg chloromethyl ketone was used to label a thiol protease involved in the conversion of proinsulin to insulin (Docherty *et al.*, 1982). In addition, the basic principles of affinity labeling based on selective reaction of chloromethyl ketone substrate analogues, initially demonstrated with proteolytic enzymes, have found ready application in a number of other areas. Receptor labeling is demonstrated by the covalent labeling of opiate receptors with chloromethyl ketone analogues of enkephalin (Venn and Barnard, 1981; Szücs *et al.*, 1987). Foucaud and Biellmann (1983) have probed the structure and function of several dehydrogenases using 3-chloroacetylpyridine adenine dinucleotide. These examples are only suggestive of the immense value this class of affinity label has had in peptide and protein biochemistry.

2.3.3 Halogenated Ketones as Transition State Analogue Inhibitors of Hydrolytic Enzymes

2.3.3.1 Introduction

Certain peptide and amino acid analogues containing the car-boxaldehyde functionality are effective inhibitors of serine and cysteine proteases (Gelb *et al.*, 1985, and references therein). NMR spectroscopy has provided convincing evidence that the mechanism of this inhibition involves formation of a hemiacetal with a serine hydroxy group present at the active site of the enzyme (Shah *et al.*, 1984). Thus, such compounds function as "transition state analogues" through formation of a tetrahedral intermediate resembling the active-site transition state (Wolfenden, 1976). Aldehydes are also effective inhibitors of metalloproteases (Galardy and Kortylewicz, 1984). In this case, data suggest that the hydrated form of the aldehyde, which resembles the tetrahedral intermediate formed during hydrolysis of the substrate, may be the active species.

The presence of only one "arm" in a carboxaldehyde for functionaliza-tion places limitations on extensions of this theme, since derivatives having recognition structures on both sides of the carbonyl group likely would show greater selectivity toward the targeted protease. However, peptides incorporating an ordinary internal ketone would be expected to exhibit low reactivity toward active-site nucleophiles, a reflection of reduced electrophilic character of the carbonyl group relative to the aldehydic carbonyl. For the same reason, hydrate formation would be diminished. On the other hand, substitution of fluorine adjacent to a carbonyl group dramatically increases the degree of solvation of that group. That the effect of the presence of even one fluorine is substantial is seen in the hydrate composition of fluoropyruvate (85–95 %) compared to pyruvate (10–14 %) (Goldstein *et al.*, 1978). Thus, α-fluorinated internal ketone analogues should show an increased reactivity toward active-site nucleophiles in serine proteases or, as hydrates, bind readily to metallo- and aspartylproteases (Gelb *et al.*, 1985).

2.3.3.2 Inhibition of Acetylcholinesterase

Brodbeck *et al.* (1979) studied the inhibition of acetylcholinesterase by a series of aldehydes and ketones containing the trifluoromethyl group, which imparts greatly increased electrophilic character to an adjacent carbonyl group. Compounds which in aqueous solution readily form hydrates showed a time-dependent reversible inhibition of acetylcholinesterase. The most potent analogue tested was *m*-(trimethylammonio)-trifluoroacetophenone (Fig. 2-5), which inhibited acetylcholinesterase with a 50 % inhibitory concentration (IC_{50}) of 1.3×10^{-8} M. A tight enzyme–inhibitor complex was formed which made the inhibition only partially reversible.

Gelb *et al.* (1985) examined fluoroketone analogues as inhibitors of three enzymes—acetylcholinesterase, carboxypeptidase A (a zinc metalloenzyme), and pepsin (an aspartylprotease). A series of acetylcholine analogues were prepared containing as a structural unit a trifluoromethyl or difluoromethylene ketone. To simplify syntheses, the trimethylammonium group of acetylcholine was represented by a *tert*-butyl group. An example of the effect of fluorine substitution is evident from a comparison of the trifluoromethyl ketone **1** and difluoromethylene ketone **2** (Fig. 2-6), potent reversible inhibitors of acetylcholinesterase ($K_i = 16$ nM and 1.6 nM, respectively), with the corresponding methyl ketone **3** (Fig. 2-6) ($K_i = 310$ μM). Although it was not determined whether the fluorinated

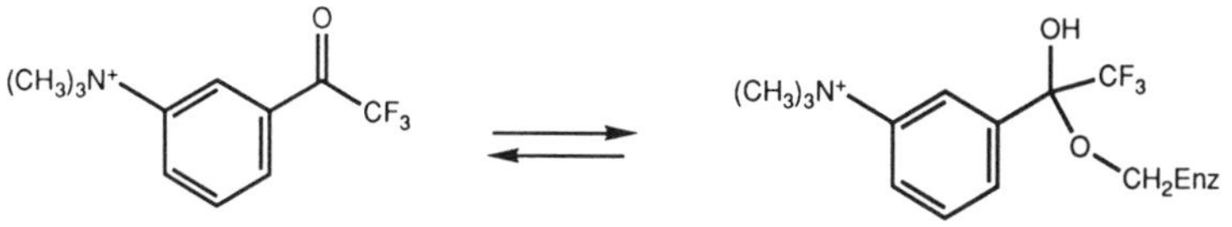

Figure 2-5. Formation of a tight enzyme–inhibitor complex between *m*-(trimethylammonio)trifluoroacetophenone and acetylcholinesterase (Brodbeck *et al.*, 1979).

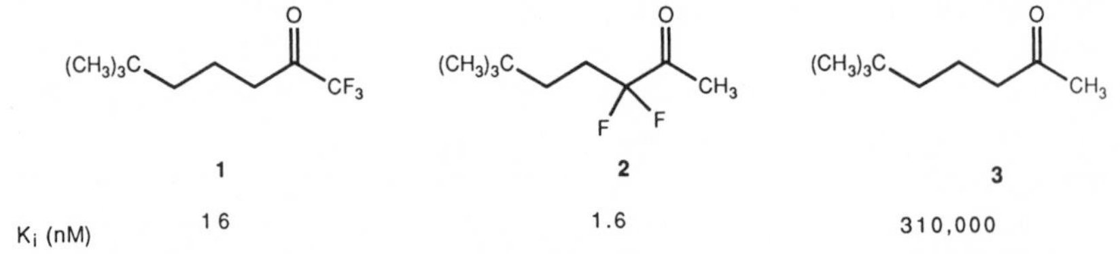

	1	2	3
K_i (nM)	16	1.6	310,000

Figure 2-6. Effect of fluorine substitution on the efficiency of inhibition of acetylcholinesterase by methyl ketones (Gelb *et al.*, 1985).

analogues reacted as the hydrated or the carbonyl form, formation of a hemiacetal with active-site serine was assumed to be the mechanism of inhibition.

2.3.3.3 Inhibition of Carboxypeptidase A

Ketomethylene substrate analogues of phenylalanine have been studied as inhibitors of carboxypeptidase A. An example is $(-)$-3-(p-methoxybenzoyl)-2-benzylpropanoic acid (**4**, Fig. 2-7) ($K_i = 100 \ \mu M$) (Sugimoto and Kaiser, 1978). The trifluoromethyl ketone analogue **5** is considerably more potent ($K_i = 0.2 \ \mu M$) (Gelb *et al.*, 1985) and is slightly more potent than the aldehydic analogue **6** ($K_i = 0.48 \ \mu M$) (Galardy and Kortylewicz, 1984). While the strong inhibition shown by **5** was attributed to the formation of a tightly bound tetrahedral adduct with the enzyme, it was not determined whether this adduct was the hydrated ketone or was formed by the addition of Glu-270 to the ketone.

2.3.3.4 Inhibition of Aspartylproteases (Pepsin and Renin)

The aspartylprotease renin, an enzyme with high substrate specificity, cleaves the circulating protein angiotensinogen at the Leu–Val peptide

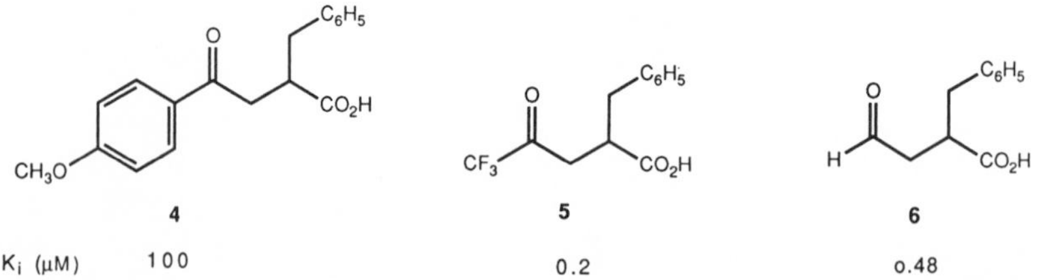

	4	5	6
K_i (μM)	100	0.2	0.48

Figure 2-7. A comparison of the effectiveness of an aryl ketone (**4**), a trifluoromethyl ketone (**5**), and an aldehyde (**6**) as inhibitors of carboxypeptidase A (see text for details and references).

bond to form angiotensin I, the biological precursor of the vasoconstrictive peptide angiotensin II. There has been much recent interest in the development of inhibitors of renin as a potential new approach for management of hypertension. A promising strategy has evolved based on the structure of statine [(3S,4S)-4-amino-3-hydroxy-6-methylheptanoic acid] (**8**; Fig. 2-8), a constituent of the microbially produced pentapeptide pepstatin (**7**; Fig. 2-8), a potent general inhibitor of aspartylproteases. A comparison of the structures of statine and the tetrahedral intermediate formed during pepsin-catalyzed hydrolysis of peptides suggested that statine-containing peptides are transition state analogue inhibitors of aspartylproteases (Marciniszyn *et al.*, 1976).

Based on this rationale, several potent aspartylprotease inhibitors have been developed in which statine is present as a structural mimic of the Leu-Val sequence cleaved by renin. Since the hydrated form of a ketone would more resemble a tetrahedral intermediate than the alcohol functionality of statine, the ketone analogue **9** (Fig. 2-9) was prepared (Rich *et al.*, 1982a). While NMR evidence indicated enzyme-catalyzed formation of a ketal-type structure (Rich *et al.*, 1982b), the ketone is a 50-fold weaker inhibitor of pepsin ($K_i = 56$ nM) than the statine-containing peptide. On the other hand, the corresponding difluoroketone analogue **10** (Fig. 2-9) was an extremely potent inhibitor of pepsin, having a K_i of only 0.06 nM (Gelb *et al.*, 1985). Maximal inhibition was obtained in less than one minute, strongly suggesting that the hydrated form of the ketone is the inhibitory species.

This work was extended to the development of renin inhibitors by the preparation of peptide analogues resembling the angiotensinogen sequence recognized by this enzyme (Fearon *et al.*, 1987). Once again, inclusion of a difluoroketone moiety in the structure resulted in potent inhibition, as

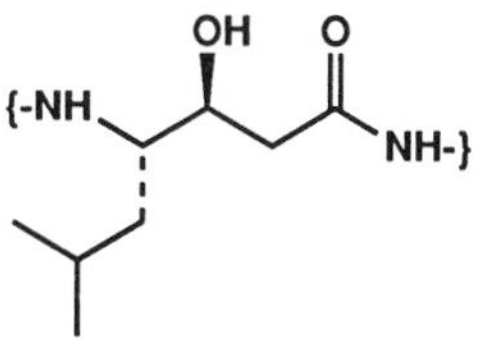

7

Pepstatin

Valeryl-Val-Val-Statyl-Ala-Statine

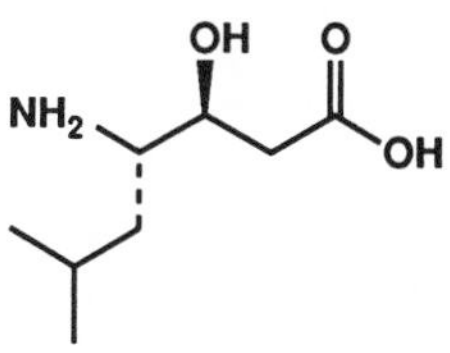

8

Statine

Figure 2-8. Pepstatin and statine.

K$_i$ (nM)

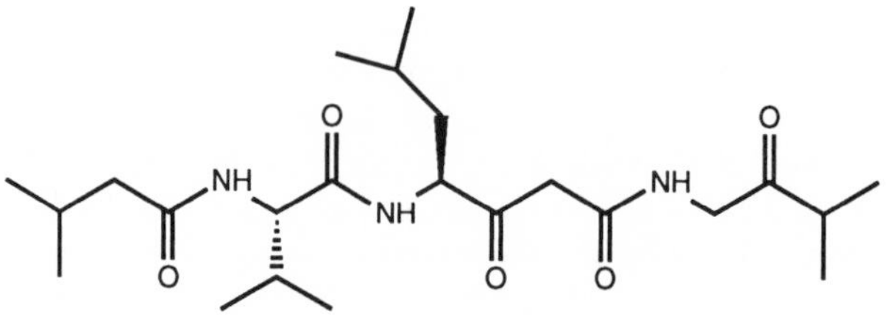

56

9

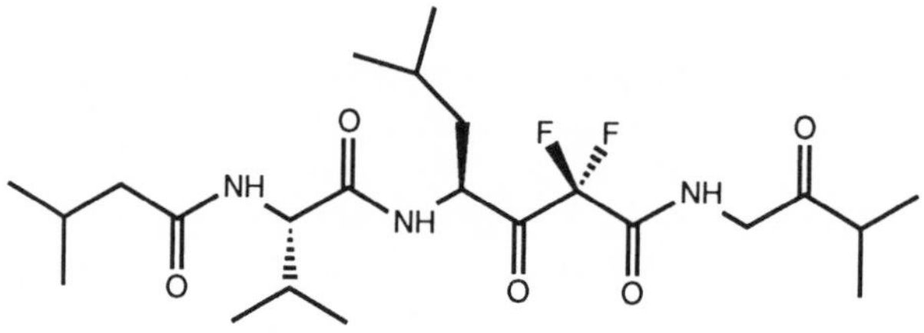

0.06

1 0

Figure 2-9. Comparison of the effectiveness of peptides containing a ketone (**9**) and a difluoroketone (**10**) mimic of the Leu-Val sequence as inhibitors of pepsin (see text for details and references).

illustrated by the comparison of the statine-containing peptide **11** ($K_i = 52$ nM), the corresponding ketone **12** ($K_i = 154$ nM), and the difluoro analogue **13** (Fig. 2-10) ($K_i = 7$ nM).

A similar study by Thaisrivongs *et al.* (1986) further illustrates the promise of this approach to the development of renin inhibitors (Fig. 2-11). Again, a statine-containing analogue, **14**, was found to be a potent renin inhibitor (IC$_{50} = 1.7$ nM). Comparison of the potency of the ketone analogue **15** (IC$_{50} = 34$ nM) and the difluoroketone analogue **16** (IC$_{50} = 0.52$ nM) shows the enhanced potency resulting from fluorine substitution. Furthermore, unlike pepstatin, **16** exhibited a high specificity for renin, relative to three other aspartylproteases (pepsin, cathepsin D, and angiotensin-converting enzyme), a finding of considerable importance for the development of therapeutic agents.

2.3.3.5 Inhibition of Phospholipase A_2

The use of the difluoroketone structural unit as a mimic of an enzyme-produced tetrahedral intermediate has been applied to phospholipase A_2, the enzyme responsible for the release of arachidonic acid from the

BocPhe-Phe-X-Leu-PheNH$_2$

| Compound | X = | K$_i$ (nM) |

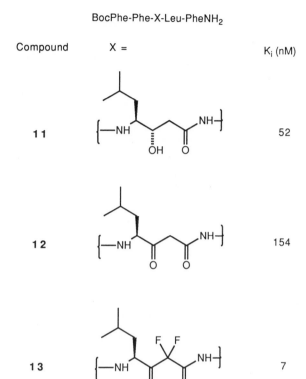

11		52
12		154
13		7

Figure 2-10. A difluoroketone-containing peptide (**13**) as an effective renin inhibitor (Fearon *et al.*, 1987).

phospholipid membrane pool (Fig. 2-12). The difluoroketone substrate analogue **17** was synthesized and shown to be a potent competitive inhibitor of phospholipase A_2 ($K_i = 50 \mu M$). This analogue was considerably more potent than previous substrate analogues incorporating amide and carbamate replacements for the ester linkage (Gelb, 1986).

2.3.3.6 Inhibition of Juvenile Hormone Esterase

Insect juvenile hormones (JH) (e.g., **18**; Fig. 2-13) are morphogenic and vitrellogenic regulators in insects. Changes in activities of the enzymes responsible for epoxide hydrolysis and ester hydrolysis control levels of juvenile hormones during development of many insects. Accordingly, there has been much interest in the development of selective inhibitors of

Boc-Phe-His-X-Ile-AMP

Compound	X =	IC$_{50}$ (nM)

1 4   1.7

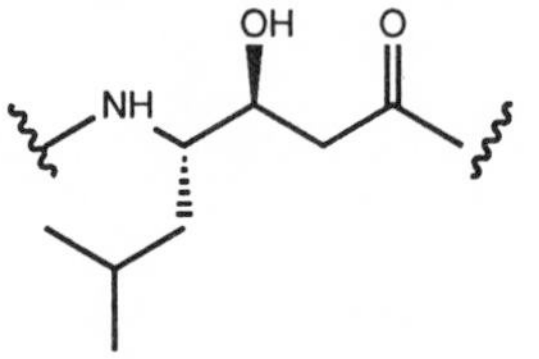

1 5   3 4

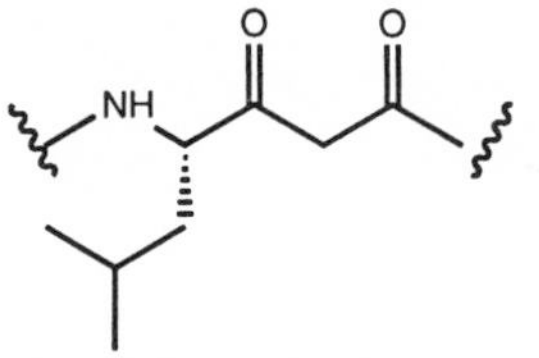

16   0.52

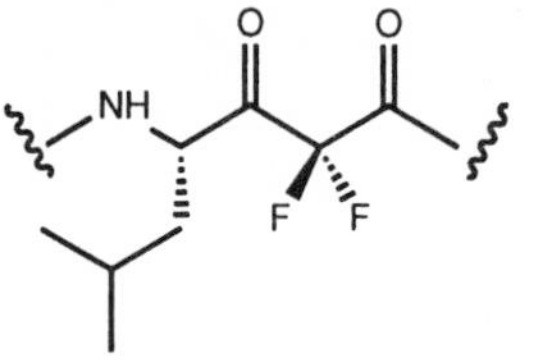

Figure 2-11. Another example of a difluoroketone-containing peptide (**16**) as an effective renin inhibitor (Thaisrivongs *et al.*, 1986).

these enzymes. The trifluoromethyl ketone analogue 1,1,1-trifluoro-3-(octylthio)propan-2-one (**19**; Fig. 2-13), wherein the sulfur 3*p* orbitals are meant to mimic the α,β-unsaturated ester π system of JH, was a slow tight-binding inhibitor of juvenile hormone esterase (JHE) (IC$_{50}$ = 2 n*M*, K_i = 0.1 n*M*) [Hammock, *et al.*, 1984; reviewed by Prestwich (1986)]. Prestwich *et al.* (1984) prepared similar sulfur- and fluorine-containing analogues having a more "juvenoid" side chain and obtained similar anti-JHE activity, but with higher selectivity. The most potent compound (**20**; Fig. 2-13) had an IC$_{50}$ of 3.2 n*M* for JHE, compared to 720 n*M* for α-naphtholacetate esterase, 62 μM for acetylcholinesterase, and $>100 \mu M$ for α-chymotrypsin.

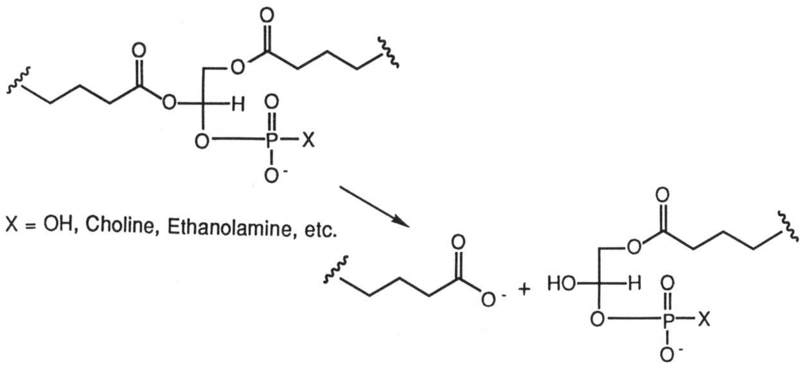

X = OH, Choline, Ethanolamine, etc.

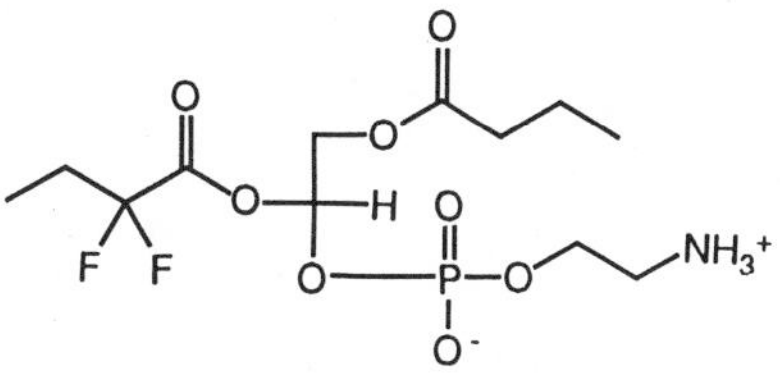

1 7

Figure 2-12. Phospholipase A_2-catalyzed hydrolysis of sn-glycero phospholipids (top) and difluoroketone analogue (**17**), as an inhibitor of phospholipase A_2 (bottom) (Gelb, 1986).

2.3.3.7 Summary

The above examples demonstrate clearly that effectiveness of ketone substrate analogues of peptides as enzyme inhibitors can be increased by introduction of fluorine adjacent to the ketone carbonyl group. Applications of this approach are certain to continue. For example, the recent

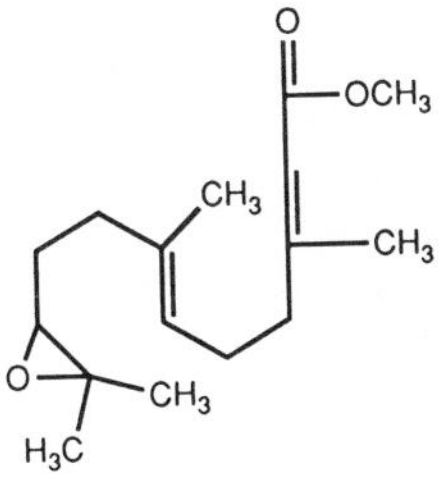

1 8 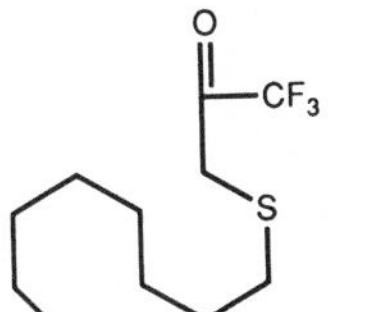   **1 9** 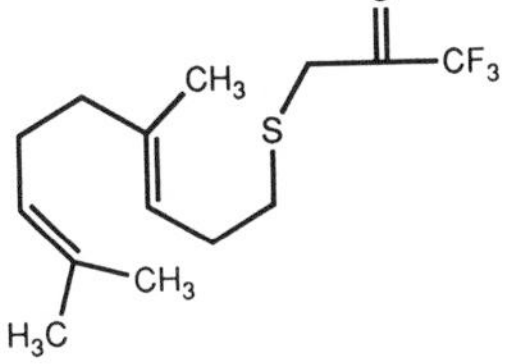   **2 0**

Figure 2-13. An insect juvenile hormone (**18**) and trifluoromethyl ketone analogue inhibitors (**19, 20**) of juvenile hormone esterase [reviewed by Prestwich (1986)].

development of effective inhibitors of HIV-1 proteases based on incorporation of a statine analogue suggests application of fluorinated analogues in
this approach to anti-AIDS drugs (Moore *et al.*, 1989).

2.4 Fluoromethylglyoxal as a Mechanistic Probe

The glyoxalase system, consisting of glyoxalase I [*S*-lactoylglutathione
methylglyoxal-lyase (isomerase)], glutathione (GSH), and glyoxalase II
(*S*-2-hydroxyacylglutathione hydrolase), converts methylglyoxal to lactic
acid. A possible role for this ubiquitous enzyme system in the regulation of
cell division has been suggested (Szent-Györgyi *et al.*, 1967). Glyoxalase I
catalyzes the rearrangement of the thiohemiacetal formed between
glutathione and methylglyoxal to the thioester of D-lactic acid, which is
subsequently hydrolyzed to lactic acid by glyoxalase II. Mechanisms considered for the initial formal internal Cannizzaro reaction involve an internal 1,2-hydride shift or, alternatively, formation of an enediol intermediate,
followed by rapid, active-site shielded proton transfer (Fig. 2-14). Kozarich
et al. (1981) synthesized 3-fluoromethylglyoxal as a probe for the
mechanism of glyoxalase I. The enediol mechanism, but not the internal
hydride shift, introduces carbanionic character at C-2, an event which
could result in elimination of fluoride (a similar approach was used to
study the mechanism of CoA carboxylase; Section 1.4.2). Incubation of
3-fluoroglyoxal with glutathione in the presence of glyoxalase I resulted in
complete fluoride release in a biphasic manner and the formation of the
glutathione thioester of pyruvic acid. In a similar experiment in the

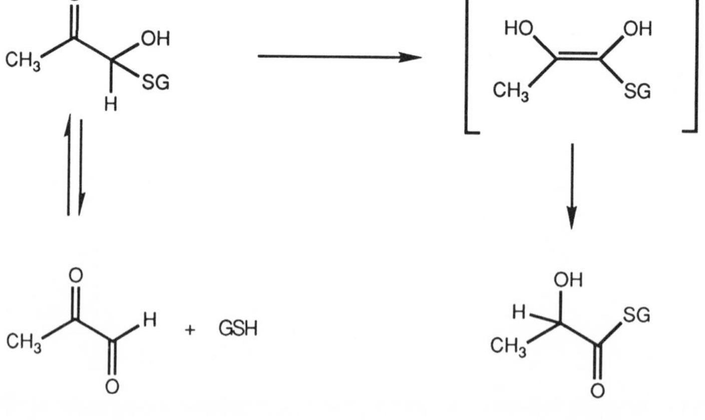

Figure 2-14. Glyoxalase-catalyzed formation of lactic acid from methylglyoxal.

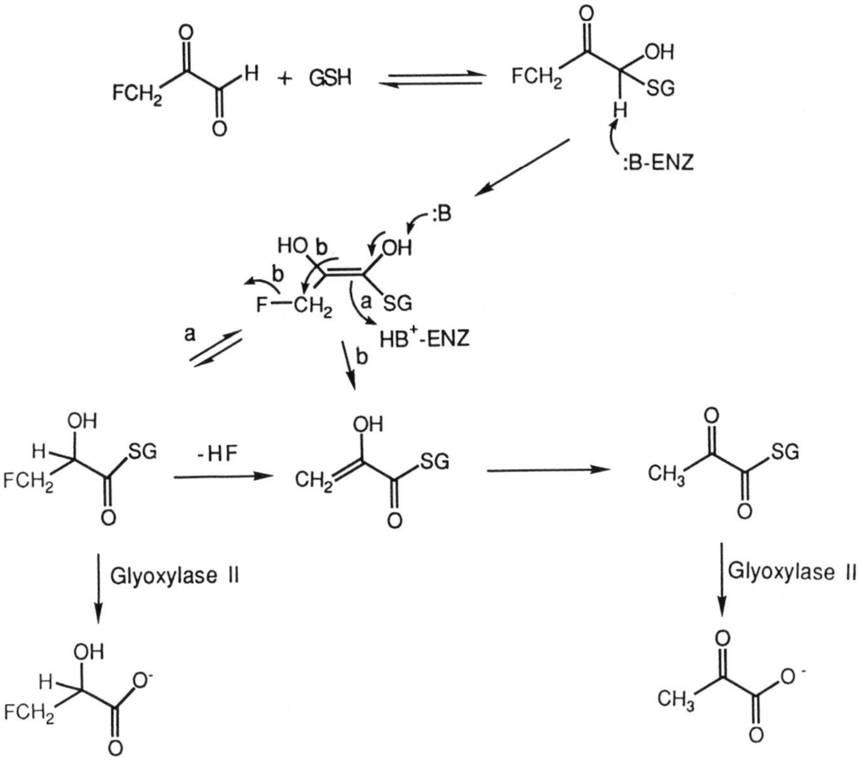

Figure 2-15. The use of 3-fluoromethylglyoxal as a substrate to demonstrate the presence of an enediol intermediate in the glyoxylase catalysis (Kozarich et al., 1981).

presence of both glyoxalase I and II, 32% fluoride release was accompanied by formation of pyruvate and fluorolactate. These data were interpreted as confirming the presence of an enediol intermediate that can partition between formation of the thioester of fluorolactic acid and formation of the thioester of pyruvate. In the absence of glyoxalase II, equilibration with the enediol and β-elimination of HF would contribute to total fluoride loss. The partitioning is revealed by glyoxalase II hydrolysis of the thioester intermediates (Fig. 2-15). Kozarich et al. noted that this sequence represents an unusual partitioning of a fluorinated substrate.

2.5 Chloral Hydrate

While the biochemistry of polyhalogenated compounds that affect central nervous system (CNS) action was discussed more thoroughly in

Chapter 8 of Vol. 9A of this series, the notoriety of chloral hydrate as a hypnotic agent prompts a brief discussion at this point. Chloral hydrate, which may be the oldest synthetic compound used in hypnotic drug therapy [reviewed by Breimer (1977)], is also a metabolite of a widely used industrial chemical trichloroethylene (see Chapter 9 in Vol. 9A of this series). The hypnotic action (induction of drowsiness with a facilitation and maintenance of sleep) of chloral hydrate, however, is caused by the principal metabolite, trichloroethanol. The popular belief that chloral hydrate and ethanol (the "Mickey Finn") have synergistic actions on the CNS apparently may be sound, the synergistic actions being a result of inhibition of ethanol metabolism by chloral as well as an ethanol-induced enhancement of trichloroethanol production (Harvey, 1985). Chloral hydrate also acts as a mutagen by disrupting mitosis. Evidence has been presented that this results from mitotic spindle collapse due to a chloral hydrate-induced increase in intracellular free calcium (Lee *et al.*, 1987). Similar increases are caused by other general anesthetics.

2.6 Summary

With the exception of Section 2.5, the topics covered in this chapter deal primarily with synthetic compounds that have been developed for probing the structure and functions of biomolecules such as nucleotides, DNA, proteins, and peptides. While selectivity of reaction of CAA and BAA with polynucleotides is determined by substrate structure—and hence is useful for the study of this structure—selectivity of haloketone affinity labels can be built into the reagent. The use of fluorinated ketone analogues is particularly effective due to the large increase in electrophilic character of the carbonyl carbon atom and also due to the minimal steric distortion produced by this substitution. For none of these topics has an exhaustive review been attempted. Instead, examples were chosen to illustrate the strategies involved, and an attempt was made to survey current research in these areas.

References

Addison, W. R., Gillam, I. C., and Tener, G. M., 1982. The nucleotide sequence of tRNA$_4^{Val}$ of *Drosophila melanogaster*. Chloroacetaldehyde modification as an aid to RNA sequencing, *J. Biol. Chem.* 257:674–677.
Barrio, J. R., Secrist, J. A., III, and Leonard, N. J., 1972a. Fluorescent adenosine and cytidine derivatives, *Biochem. Biophys. Res. Commun.* 46:597–604.

Barrio, J. R., Secrist, J. A., III, and Leonard, N. J., 1972b. A fluorescent analog of nicotinamide adenine dinucleotide, *Proc. Natl. Acad. Sci. USA* 69:2039–2042.

Bode, J., Pucher, H.-J., and Maass, K., 1986. Chromatin structure and induction-dependent conformational changes of human interferon-β genes in a mouse host cell, *Eur. J. Biochem.* 158:393–401.

Breimer, D. D., 1977. Clinical pharmacokinetics of hypnotics, *Clin. Pharmacokinet.* 2:93–109.

Brodbeck, U., Schweikert, K., Gentinetta, R., and Rottenberg, M., 1979. Fluorinated aldehydes and ketones acting as quasi-substrate inhibitors of acetylcholinesterase, *Biochim. Biophys. Acta* 567:357–369.

Chirikjian, J. G., and Papas, T. S., 1974. Inhibition of AMV DNA polymerase by polyriboadenylic acid containing ε-adenosine residues, *Biochem. Biophys. Res. Commun.* 59:489–495.

Collier, D. A., Griffin, J. A., and Wells, R. D., 1988. Non-B right-handed DNA conformations of homopurine homopyridine sequences in the murine immunoglobulin C_α switch region, *J. Biol. Chem.* 263:7397–7405.

Docherty, K., Carroll, R. J., and Steiner, D. F., 1982. Conversion of proinsulin to insulin: Involvement of a 31,500 molecular weight thiol protease, *Proc. Natl. Acad. Sci. USA* 79:4613–4617.

Fearon, K., Spaltenstein, A., Hopkins, P. B., and Gelb, M. H., 1987. Fluoro ketone containing peptides as inhibitors of human renin, *J. Med. Chem.* 30:1617–1622.

Foucaud, B., and Biellmann, J.-F., 1983. Properties of horse-liver alcohol dehydrogenase modified by the affinity label 3-chloroacetylpyridine-adenine dinucleotide, *Biochim. Biophys. Acta* 748:362–366.

Furlong, J. C., Sullivan, K. M., Murchie, A. I. H., Gough, G. W., and Lilley, D. M. J., 1989. Localized chemical hyperreactivity in supercoiled DNA: Evidence for base unpairing in sequences that induce low-salt cruciform extrusion, *Biochemistry* 28:2009–2017.

Galardy, R. E., and Kortylewicz, Z. P., 1984. Inhibition of carboxypeptidase A by aldehyde and ketone substrate analogues, *Biochemistry* 23:2083–2087.

Gelb, M. H., 1986. Fluoro ketone phospholipid analogues: New inhibitors of phospholipase A_2, *J. Am. Chem. Soc.* 108:3146–3147.

Gelb, M. H., Svaren, J. P., and Abeles, R. H., 1985. Fluoro ketone inhibitors of hydrolytic enzymes, *Biochemistry* 24:1813–1817.

Goldstein, J. A., Cheung, Y.-F., Marletta, M. A., and Walsh, C., 1978. Fluorinated substrate analogues as stereochemical probes of enzymatic reaction mechanisms, *Biochemistry* 17:5567–5575.

Hammock, B. D., Abdel-aal, Y. A. I., Mullin, C. A., Hanzlik, T. N., and Roe, R. M., 1984. Substituted thiotrifluoropropanones as potent selective inhibitors of juvenile hormone esterase, *Pest. Biochem. Physiol.* 22:209–223.

Harvey, S. C., 1985. Hypnotics and sedatives, in *The Pharmacological Basis of Therapeutics*, 7th ed. (A. G. Gilman, L. S. Goodman, T. W. Rall, and F. Murad, eds.), Macmillan, New York, pp. 339–371.

Kimura, K., Nakanishi, M., Yamamoto, T., and Tsuboi, M., 1977. A correlation between the secondary structure of DNA and the reactivity of adenine residues with chloro-acetaldehyde, *Biochem. J.* 81:1699–1703.

Kochetkov, N. K., Shibaev, V. N., and Kost, A. A., 1971. New reaction of adenine and cytosine derivatives, potentially useful for nucleic acid modification, *Tetrahedron Lett.* 1971:1993–1996.

Kohwi-Shigematsu, T., Enomoto, T., Yamada, M.-A., Nakanishi, M., and Tsuboi, M., 1978. Exposure of DNA bases induced by the interaction of DNA and calf thymus DNA helix-destabilizing protein, *Proc. Natl. Acad. Sci. USA* 75:4689–4693.

Kohwi-Shigematsu, T., Gelinas, R., and Weintraub, H., 1983. Detection of an altered DNA conformation at specific sites in chromatin and supercoiled DNA, *Proc. Natl. Acad. Sci. USA* 80:4389–4393.

Kohwi-Shigematsu, T., Scribner, N., and Kohwi, Y., 1988. An ultimate chemical carcinogen, *N*-acetoxy-2-acetylaminofluorene, detects non-B DNA structures that are reactive with chloroacetaldehyde in supercoiled plasmid DNA, *Carcinogenesis* 9:457–461.

Kozarich, J. W., Chari, R. V. J., Wu, J. C., and Lawrence, T. L., 1981. Fluoromethylglyoxal: Synthesis and glyoxalase I catalyzed product partitioning via a presumed enediol intermediate, *J. Am. Chem. Soc.* 103:4593–4595.

Krzyzosiak, W. J., Biernat, J., Ciesiolka, J., Gulewicz, K., and Wiewiorowski, M., 1981. The reactions of adenine and cytosine residues in tRNA with chloroacetaldehyde, *Nucleic Acids Res.* 9:2841–2851.

Kuśmierek, J. T., and Singer, B., 1982. Chloroacetaldehyde-treated ribo- and deoxy-ribopolynucleotides. 2. Errors in transciption by different polymerases resulting from ethenocytosine and its hydrated intermediate, *Biochemistry* 21:5723–5728.

Lee, G. M., Diguiseppi, J., Gawdi, G. M., and Herman, B., 1987. Chloral hydrate disrupts mitosis by increasing intracellular free calcium, *J. Cell. Sci.* 88:603–612.

Leonard, N. J., 1984. Etheno-substituted nucleotides and coenzymes: Fluorescence and biological activity, *CRC Crit. Rev. Biochem.* 15:125–199.

Marciniszyn, J., Jr., Hartsuck, J. A., and Tang, J., 1976. Mode of inhibition of acid proteases by pepstatin, *J. Biol. Chem.* 251:7088–7094.

McCray, J. W., and Weil, R., 1982. Inactivation of interferons: Halomethyl ketone derivatives of phenylalanine as affinity labels, *Proc. Natl. Acad. Sci. USA* 79:4829–4833.

McLean, M. J., and Wells, R. D., 1988. The role of DNA sequence in the formation of Z-DNA versus cruciforms in plasmids, *J. Biol. Chem.* 263:7370–7377.

McMurray, J. S., and Dyckes, D. F., 1987. Evidence for hemiketals as intermediates in the inactivation of serine proteinases with halomethyl ketones, *Biochemistry* 25:2298–2301.

Meyer, R. B., Shuman, D. A., Robins, R. K., Miller, J. P., and Simon, L. N., 1973. Synthesis and enzymic studies of 5-aminoimidazole and N-1- and N⁶-substituted adenine ribonucleoside cyclic 3′,5′-phosphates prepared from adenosine-cyclic-3′,5′-phosphate, *J. Med. Chem.* 16:1319–1323.

Moore, M. L., Bryan, W. M., Fakhoury, S. A., Magaard, V. W., Huffman, W. F., Dayton, B. D., Meek, T. D., Hyland, L., Dreyer, G. B., Metcalf, B. W., Strickler, J. E., Gorniak, J. G., and Debouck, C., 1989. Peptide substrates and inhibitors of the HIV-1 protease, *Biochem. Biophys. Res. Commun.* 159:420–425.

Powers, J. C., 1977. Reaction of serine proteases with halomethyl ketones, *Methods Enzymol.* 46:197–208.

Prestwich, G. D., 1986. Fluorinated sterols, hormones and pheromones: Enzyme-targeted disruptants in insects, *Pestic. Sci.* 37:430–440.

Prestwich, G. D., Eng, W.-S., Roe, R. M., and Hammock, B. D., 1984. Synthesis and bioassay of isoprenoid 3-alkylthio-1,1,1-trifluoro-2-propanones: Potent, selective inhibitors of juvenile hormone esterases, *Arch. Biochem. Biophys.* 228:639–645.

Rich, D. H., Boparai, A. S., and Bernatowicz, M. S., 1982a. Synthesis of a 3-oxo-4(*S*)-amino acid analog of pepstatin. A new inhibitor of carboxyl (acid) proteases, *Biochem. Biophys. Res. Commun.* 104:1127–1133.

Rich, D. H., Bernatowicz, M. S., and Schmidt, P. G., 1982b. Direct ¹³C NMR evidence for a tetrahedral intermediate in the binding of a pepstatin analogue to porcine pepsin, *J. Am. Chem. Soc.* 104:3535–3536.

Rich, A., Nordheim, A., and Wang, A. H.-J., 1984. The chemistry and biology of left-handed DNA, *Annu. Rev. Biochem.* 53:791–846.

Schoellman, G., and Shaw, E., 1963. Direct evidence for the presence of histidine in the active center of chymotrypsin, *Biochemistry* 2:252–255.

Schulman, L. H., and Pelka, H., 1976. Location of accessible bases in *Escherichia coli* formylmethionine transfer RNA as determined by chemical modification, *Biochemistry* 15:5769–5775.

Secrist, J. A., III, Barrio, J. R., Leonard, N. J., and Weber, G., 1972a. Fluorescent modification of adenosine-containing coenzymes. Biological activities and spectroscopic properties, *Biochemistry* 11:3499–3506.

Secrist, J. A., III, Barrio, J. R., Leonard, N. J., Villar-Palasi, C., and Gilman, A. G., 1972b. Fluorescent modification of adenosine 3',5'-monophosphate: Spectroscopic properties and activity in enzyme systems, *Science* 177:279–280.

Shah, D. O., Lai, K., and Gorenstein, D. G., 1984. ^{13}C NMR spectroscopy of "transition-state analogue" complexes of *N*-acetyl-L-phenylalinal and α-chymotrypsin, *J. Am. Chem. Soc.* 106:4272–4273.

Shaw, E., 1970. Chemical modification by active-site-directed reagents, in *The Enzymes*, Vol. I, 3rd ed. (P. D. Boyer, ed.), Academic Press, New York, pp. 91–146.

Sugimoto, T., and Kaiser, E. T., 1978. Carboxypeptidase A catalyzed enolization of a ketonic substrate. A new stereochemical probe for an enzyme-bound nucleophile, *J. Am. Chem. Soc.* 100:7750–7751.

Szent-Györgyi, A., Együd, L. G., and McLaughlin, J. A., 1967. Keto-aldehydes and cell division, *Science* 155:539–541.

Szücs, M., Belcheva, M., Simon, J., Benyhe, S., Toth, G., Hepp, J., Wollemann, M., and Medzihradszky, K., 1987. Covalent labeling of opioid receptors with ^{3}H-D-Ala^2-Leu^5-enkephalin chloromethyl ketone I. Binding characteristics in rat brain membrane, *Life Sci.* 41:177–184.

Thaisrivongs, S., Pals, D. T., Kati, W. M., Turner, S. R., Thomasco, L. M., and Watt, W., 1986. Design and synthesis of potent and specific renin inhibitors containing difluorostatine, difluorostatone, and related analogues, *J. Med. Chem.* 29:2080–2087.

Venn, R. F., and Barnard, E. A., 1981. A potent peptide affinity reagent for the opiate receptor, *J. Biol. Chem.* 256:1529–1532.

Vogt, N., Marrot, L., Rousseau, N., Malfoy, B., and Leng, M., 1988. Chloroacetaldehyde reacts with Z-DNA, *J. Mol. Biol.* 201:773–776.

Wells, R. D., 1988. Unusual DNA structures, *J. Biol. Chem.* 263:1095–1098.

Wolfenden, R., 1976. Transition state analog inhibitors and enzyme catalysis, *Annu. Rev. Biophys. Bioeng.* 5:271–306.

Biochemistry of Halogenated
Analogues of Steroids,
Isoprenyl Derivatives,
and Other Terpenoids

3.1 Biological Properties of Halogenated Steroids

The serendipitous discovery by Fried and Sabo (1953, 1954) that 9α-halocortisones have enhanced glucocorticoid activity marked the genesis of a major field of medicinal chemistry. The systematic study of halogenated, and especially fluorinated, steroids that followed has produced an enormous number of new analogues and has resulted in many useful pharmaceutical and medicinal agents. In this section, a brief review of the development of halogenated glucocorticoids will be given, followed by comments on other biomedical applications of halogenated steroids. The use of halogenated steroids as biological tracers and mechanistic probes will be discussed, with special attention given to recent research on the development of radiohalogenated steroids as steroid receptor-based imaging agents.

3.1.1 Biosynthesis of Steroid Hormones

In the final stages of the biosynthesis of steroid hormones, the cholesterol side chain is oxidatively cleaved to give pregnenolone, the precursor of all steroid hormones (Fig. 3-1). Double-bond migration and oxidation of the 3-hydroxyl group give progesterone, a gestagen. Hydroxylation of progesterone at C-11, C-17, and C-21 produces the major glucocorticoid hormone, cortisol. Aldosterone, the major mineralocorticoid, is produced from progesterone by hydroxylation at C-21, followed by C-11 hydroxylation and oxidation of the C-18 angular methyl group. The androgens, androstanedione and testosterone, are derived from progesterone by oxidative cleavage of the side chain at C-21. The estrogens, estrone and estradiol, are derived from androstanedione and testosterone, respectively, by loss of the C-6 methyl group and aromatization of ring A (Fig. 3-1).

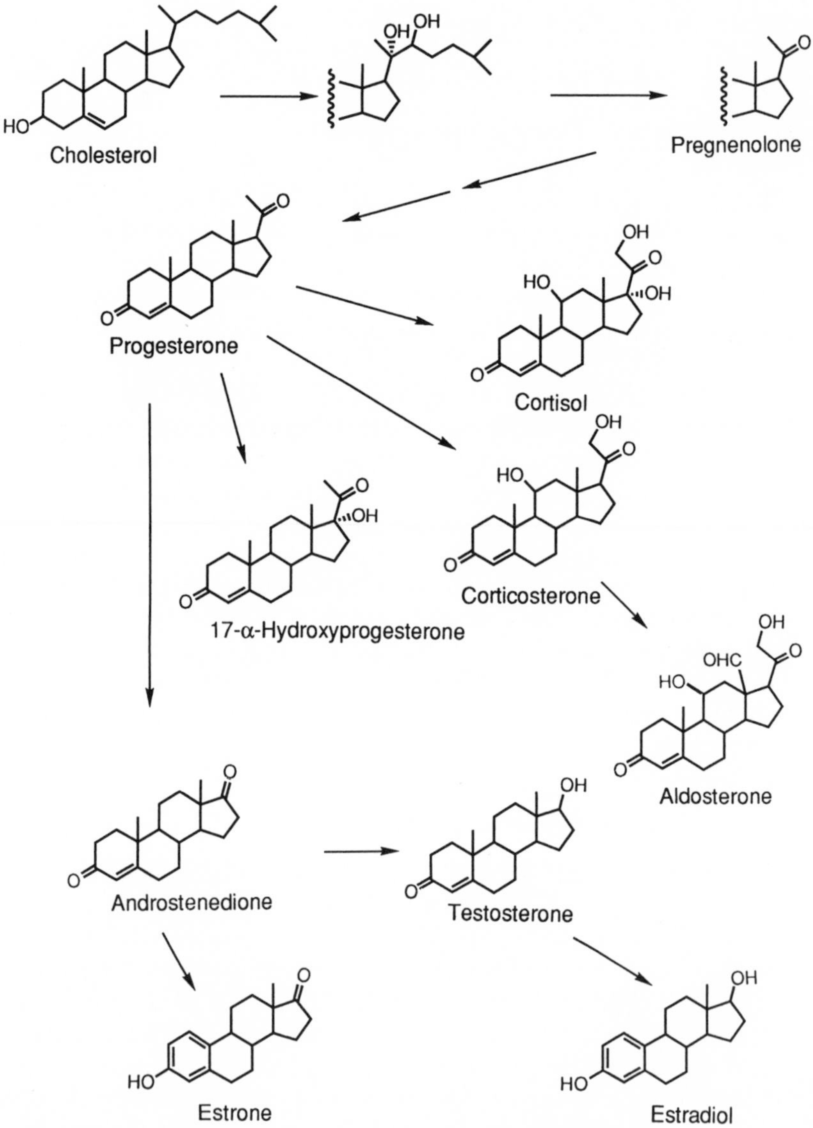

Figure 3-1. Biosynthesis of steroid hormones.

3.1.2 Biological Functions of Corticosteroids

The natural corticosteroids, elaborated in the adrenal gland, include cortisol, cortisone, corticosterone, 11-desoxycorticosterone, and aldosterone (Fig. 3-1). These hormones stimulate liver glycogen deposition (glucocorticoid activity) and retention of sodium, chloride, and bicarbonate by the kidney (mineralocorticoid activity) and have anti-inflammatory activity. Cortisol and cortisone affect all three activities but have relatively weak effects on electrolyte metabolism. Nonetheless, changes in electrolyte balance are serious side effects that often accompany doses of cortisone and cortisol sufficient to alleviate the inflammatory symptoms of rheumatoid arthritis. Aldosterone is the principal mineralocorticoid [for a review of adrenocortical steroid function, see Haynes and Murad (1985)]. Because adrenocorticosteroids have lympholytic activity and suppress mitosis of lymphocytes, these agents have received much attention as antineoplastic drugs, especially for the treatment of acute childhood leukemias and lymphomas.

3.1.3 Effects of Halogenation on the Biological Properties of Corticosteroids

3.1.3.1 9α-Fluoro-11β-hydroxy Substitution

Fried and Sabo (1953) used a liver-glycogen assay for glucocorticoid activity to confirm the structure of 9α-bromohydrocortisone (9α-bromo-17α-hydroxycorticosterone; Fig. 3-2), to be used as a synthetic intermediate. The finding that the activity of this analogue was increased relative to that of the parent was a revolutionary event, contradicting many previous indications that any modification of the cortisol structure leads to diminished activity. A study of other 9α-halohydrocortisones showed that glucocorticoid, anti-inflammatory, and mineralocorticoid activity all increased with decreasing atomic weight of the halogen. Unfortunately, the dramatic increase in anti-inflammatory and glucocorticoid activity of 9α-fluorohydrocortisone was accompanied by an even greater increase in mineralocorticoid activity (Fried, 1957; Fried and Borman, 1958).

Subsequent studies showed that increased hormonal activity was present not only in 9α-fluorohydrocortisone, but also in any series wherein the 9α-fluoro-11β-hydroxyl functionality was present. The 6α-fluoro-11β-hydroxyl series also had increased activity. Thus, it was possible to introduce structural variations in other parts of the steroid nucleus designed to attenuate undesired biological properties. For example, dehydrogenation of the 1–2 bond increases anti-inflammatory activity whereas addition of a

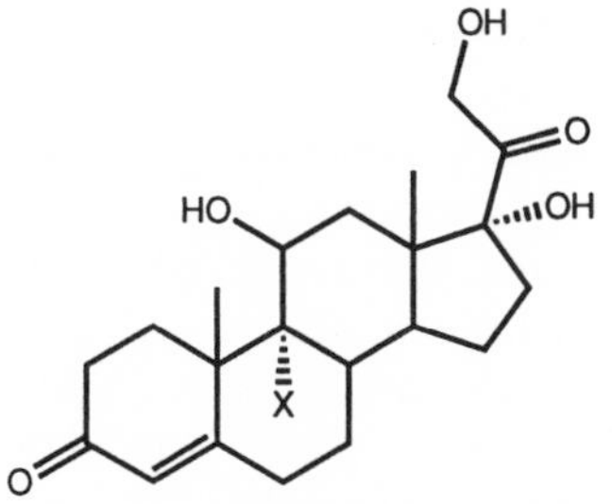

9α-Halo-17α-hydroxycorticosterone
(X = F, Cl, Br, I)

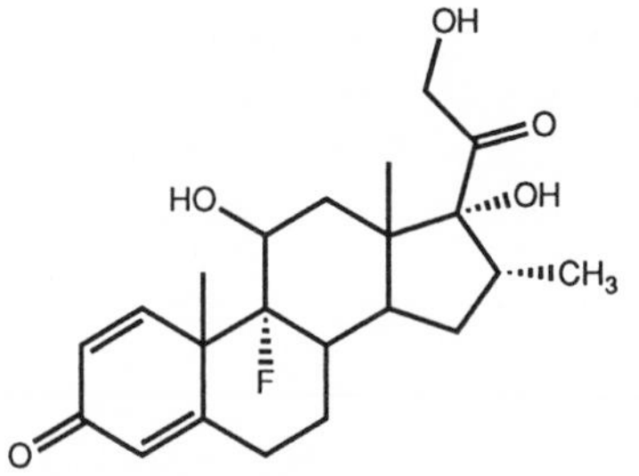

Dexamethasone 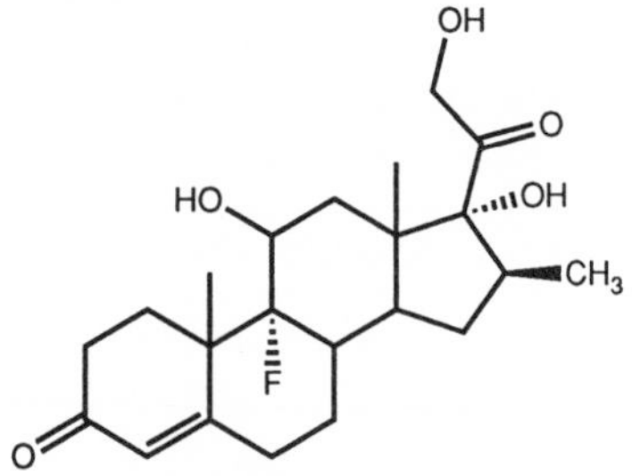   Betamethasone

Figure 3-2. Examples of 9α-halo-11β-hydroxy steroids.

16α-hydroxyl or methyl group greatly diminishes mineralocorticoid activity without affecting the anti-inflammatory response. By such structural adjustments, several potent anti-inflammatory agents have been developed, including such well-known drugs as dexamethasone and betamethasone (Fig. 3-2). In his review, Filler (1979) compiled a list of commercially significant fluorine-containing corticosteroids, illustrating the utility of these analogues as medicinal agents. For a comprehensive review of the development of fluorinated steroids as anti-inflammatory agents, the review by Chen and Borrevang (1970) is also recommended.

3.1.3.2 Mechanism of Halogen-Induced Increased Glucocorticoid Activity

The clear correlation between glucocorticoid activity and the electro-negativity of the 9α-substituent prompted Fried to propose that increased activity is a function of the effectiveness of the 11β-hydroxyl group to serve as a hydrogen bond donor (Fried and Borman, 1958). However, although the 11β-hydroxyl group is important for activity, results of studies to quantitate the effect of 9α-substitution on corticosteroid activity indicate that additional factors may be involved. Molecular orbital calculations support a fluorine-induced increase in acidity of the 11β-hydroxyl group, but the modest nature of this increase casts doubt on this as the complete cause of increased biological activity (Kollman *et al.*, 1973). Conformational effects may contribute, since X-ray analysis has shown that the presence of a 9α-fluoro substituent causes a significant distortion of the A ring (Weeks *et al.*, 1973), and a similar distortion in solution has been suggested by NMR studies (Wong *et al.*, 1984). Increased biological activity may also result from a decreased rate of oxidation of the 11β-hydroxyl to the ketone in the fluorinated compound (Ringold *et al.*, 1964). Wolff and Hansch (1973) have correlated steric, inductive, and hydrogen bonding properties of a series of 9α-substituents.

3.1.3.3 Fluorinated Antimineralocorticoids

While a major impact of halogenated corticosteroids has been in the development of anti-inflammatory agents, halogenated analogues have been investigated in virtually every aspect of the biological behavior of steroids. For example, the aldosterone antagonist spironolactone (Fig. 3-3) is used clinically in antimineralocorticoid therapy. However, adverse side affects result from antiandrogenic and progestational properties of this drug. The 9α-fluorinated spirolactone (Fig. 3-3) is a more potent aldosterone antagonist than spironolactone and, moreover, has low to

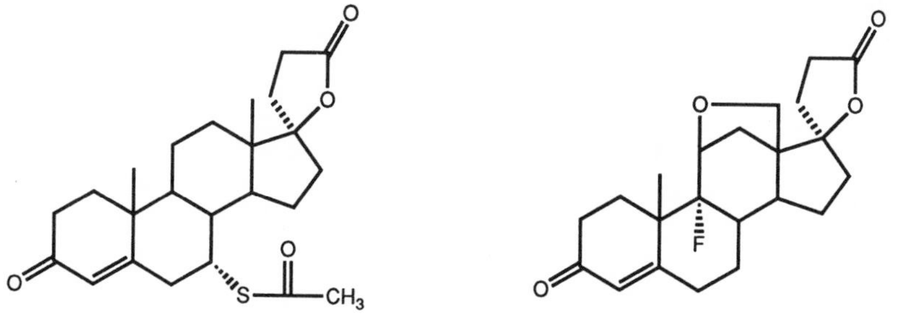

Spironolactone

11β, 18-Epoxy-3-(9α-fluoro-17β-hydroxy-3-oxoandrost-4-en-17α-yl)
propionic acid γ-lactone

Figure 3-3. The aldosterone antagonists spironolactone and the more potent and more selective 9α-fluorinated spirolactone analogue.

negligible affinity for androgen, progestin, estrogen, and glucocorticoid receptors. In this work, activity attributable to 9α-fluorination is seen in the absence of the 11β-hydroxyl group (Kamata *et al.*, 1985).

3.1.4 Halogenated Analogues of Steroidal Sex Hormones

The potential utility of steroid analogues as agents to alter endocrine functions has resulted in the synthesis and testing of an enormous number of new analogues, including many containing halogen. Chen and Borrevang (1970) have reviewed the development of fluorinated steroids with respect to gestagenic, androgenic, antiandrogenic, antiovulatory, and antigonadotropic effects. The development of several radiohalogenated steroids as steroid receptor-based imaging agents for the estrogen, progesterone, and androgen receptors will be discussed below. Among the medical applications of steroid sex hormones are their use in the treatment of several hormone-dependent cancers, especially cancer of the breast and prostate [for a review of concepts and strategies used in anticancer hormone therapies, see Robinson and Jordan (1988)]. An example of halogenated analogues that have shown promise as anticancer agents is the fluorinated androgen fluoxymesterone (Fig. 3-4), studied for the treatment

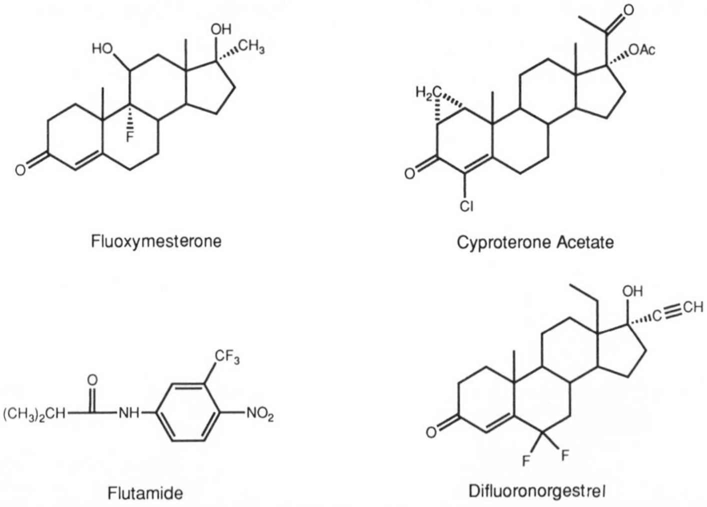

Fluoxymesterone

Cyproterone Acetate

Flutamide

Difluoronorgestrel

Figure 3-4. Examples of fluorinated analogues of sex hormones: the fluorinated androgen fluoxymesterone, active against breast cancer; the potent antiandrogen cyproterone acetate, used in the treatment of prostate cancer; the fluorinated nonsteroidal antiandrogen flutamide, also effective against prostate cancer; and the potent progestin difluoronorgestrel, effective as a contraceptive agent.

of advanced breast cancer (e.g., Swain *et al.*, 1988). The potent antiandrogen cyproterone acetate (Fig. 3-4) is used in the treatment of prostate cancer. A "pure" antiandrogen having a nonsteroidal structure, flutamide (Fig. 3-4), has shown particular promise against prostate cancer (Labrie *et al.*, 1988). Synthetic progestins have been developed for a number of applications, including use as oral contraceptives. 6,6-Difluoronorgestrel (Fig. 3-4) is an example of a potent progestin that has been used extensively, particularly in veterinary science.

3.1.5 Radiohalogenated Steroids as Biological Tracers

The ready availability of several radioactive halogen nuclides, some with particularly useful decay behavior, has made radiohalogenation an attractive approach to the preparation of radiopharmaceuticals for a wide range of applications. A brief overview of the radiochemistry of halogen nuclides follows.

3.1.5.1 Radioisotopes of Halogens Used in Nuclear Medicine

Reactor-produced ^{131}I has been used extensively in clinical diagnoses because of its convenient half-life (8.04 days) and low cost and because the 364-keV γ radiation is readily detected. A disadvantage is the high radiation dose from the accompanying β emission (Kowalsky and Perry, 1987). Cyclotron-produced ^{123}I, because of the short half-life (13.1 h) and lack of β radiation, gives a much lower internal radiation dose to the patient and thus has major advantages over ^{131}I. This neutron-deficient nuclide decays by electron capture (EC) and produces readily detected γ radiation (159 keV). The neutron-deficient ^{125}I decays by EC, with a half-life of 59.9 h, to give ^{125}Te. Although only a small amount of γ radiation is produced by this decay, ^{125}Te characteristic X rays, which can be used for detection, are produced efficiently. Iodine-125 labeling is used extensively in a multitude of biochemical applications, such as receptor binding assays and radioimmunoassays (Kowalsky and Perry, 1987; Nozaki, 1983).

Fluorine-18, the longest-lived radiofluorine (109-min half-life), can be made by reactor radiation and by cyclotron production. Radioactive decay of ^{18}F produces β^+ particles which are annihilated by electrons to give two 551-keV photons emitted at 180° from each other, such that the three-dimensional origin of this emission can be determined by coincidence detection. The short half-life of ^{18}F and low β^+ energy make this nuclide ideal for positron emission tomography (PET) imaging, as reflected in the extensive research activity in this area.

Cyclotron-produced ^{39}Cl (a β^- emitter, 56-min half-life) can be used as a tracer, but production is inefficient. On the other hand, reactor-produced ^{36}Cl (β^- emission, 3×10^5-year half-life) has many applications in chemistry and biology, an example of which will be given in Chapter 7 (see Section 7.3.1.2).

The β^+ emitter ^{75}Br (96-min half-life) can be used for PET imaging, and ^{77}Br (56-h half-life) emits γ rays suitable for diagnostic purposes. The relative ease of bromination reactions relative to fluorination procedures and the greater strength of the $C-Br$ bond relative to the $C-I$ bond are advantages offered by these bromine isotopes (Nozaki, 1983).

3.1.5.2 Radiohalogenated Cholesterol Analogues as Adrenal Imaging Agents

The adrenal cortex is a major storage area of cholesterol, the principal precursor of adrenocortical steroids (two other important storage sites are the corpora lutea cells of the ovary and the Sertoli cells of the testis) (Beierwaltes et al., 1978). 19-[^{131}I]Iodocholesterol (Fig. 3-5), shown to be concentrated efficiently in adrenal glands of animals, was used in the first successful imaging of adrenal glands of humans *in vivo* by scintillation scanning (Thrall et al., 1978, and references therein). 6β-[^{131}I]Iodomethyl-19-norcholes-5(10)-en-3β-ol (NP-59; Fig. 3-5) was subsequently found to have significantly higher uptake into the adrenal cortex and to be metabolically more stable and is now the reagent of choice for adrenal scintigraphy. Adrenal imaging now has a wide range of diagnostic applications related to adrenal disorders and to the evaluation of adrenal function in general.

3.1.5.3 Radiohalogenated Estrogen Receptor Ligands

In recent years, there has been much interest in the development of steroid receptor-based imaging agents, particularly for applications related

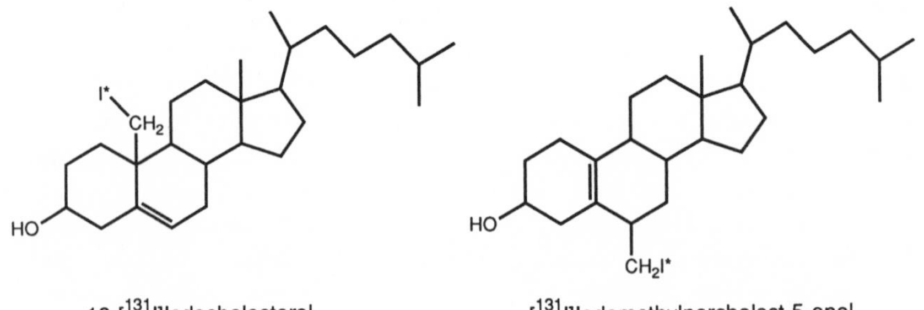

19-[^{131}I]Iodocholesterol   [^{131}I]Iodomethylnorcholest-5-enol

Figure 3-5. Radiohalogenated cholesterol analogues used as adrenal scanning agents.

to diagnosis and treatment of cancer of the breast, ovary, and prostate. While ^{3}H-labeled estradiol of high specific activity was used initially for detection and assay of the estrogen receptor, much of the recent research has focused on the preparation of steroid analogues substituted with radioactive halogen nuclides. Thus, the recognition that the response of breast tumors to various forms of hormonal therapy could be correlated to the presence of functional estrogen receptors prompted interest in development of estrogens labeled with high-energy isotopes, suitable for external detection. For such an approach to be successful, three criteria must be satisfied: (1) since steroid hormone receptors are present in low concentrations, the agent must have high specific activity; (2) high binding selectivity is required; and (3) the agent must have appropriate metabolic characteristics—for example, it must possess sufficient metabolic stability to permit imaging, and it must be cleared rapidly from background tissue [for a more thorough discussion of these factors, see the review by Brandes and Katzenellenbogen (1988)]. Indeed, initial efforts to develop estrogen receptor imaging agents met with limited success. For example, although 3-iodohexestrol (Fig. 3-6) had significant receptor binding affinity, the increased lipophilicity caused by iodine substitution resulted in increased binding to nonreceptor proteins, and selective, estrogen-specific uptake by the uterus could not be demonstrated (Katzenellenbogen *et al.*, 1975). In another example that demonstrates the importance of metabolic stability, rapid deiodination of 17α-[^{125}I]iodoethynylestradiol and 17α-[^{125}I]iodoethynyl-11β-methoxyestradiol (Fig. 3-6) occurred in the presence of protein, precluding their use *in vivo* or *in vitro* (Mazaitis *et al.*, 1980). In 1979, 16α-[^{125}I]iodoestradiol (Fig. 3-6), a potent γ-emitting estrogen ligand, was prepared (Hochberg, 1979) and was shown to have selective binding to the estrogen receptor comparable to that of [^{3}H]estradiol (Hochberg and Rosner, 1980). While 16α-[^{125}I]iodoestradiol was shown to be concentrated *in vivo* in human ovarian tumors in proportion to estrogen receptor concentration, rapid metabolism apparently will preclude external imaging using the corresponding ^{123}I-labeled ligand (Hochberg *et al.*, 1985a). The closely related analogue 16α-[[77]Br]bromoestradiol-17β had uptake behavior comparable to that of iodoestradiol, except that uterine uptake of the bromo analogue was significantly higher at longer time periods (McElvany *et al.*, 1982). 16α-[[77]Br]Bromo-11β-methoxyestradiol-17β (Fig. 3-6) had an even more selective uptake and longer retention in estrogen target tissues than did either 16α-iodo- or 16α-bromoestradiol (Katzenellenbogen *et al.*, 1982).

Several ^{18}F-labeled estrogen receptor ligands have been prepared as potential PET scanning agents for imaging of estrogen receptor-rich tissue, especially tumors of the breast. 16α-[^{18}F]Fluoroestradiol-17β (Fig. 3-7)

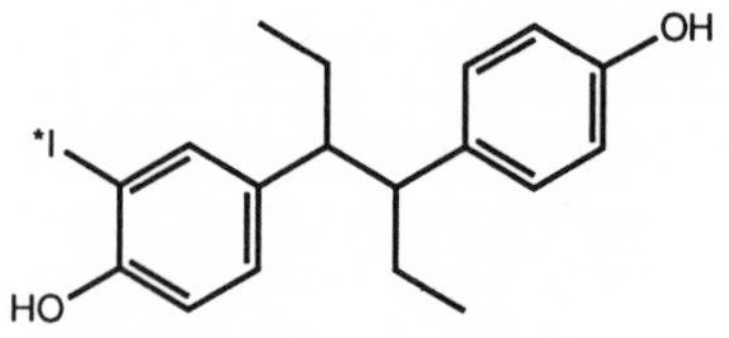

3-[^{131}I]Iodohexestrol

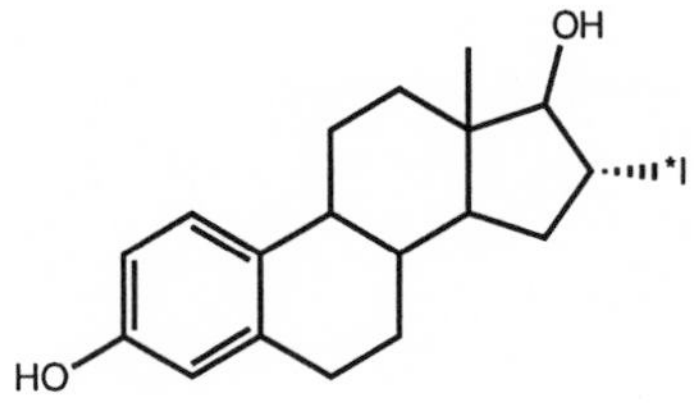

16α-[^{125}I]Iodoestradiol

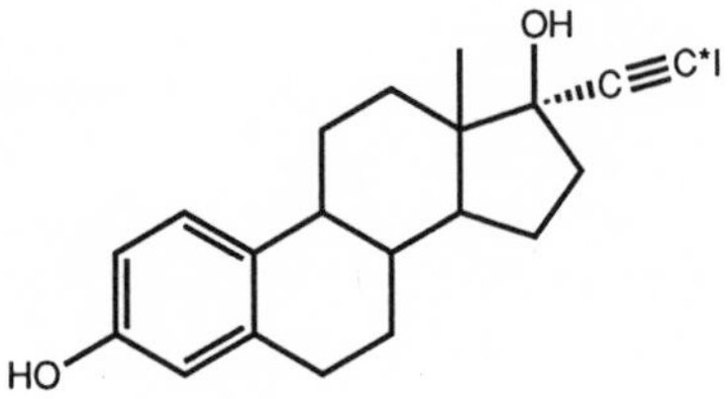

17α-[^{125}I]Iodoethynylestradiol

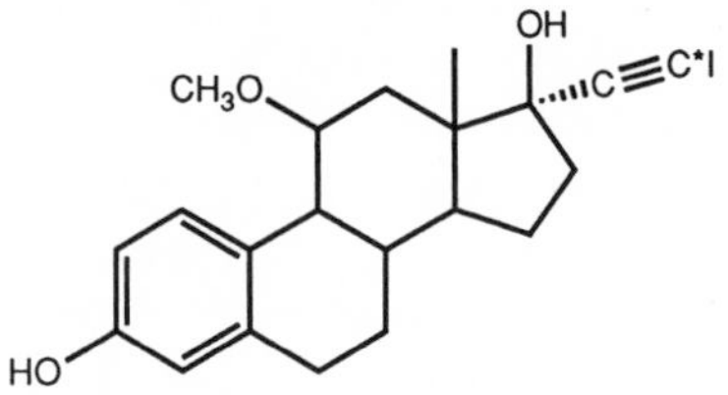

17α-[^{125}I]Iodoethynyl-11β-methoxy-
estradiol

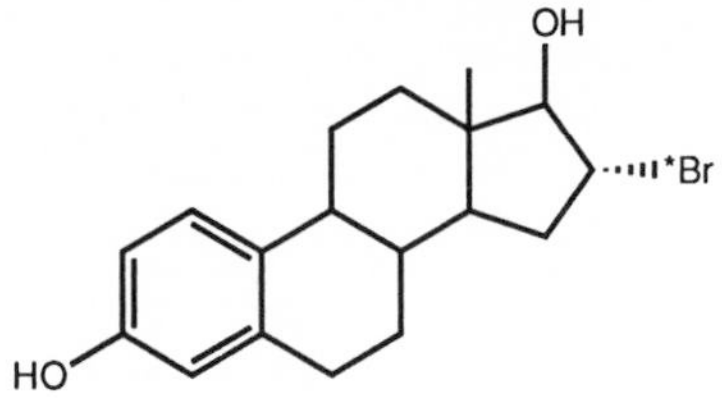

16α-[^{77}Br]Bromoestradiol

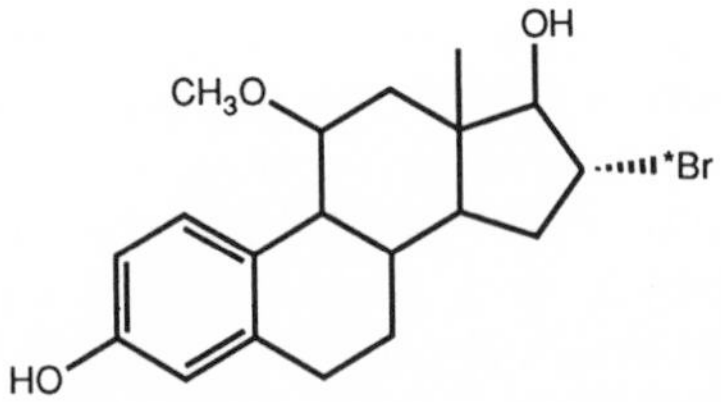

16α-[^{77}Br]Bromo-11β-methoxyestradiol

Figure 3-6. Examples of estrogen receptor imaging agents.

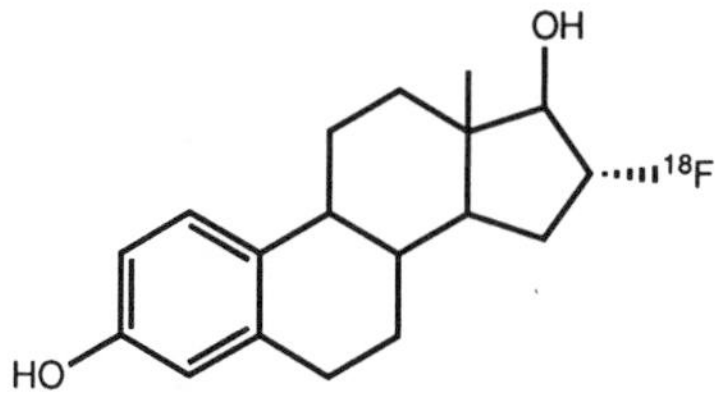

16α-[^{18}F]Fluoroestradiol

Figure 3-7. An effective PET imaging agent for the estrogen receptor.

was the most promising of a series of analogues that were taken up with very high selectivity into target tissues (Kiesewetter *et al.*, 1984). This analogue subsequently was selected as the labeled estrogen to be used in the first attempt to image human breast carcinoma with PET. In this study, an excellent correlation was found between uptake of the analogue at sites of primary carcinomas and foci of metastases, as measured by PET imaging, and the tumor estrogen receptor concentration measured *in vitro* after excision. Thus, this procedure shows promise for confirmation of diagnosis of primary and metastatic lesions of breast carcinoma, as well for the development of a noninvasive technique for prediction of hormone therapy responsiveness (Mintun *et al.*, 1988).

3.1.5.4 Radiolabeled Progestins

From the above examples, it is apparent that much progress has been made in the development of estrogen receptor imaging agents based on radiohalogenated estrogens. However, for a number of reasons, much recent attention has shifted to the development of radioligands with selective affinity for the progesterone receptor. Response to hormone therapy has been shown to correlate more precisely with progesterone receptor concentration than with estrogen receptor concentration. The presence of defective estrogen receptors, capable of binding ligands, but otherwise nonfunctional, has been invoked as one explanation for lower estrogen receptor–response correlation. On the other hand, progesterone receptor levels are low unless expressed by a functioning estrogen receptor system, and thus such false-positive responses would not be evident in progesterone receptor assays. In addition, since hormone therapy would often result in saturation of the estrogen receptor [for example, the frequently used antiestrogen tamoxifen (Fig. 3-8) would be expected to have high occupancy of available estrogen receptors], making estrogen receptor

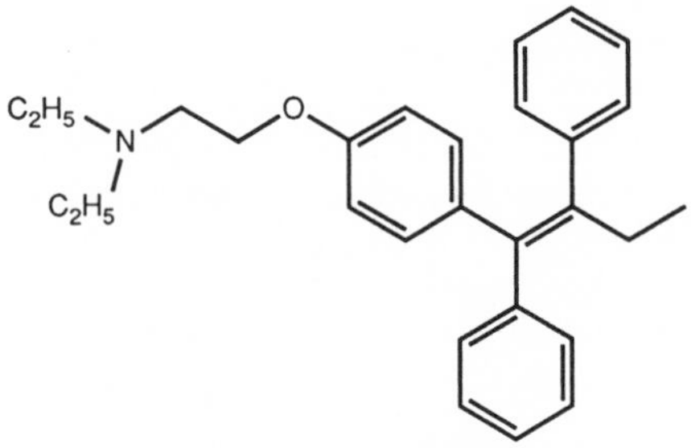

Figure 3-8. Tamoxifen.

assays difficult during therapy. Again, progesterone receptor concentration would reflect the concentration of functioning estrogen receptors and the estrogen or antiestrogen used in therapy would not interfere with the assay (Brandes and Katzenellenbogen, 1988, and references therein).

In one approach to the development of progesterone receptor radioligands, several iodinated analogues of 19-nortestosterone and of aromatic steroids were prepared, and their affinities for the progesterone receptor were determined (Fig. 3-9). 16α-Iodo-19-nortestosterone and (E)- and (Z)-17α-(2-iodovinyl)-19-nortestosterone were found to compete effectively for the progesterone receptor with R5020 (a tritiated ligand selective for this receptor), suggesting potential applications of these analogues as progesterone receptor ligands (Hoyte et al., 1985). Indeed, (E)-17α-(2-[^{125}I]iodovinyl)-19-nortestosterone was found to behave comparably to R5020 and appears thus to be a sensitive and specific γ-emitting probe for the progesterone receptor (Hochberg et al., 1985b).

Katzenellenbogen and co-workers have explored the use of ^{18}F-fluorinated androgens and progestins as positron-emitting probes for the androgen and progesterone receptor, respectively. Receptor binding and in

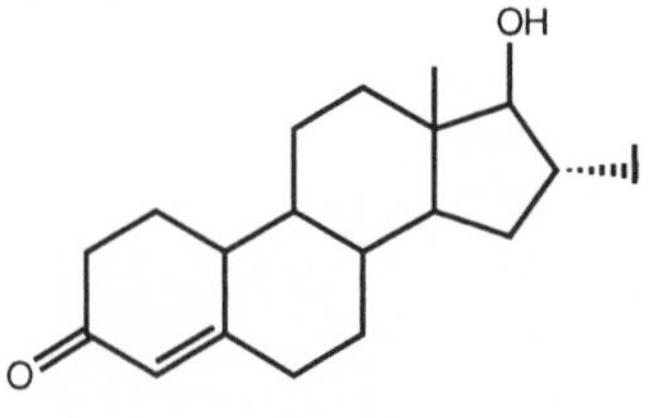

16α-Iodo-19-nortestosterone

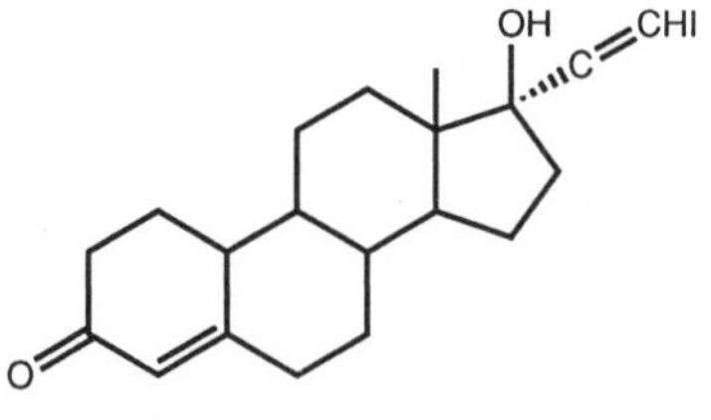

E- and Z-17α-2-(Iodovinyl)-19-nortestosterone

Figure 3-9. Progesterone receptor imaging agents.

vivo uptake studies had revealed that suitable imaging would be achieved only with ligands having significantly greater affinity than progesterone itself. To facilitate the search for such ligands, the structural features which affect binding in a series of steroids were analyzed systematically to determine which positions could be functionalized with the expectation of high and selective receptor affinity (Brandes and Katzenellenbogen, 1988). Of several analogues prepared in one study, 17α-(3-fluoro-1-propynyl)nortestosterone (Fig. 3-10) had the highest affinity and selectivity for the progesterone receptor (Brandes and Katzenellenbogen, 1987). In another report, a facile and rapid route to 21-[^{18}F]fluoro-16α-ethyl-19-norprogesterone (Fig. 3-10) was described. This compound was shown to have highly selective uptake into target tissue of estrogen-primed immature rats and appears to be highly promising as a PET imaging agent with potential clinical applications (Pomper *et al.*, 1988).

3.1.5.5 Radiolabeled Androgens

The potential prognostic and diagnostic value in the assay of androgen receptor concentrations in prostate tumors similarly has stimulated research toward the development of radiolabeled androgens [reviewed by Counsell *et al.* (1981) and Brandes and Katzenellenbogen (1988)]. A recent approach was based on the fact that 16α-iodoestradiol has good affinity for the estrogen receptor and that D-ring requirements for ligands to the estrogen and androgen receptors were considered to be comparable. Accordingly, 16α-iodo-, 16α-bromo-, and 16β-bromo-5α-dihydrotestosterone (Fig. 3-11) were synthesized, and their affinities for the androgen receptor were evaluated. In contrast to the estrogen series, the halogenated

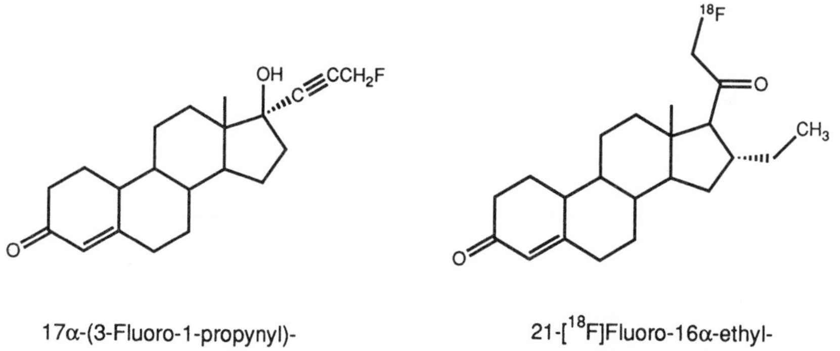

17α-(3-Fluoro-1-propynyl)-
nortestosterone

21-[^{18}F]Fluoro-16α-ethyl-
19-norprogesterone

Figure 3-10. PET imaging agents for the progesterone receptor.

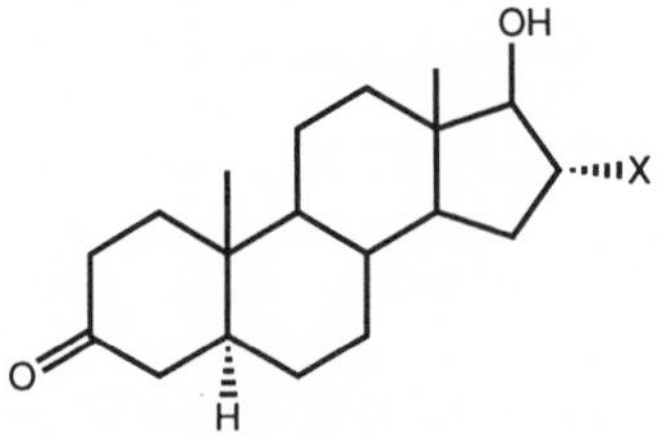

16α-Halo-5α-dihydrotestosterone   16β-Bromo-5α-dihydrotestosterone
(X = Br, I)

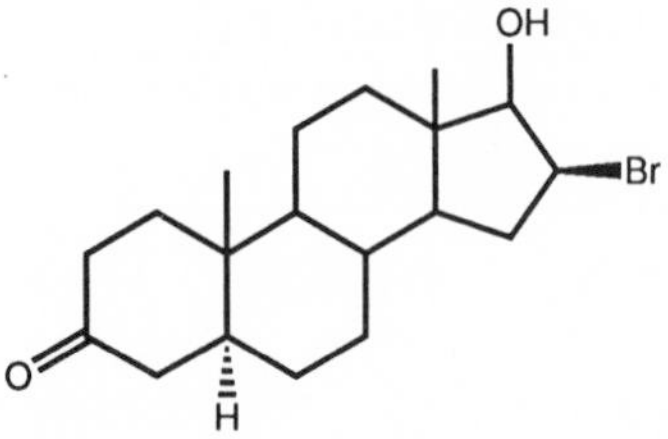

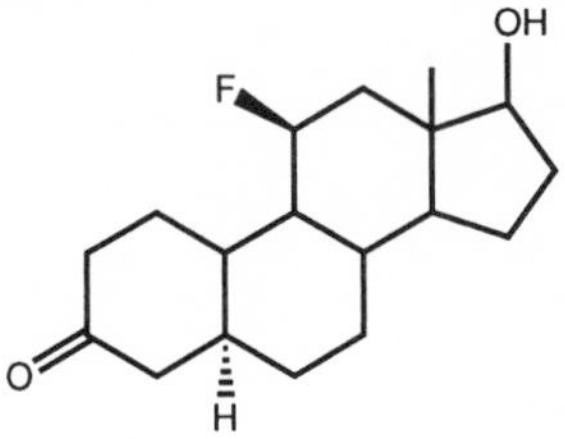

11β-Fluoronordihydrotestosterone

Figure 3-11. Androgen receptor imaging agents.

testosterone analogues had significantly lower receptor affinity than did dihydrotestosterone, and no specific binding of the 16α-iodo analogue was detected (Hoyte *et al.*, 1982).

Brandes and Katzenellenbogen (1987) found that, in a series of fluorinated androgens evaluated, 11β-fluoronordihydrotestosterone had the highest affinity for the androgen receptor (Fig. 3-11). Brandes and Katzenellenbogen (1988) also have presented an extensive review of the effects of structural modification on the binding of androgens to the androgen receptor in order to facilitate the rational design of PET imaging agents based on ^{18}F substitution.

3.1.5.6 Ecdysone Receptor Radioligands

Ecdysone (Fig. 3-12) regulates the timing of development and metamorphosis of arthropods. Studies on ecdysone receptors have been hampered by the lack of a suitable highly radioactive ligand with high affinity for the receptor. Cherbas *et al.* (1988) prepared 26-iodoponasterone (Fig. 3-12) and found it to be one of the most potent ecdysones known. The radiolabeled form was prepared using carrier-free [125]I. This analogue should be useful as a specific radioligand for the detection of ecdysone receptors.

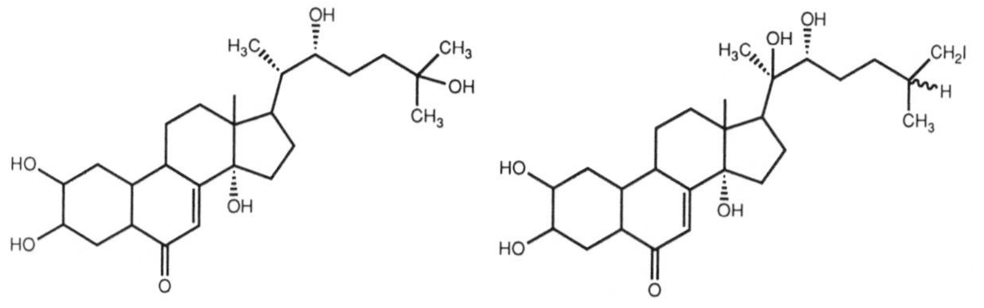

Ecdysone   26-Iodoponasterone

Figure 3-12. Ecdysone and the potent iodinated analogue potentially useful, in radioiodinated form, as an ecdysone receptor imaging agent.

3.1.6 Halogenated Steroids as Mechanistic Probes

3.1.6.1 Haloacetyl Steroids as Electrophilic Affinity Labels

Haloacetyl analogues of steroids have been used effectively to label steroid receptors as well as steroid processing enzymes [for reviews, see Gronemeyer and Govindan (1986), Katzenellenbogen and Katzenellenbogen (1984), Benisek et al. (1982), and Katzenellenbogen (1977)]. The extensive studies on estradiol 17β-dehydrogenase using a series of bromoacetoxy-labeled estrogen analogues illustrate the potential of this approach (Chin et al., 1982, and references therein). In another example, the active site of the C_{21} steroid side-chain cleavage cytochrome P-450 enzyme (responsible for the production of C_{19} androgenic steroids from the C_{21} steroids pregnenolone and progesterone) was labeled with 17-(bromoacetoxy)progesterone (Onoda et al., 1987). Many other examples of enzyme and receptor labeling with haloacetyl steroid analogues could be cited. The use of Δ^9-[16α-^{125}I]iodo-19-nortestosterone (Fig. 3-13) as a

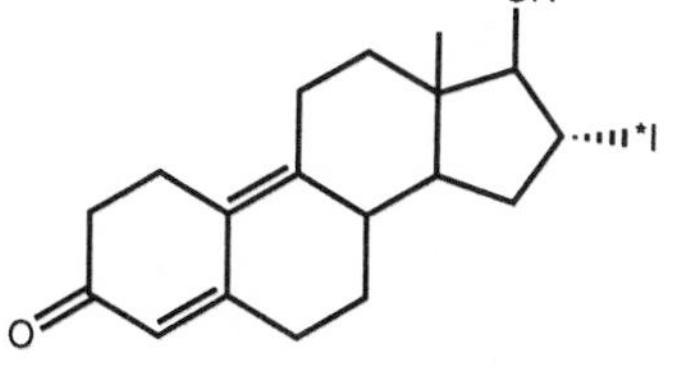

Δ^9-[16α-^{125}I]Iodo-19-nortestosterone

Figure 3-13. A photoaffinity label for the progesterone receptor.

γ-emitting photoaffinity label for the progesterone receptor has been reported (Lamb *et al.*, 1988). The dienone moiety functions as the photosensitive group.

3.1.6.2 A Mechanism-Based Aromatase Inhibitor

Marcotte and Robinson (1982) prepared 19,19-difluoroandrost-4-ene-3,17-dione (Fig. 3-14) as an irreversible inhibitor of estrogen synthase (aromatase). This cytochrome P-450 monooxygenase has the very important function of converting androgens into estrogens, and agents capable of modulating the activity of this enzyme have obvious clinical applications. The difluoro analogue is a potent irreversible inhibitor of estrogen synthase, causing a time-dependent inactivation of the enzyme ($K_i = 1$ μM, $k_{inact} = 0.023$ min^{-1}). An intermediate reactive acyl fluoride was suggested as the inactivating species (Fig. 3-14).

3.1.7 Fluorinated Steroids as Pro-Insecticides

Prestwich *et al.* (1984) have developed 29-fluorophytosterols (Fig. 3-15) as pro-insecticides. The toxicity of these analogues is produced by the *in vivo* formation of *erythro*-2-fluorocitrate. Enzymatic removal of the

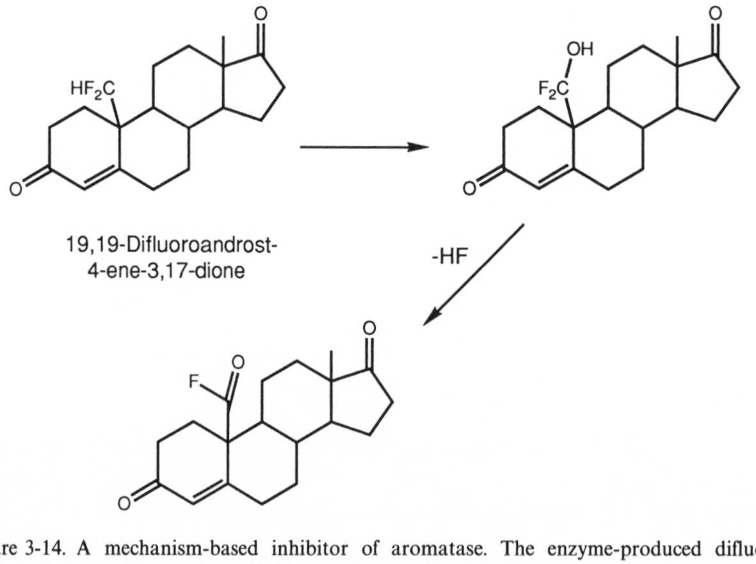

Figure 3-14. A mechanism-based inhibitor of aromatase. The enzyme-produced difluoroalcohol eliminates HF to give a highly reactive acyl fluoride (Marcotte and Robinson, 1982).

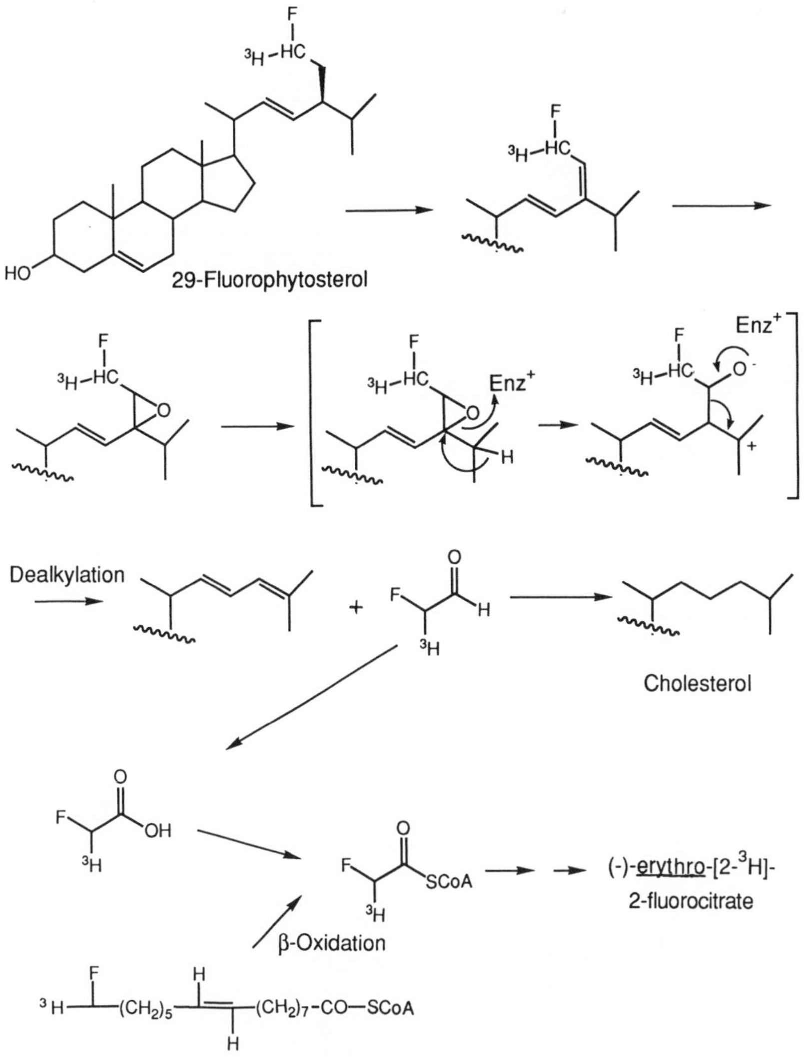

Figure 3-15. 29-Fluorophytosterol as a prodrug for the *in vivo* production of fluorocitrate. See text for details (Prestwich *et al.*, 1984).

fluoroethyl side chain, formally as fluoroacetaldehyde, presumably leads to the formation of fluoroacetyl-CoA, which then is processed by the insect to fluorocitrate. This and related work exploiting fluorinated sterols, hormones, and pheromones for the disruption of insect growth, development, or communication has been reviewed by Prestwich (1986).

3.1.8 Halogenated Steroids—Summary

Probably in no other field of medicinal chemistry has halogen substitution been exploited to the extent that it has in the development of steroid analogues. Halogenation, together with other systematic structural variations, has produced a huge number of useful analogues, as well as an enormous volume of literature. In the preceding sections, an attempt has been made to give an overview of certain aspects of this research. For more thorough discussions of these areas, the reviews and recent literature cited should be consulted.

3.2 Halogenated Analogues of Vitamin D

The role of vitamin D in calcium metabolism was discussed briefly in Chapter 2 in Vol. 9A of this series. Specific fluorination of vitamin D_3 and its functional metabolites has received much recent attention as a method for producing analogues which could have more potent or more specific activity, which could serve as inhibitors of vitamin D_3 metabolism, or which could act as vitamin D_3 antagonists. Kobayashi and Taguchi (1982) and Welch (1987) have reviewed this subject.

3.2.1 Biosynthesis and Metabolism of Vitamin D_3

The biosynthesis of vitamin D_3 (D_3) and its functional metabolism are shown in Fig. 3-16. Vitamin D_3, either produced in the skin from cholesterol or adsorbed through the small intestine, is hydroxylated in the liver to give $25\text{-OH-}D_3$. As the major circulating metabolite of D_3, $25\text{-OH-}D_3$ must undergo further hydroxylation in the kidney at carbon-1 or carbon-24 to give $1\alpha,25\text{-}(OH)_2\text{-}D_3$ or $24(R),25\text{-}(OH)_2\text{-}D_3$, respectively. The former is the most active form of vitamin D_3 and induces bone calcium and phosphate mobilization and bone calcification.

3.2.2 Introduction of Fluorine to Block C-25 or 1α-Hydroxylation as an Approach to D_3 Antagonists

$25\text{-F-}D_3$ (Fig. 3-17) has been prepared in an attempt to produce an analogue in which the obligatory hydroxylation of C-25 would be blocked. Although this analogue was an inhibitor of 25-hydroxylase *in vivo*, it did not block *in vivo* vitamin D activity. In a similar study, $1\alpha\text{-OH-25-F-}D_3$

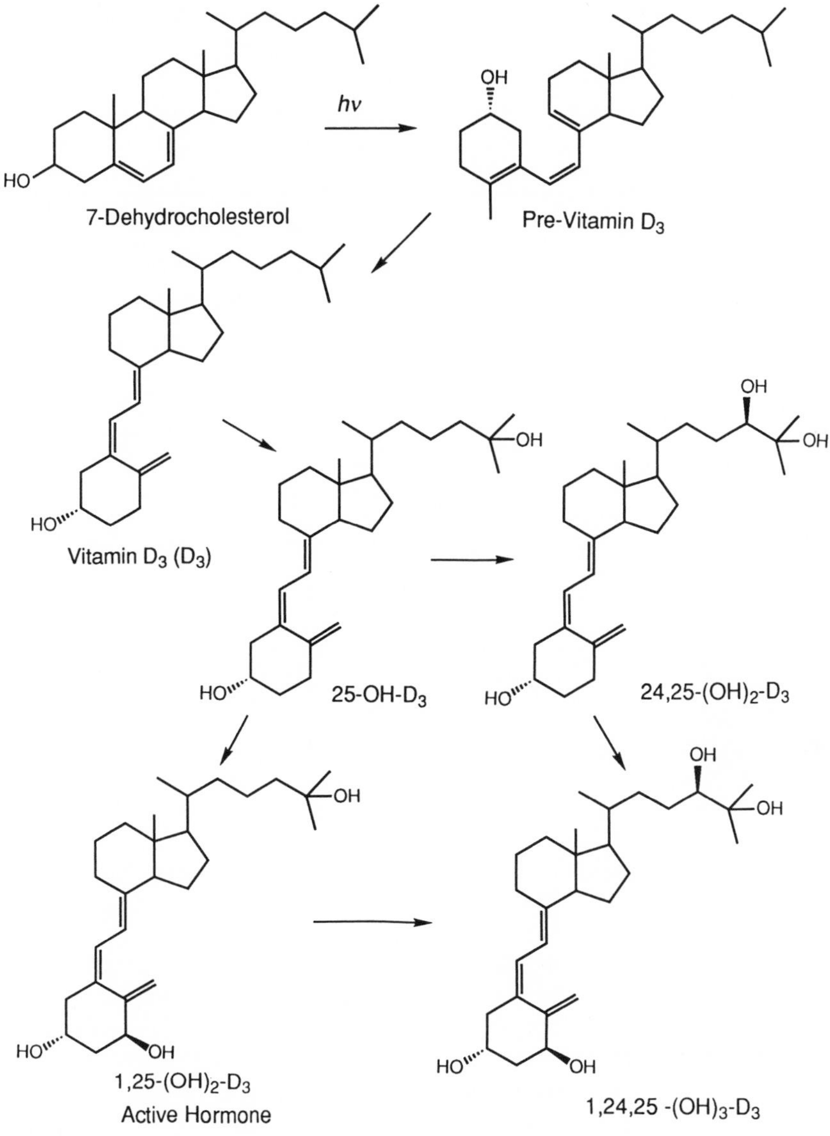

Figure 3-16. Biosynthesis of vitamin D_3 and related hormones.

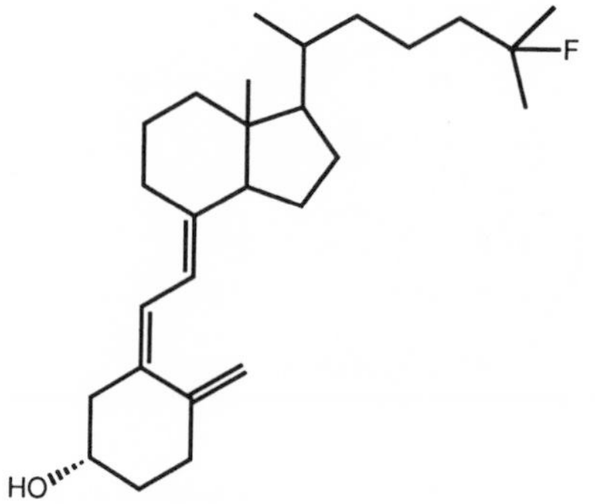

25-F-D$_3$ 1α,24R-(OH)$_2$-25-F-D$_3$

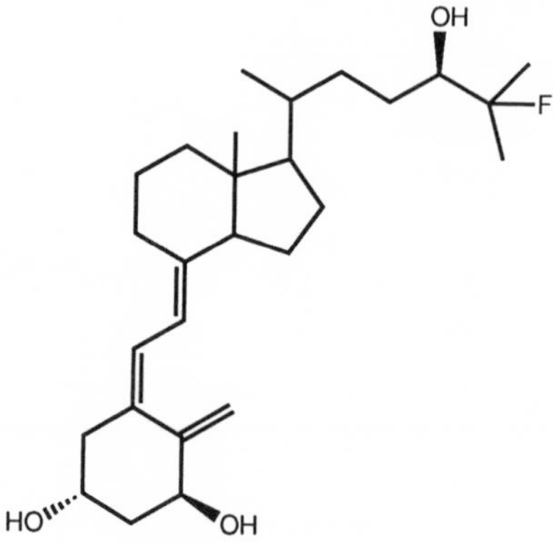

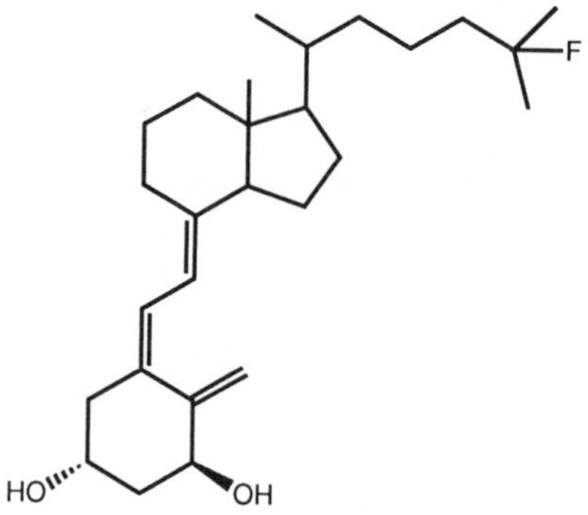

1α-OH-25-F-D$_3$

Figure 3-17. Examples of C-25-fluorinated D$_3$ analogues as potential vitamin D$_3$ antagonists.

(Fig. 3-17) was shown to be 50 times less active than 1α,25-(OH)$_2$-D$_3$ with respect to stimulation of bone calcium mobilization and intestinal calcium transport. Partial separation of vitamin D$_3$ functions was achieved with 1α,24(R)-(OH)$_2$-25-F-D$_3$ (Fig. 3-17). In sum, the results obtained with these and other 25-F analogues suggest that the 25-hydroxyl group is not obligatory for activity, but can be replaced by the 24-hydroxyl functionality. In these analogues, fluorine appears to function as a replacement for hydrogen rather than for a hydroxyl group [reviewed by Kobayashi and Taguchi (1982)].

Blockade of the critical 1α-hydroxylation in 1α-F-25-OH-D$_3$ (Fig. 3-18)

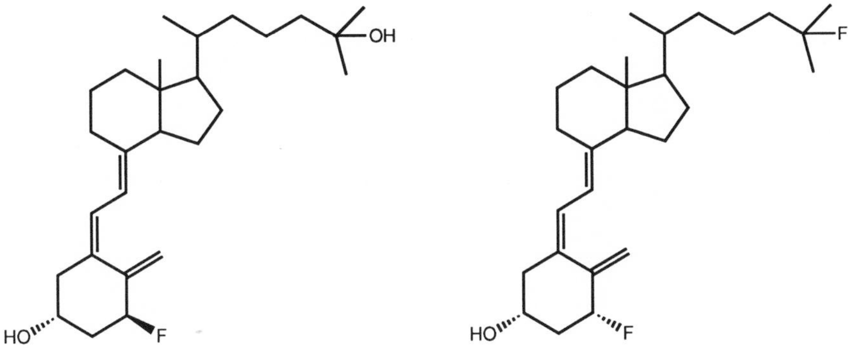

1α-F-25-OH-D$_3$ 1β,25-F$_2$-D$_3$

Figure 3-18. Potential D$_3$ antagonists based on C-1 fluorination.

results in an analogue which has no vitamin D$_3$ activity. However, this analogue had a 30-fold greater affinity for the chick intestinal D$_3$ receptor than did 25-OH-D$_3$ (Ohshima *et al.*, 1984). 1β,25-F$_2$-D$_3$ (Fig. 3-18) also was inactive and showed no affinity for 1α-hydroxylase (Paaren *et al.*, 1981).

3.2.3 Introduction of Fluorine to Increase D$_3$ Activity

Two approaches to produce active potent analogues through selective fluorination have been used. One strategy was based on the precedent of

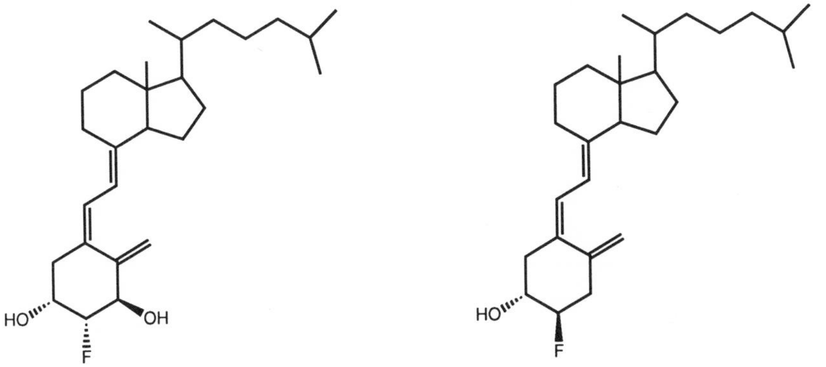

2β-F-1α-OH-D$_3$ 2α-F-D$_3$

Figure 3-19. Incorporation of a trans-β-fluorohydrin structural unit as an approach to the development of more potent D$_3$ analogues.

increased potency caused by the presence of the *trans*-fluorohydrin moiety in 9α-fluorocorticosteroids. Thus, 2β-F-1α-OH-D$_3$ (Fig. 3-19) was prepared and found to be more active than D$_3$ itself (Oshida *et al.*, 1980). In contrast, 2α-F-D$_3$ was found to be equivalent to D$_3$ itself (Kobayashi *et al.*, 1986). Whereas the 9α-fluoro substituent is *trans* to the 11β-hydroxyl group in the potent 9α-fluorocorticosteroids, hydroxylation of 2α-F-D$_3$ to produce the active hormone would place the critical 1α-hydroxyl group in a *cis* relation to fluorine.

Increased activity can result from the ability of a fluorine substituent to block degradation. Whereas fluorine at C-25 behaves like hydrogen,

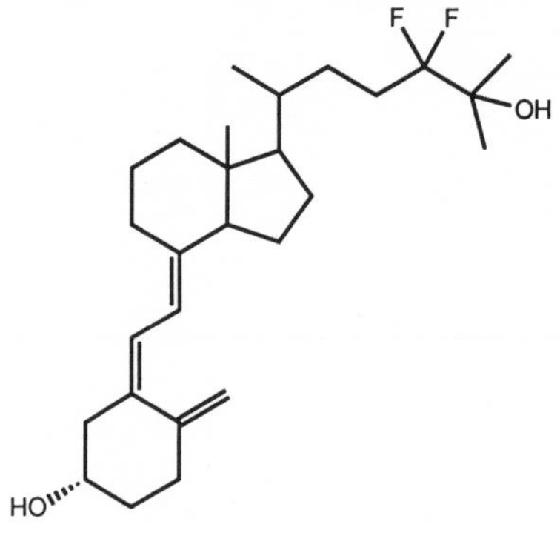

24,24-F$_2$-25-OH-D$_3$ 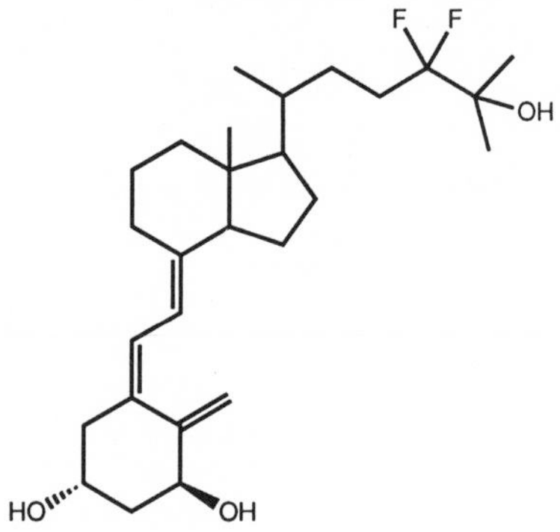   24,24-F$_2$-1α,25-(OH)$_2$-D$_3$

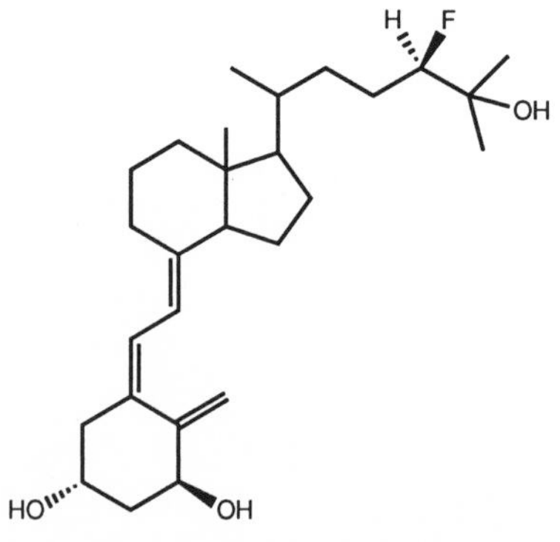

24(R)-F-1α,25-(OH)$_2$-D$_3$

Figure 3-20. Examples of analogues that have fluorine substituents situated at positions critical to hydroxylative deactivation as a strategy to produce more active hormones.

based on the effects on biological behavior, more dramatic results were found in analogues having multiple fluorine substituents in other positions of the side chain. In particular, $24,24\text{-}F_2\text{-}25\text{-}OH\text{-}D_3$ (Fig. 3-20) was found to be seven times more potent than $25\text{-}OH\text{-}D_3$ in bone resorptive activity *in vitro*, although *in vivo* calcium-mobilizing activity was unaffected (Stern *et al.*, 1981). More interesting were the *in vivo* properties of $24,24\text{-}F_2\text{-}1\alpha,25\text{-}(OH)_2\text{-}D_3$ (Fig. 3-20). While *in vitro* bone-resorbing effects and receptor binding were comparable to those of $1\alpha,25\text{-}(OH)_2\text{-}D_3$ (Stern *et al.*, 1981), intestinal Ca^{2+} transport, P_i and Ca^{2+} mobilization, and antirachitic activity were increased 5 to 10 times. The comparable binding characteristics suggest that better target organ response is not the cause for the greater potency, but, rather, blockade of hydroxylation at C-24 may decrease the rate of inactivation (Okamoto *et al.*, 1983). $24(R)\text{-}F\text{-}1\alpha,25\text{-}(OH)_2\text{-}D_3$ (Fig. 3-20), having fluorine specifically located at the $24R$ position, has a significantly longer half-life than $1\alpha,25\text{-}(OH)_2\text{-}D_3$, a property attributed to blockade of hydroxylation at the $24R$ position (Shiuey *et al.*, 1988). This analogue had potent *in vivo* antirachitogenic activity. Both $1\alpha,25\text{-}(OH)_2\text{-}D_3$ and $24,24\text{-}F_2\text{-}1\alpha,25\text{-}(OH)_2\text{-}D_3$ at $1\text{–}100 \text{ n}M$ were effective in induction of macrophage differentiation of normal and leukemic myeloid stem cells (Koeffler *et al.*, 1984).

Examples of other side-chain-fluorinated analogues include 23,23-difluorinated analogues (Nakada *et al.*, 1985) and analogues with trifluoromethyl groups replacing the isopropyl methyl groups (Stern *et al.*, 1984) (Fig. 3-21). In the $25\text{-}OH\text{-}D_3$ series, the hexafluoro analogue was 40-fold more potent than the parent in the *in vitro* bone resorption assay.

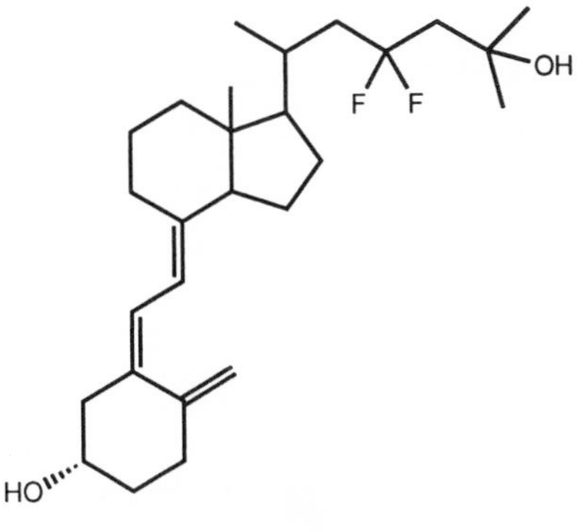

23,23-F_2-25-OH-D_3 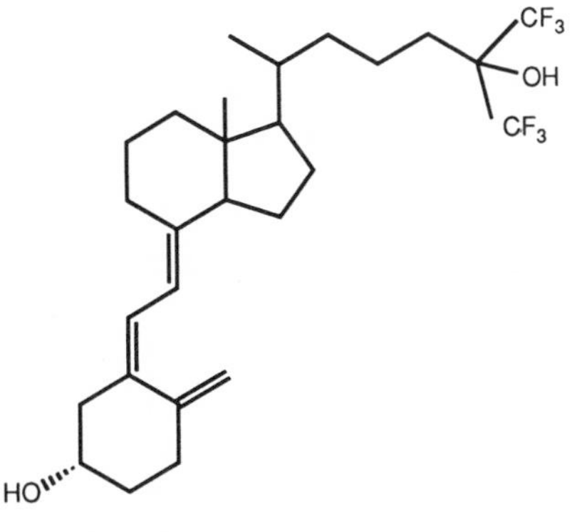   26,26,26,27,27,27-F_6-25-OH-D_3

Figure 3-21. Further examples of side-chain fluorinated D_3 analogues.

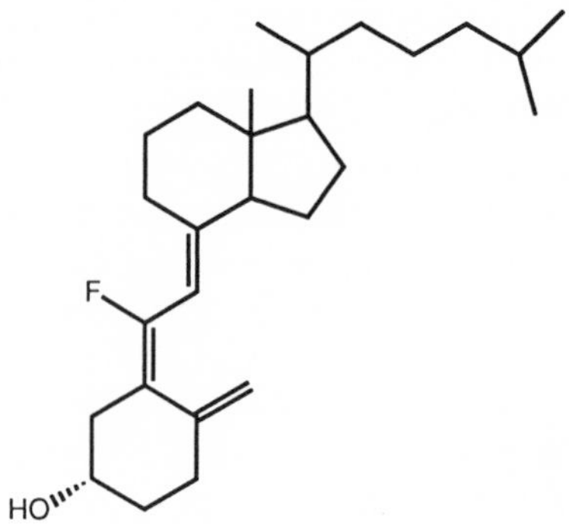

6-F-D$_3$

Figure 3-22. A fluorinated D$_3$ antagonist.

The previous studies have concentrated on positions where the presence of fluorine may affect the physicochemical properties of hydroxyl groups or may alter metabolism by blocking hydroxylation. Wilhelm *et al.* (1984) have reported that 6-F-D$_3$ (Fig. 3-22) had no D$_3$ activity, but competed with 1α,25-(OH)$_2$-D$_3$ and 25-OH-D$_3$ for binding to the 1α,25-(OH)$_2$-D$_3$-specific intestinal receptor with an affinity slightly less than that of the natural ligand, but greater than that of 25-OH-D$_3$. Several aspects of this result are notable. First, this shows that the triene portion of the molecule has an important role in receptor interaction. Second, since D$_3$ itself has no capacity to interact with the receptor, the presence of fluorine at C-6 serves the functional role, with regard to receptor binding, of C-1 or C-25 hydroxylation. Finally, this is the first vitamin D$_3$ antagonist reported to antagonize binding to the intestinal 1α,25-(OH)$_2$-D$_3$ receptor and to effect *in vivo* inhibition of intestinal Ca^{2+} absorption.

3.3 Biochemistry of Halogenated Retinoids

In a process central to the biochemistry of vision, 11-*cis*-retinal becomes covalently attached to opsin as a protonated Schiff base formed with a lysine ε-amino group of the protein. This attachment forms the visual pigment rhodopsin. The photochemically induced isomerization of rhodopsin at C-11 produces all-*trans* rhodopsin (bathorhodopsin), generally considered the primary photoproduct of vision (Fig. 3-23). Recent research by Nakanishi and co-workers, Liu and co-workers, and

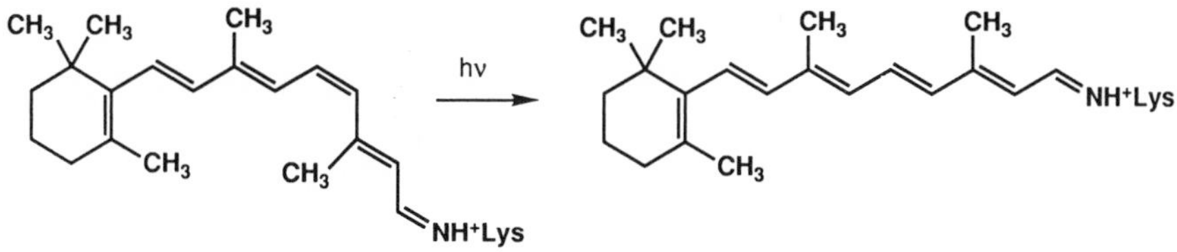

Rhodopsin (11-cis)   Bathorhodopsin (all trans)

Figure 3-23. The photochemically induced isomerization of rhodopsin at C-11 to produce all-*trans* rhodopsin (bathorhodopsin) in the primary photochemical process of vision.

others has produced a detailed understanding of the molecular events involved in these processes [for a recent review, see Liu and Browne (1986)].

Liu and co-workers have prepared a series of fluorinated retinals and rhodopsins as mechanistic probes in their research on the biochemistry of vision, one goal being to develop a sensitive ^{19}F-NMR probe to follow the photochemical and subsequent dark reactions (Shichida *et al.*, 1987, and references therein). 10-Fluoro- and 14-fluororetinals and fluororhodopsins were the first to be synthesized (Asato *et al.*, 1978). 11-*cis*-12-Fluororetinal and 12-fluororhodopsin (the structures of the retinals are given in Fig. 3-24), with fluorine situated on the configurationally significant 11,12 double bond, appeared more promising for ^{19}F-NMR studies. 12-Fluororhodopsin resembled rhodopsin with respect to many physical and chemical properties, although line broadening complicated interpretation of NMR spectra (Liu *et al.*, 1981). Anomalous photochemical behavior of 10-fluororhodopsin made this analogue less attractive than the 12-fluoro analogue as a model for natural visual pigments, although this unusual behavior has in itself proven quite useful (see below) (Liu *et al.*, 1986). 19,19,19- and 20,20,20-Trifluororetinals (Fig. 3-24) (Hanzawa *et al.*, 1985; Mead *et al.*, 1985, and references therein) have been synthesized to serve as more sensitive NMR probes. A vicinal difluororetinal—expected to be particularly useful for studying *cis–trans* isomerization of the corresponding rhodopsins because of expected large differences in coupling between *cis* and *trans* vicinal olefinic fluorines—has been reported (Fig. 3-24) (Asato and Liu, 1986).

The behavior of 9-*cis*-10-fluororhodopsin has been important in a study of the precise mechanism by which bathorhodopsin is formed. A *cis–trans* isomerization involving rotation of a single carbon–carbon bond to give s-*trans* bathorhodopsin, as formulated originally, is a volume-demanding process, seemingly inconsistent with the rapid primary photochemical process observed. An alternative mode of isomerization has

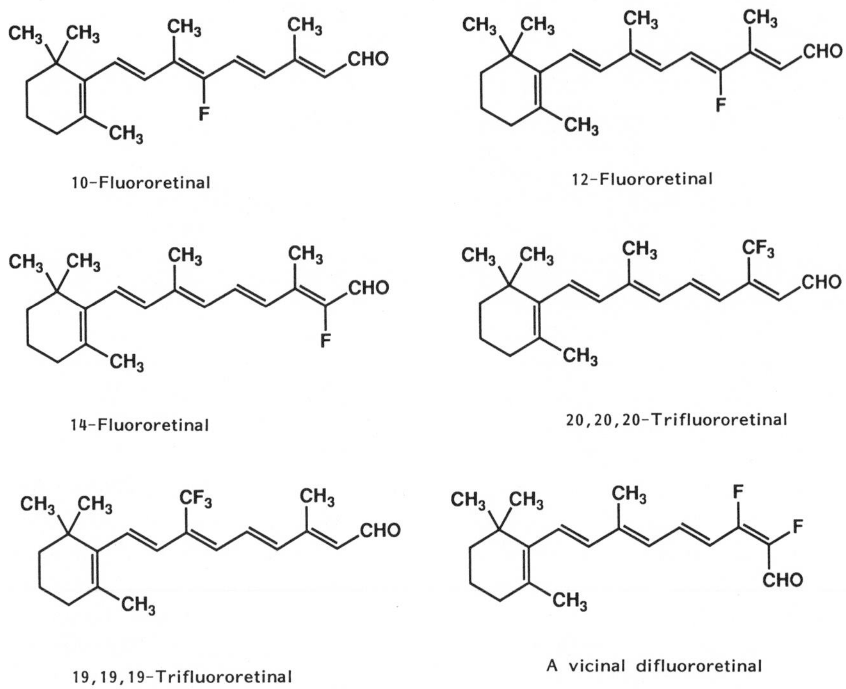

Figure 3-24. Examples of fluorinated analogues of retinal.

been proposed which involves rotation of two adjacent bonds flanking a carbon center "*n*." This "hula-twist at center *n*" (H.T.-*n*), producing 10-*s*-*cis* all-*trans* bathorhodopsin (Fig. 3-25) as the primary photochemical product, is much less volume-demanding and, further, is consistent with spectral data for the process (Liu and Browne, 1986). Investigation of the cause of the anomalous behavior of 10-fluororhodopsin has revealed specific sites of interaction between opsin and the chromophore and has provided support for the H.T.-*n* model for the photochemical isomerization. A dramatic decrease in quantum yield unique in the photoisomerization of 9-*cis*-10-fluororhodopsin has been attributed to hydrogen bonding of fluorine at C-10 to a carboxyl group on the protein. Isomerization of the 9-*cis* isomer by a hula twist at C-10 (H.T.-10) places the 10-substituent at the reaction center (Fig. 3-26). By the same reasoning, a 10-substituent in the 11-*cis* analogue does not directly coincide with the reaction center in the H.T.-11 process, and photoisomerization occurs efficiently (Liu *et al.*, 1986). Use of 9-*cis*-10-fluororhodopsin in a dark exchange reaction with

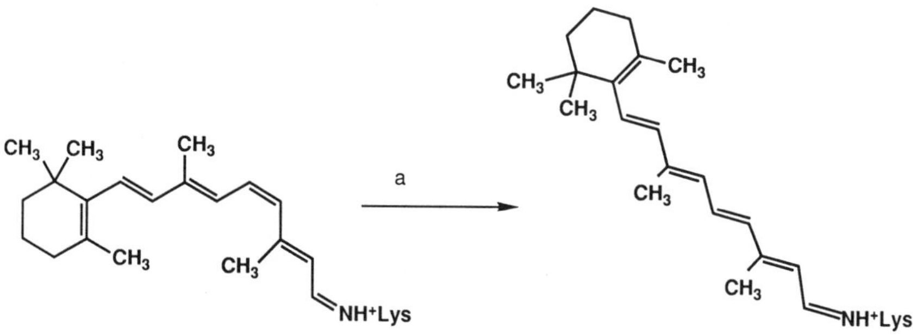

Rhodopsin (11-cis)

(All trans) Bathorhodopsin
(s-trans)

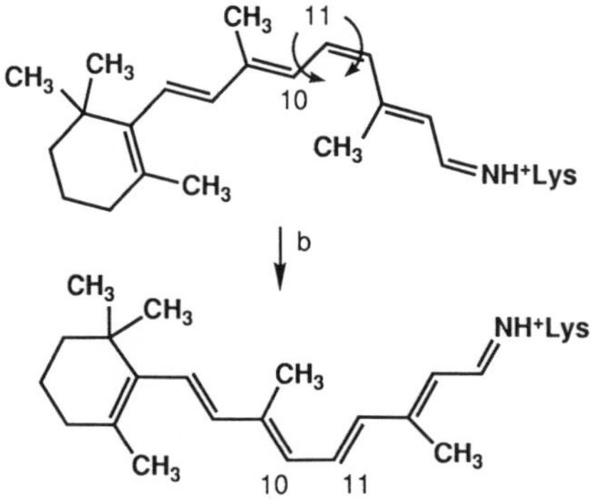

(All-trans) 10-s-cis-bathorhodopsin

Figure 3-25. Two mechanisms for photoisomerization of 11-cis-rhodopsin: (a) cis–trans isomerization involving rotation of a single carbon–carbon (here, C-10−C-11) bond to give s-trans bathorhodopsin; and (b) rotation of two adjacent bonds flanking a carbon center "n" (here, C-11) to give 10-s-cis all-trans bathorhodopsin (Liu and Browne, 1986).

the 11-cis natural chromophore facilitated the study of this process because of the low photosensitivity of the analogue (Crescitelli, 1988).

While the presence of fluorine has no effect on the photoisomerization of 11-cis-10-fluororhodopsin to all-trans bathorhodopsin (fluorine on C-10 in an H.T.-11 process), the fate of the bathoproduct (all-trans) is altered by fluorine at C-10. Natural bathorhodopsin and batho-12-fluororhodopsin (both all-trans) are photoisomerized by blue light to a mixture of C-11-cis and C-9-cis isomers. In contrast, batho-10-fluororhodopsin, under the same conditions, gives only the original C-11-cis structure, showing that isomerization of the 9,10 double bond is inhibited. These data again are consistent with anchoring of C-10 to the protein through an electrostatic

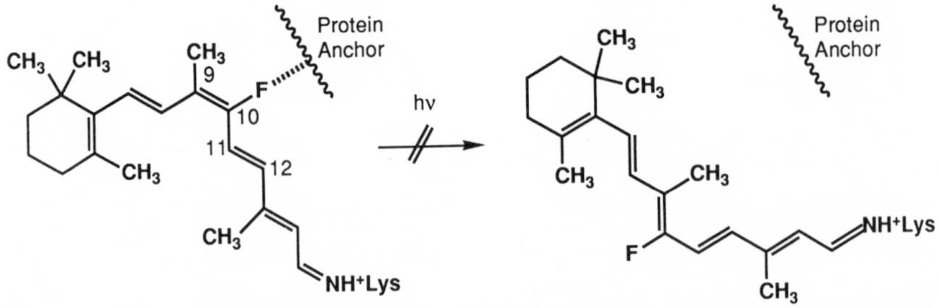

Figure 3-26. A decrease in quantum yield in the photoisomerization of 9-*cis*-10-fluororhodopsin is explained by hydrogen bonding of fluorine at C-10 to an "anchor" on the protein, an interaction that inhibits "hula" rotation (Liu *et al.*, 1986).

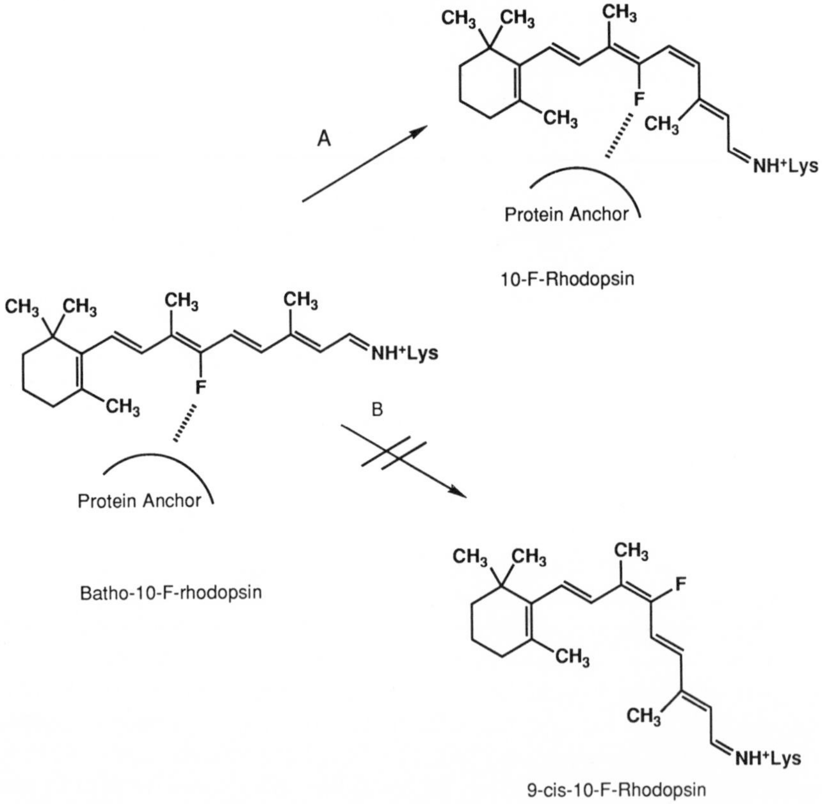

Figure 3-27. An electrostatic interaction of C-10 fluorine of batho-10-fluororhodopsin with the protein inhibits photoisomerization of the 9,10 double bond but has no effect on isomerization at C-11 (Shichida *et al.*, 1987).

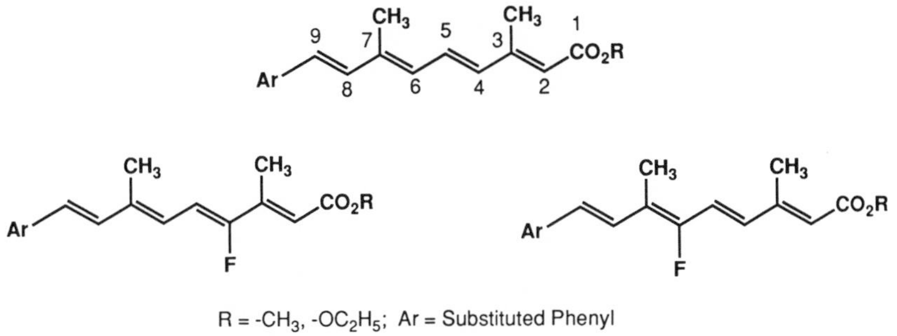

R = -CH₃, -OC₂H₅; Ar = Substituted Phenyl

Figure 3-28. Examples of fluorinated retenoic acids analogues developed as anticancer agents.

interaction of the fluorine with the protein (Fig. 3-27), as discussed above (Shichida *et al.*, 1987).

The preceding discussion offers a sampling of extensive ongoing research in which the behavior of retinal analogues has been exploited to study the restricted cavity of the chromophore binding site of opsin. In a recent additional example, photopigments of several analogues of 11-*cis*- and 9-*cis*-retinal, including the fluorinated and chlorinated analogues, were generated from two opsins from different sources, and the spectral properties were compared (Crescitelli and Liu, 1988).

Several fluorinated retinoic acids have been investigated in research designed to improve the action of retinoids against papillomas and carcinomas in mice. From testing of many analogues, including those with replacement of the β-ionone ring with an aromatic ring, substitution of fluorine at C-4 and C-6 (Fig. 3-28) showed the most promise for increasing effectiveness (Lovey and Pawson, 1982).

3.4 Fluorinated Analogues of Terpenoid Biosynthetic Intermediates: Mevalonic Acid, Isopentenyl Pyrophosphate, and Dimethylallyl Pyrophosphate

Mevalonic acid (Mev) (Fig. 3-29) is a key intermediate in the terpene biosynthetic pathway and thus is a precursor of such diverse natural products as steroids, carotenoids, and plant terpenoids. Early steps in terpene biosynthesis involve phosphorylation and decarboxylation of Mev to produce isopentenyl pyrophosphate (IPP), isomerization of IPP to dimethylallyl pyrophosphate (DMAPP), and condensation between C-1 of DMAPP and C-4 of IPP to give geranyl pyrophosphate (Fig. 3-29).

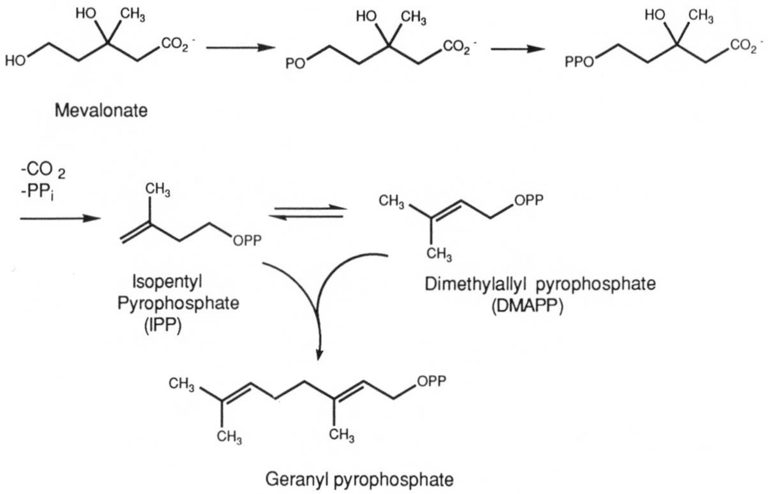

Figure 3-29. Early steps in the biosynthesis of terpenes from mevalonic acid.

Fluorinated analogues of Mev, IPP, and DMAPP have been developed as potential inhibitors and as mechanistic probes for the enzymes responsible for these conversions.

3.4.1 Fluoromevalonic Acids as Inhibitors of Juvenile Hormone and Cholesterol Biosynthesis

Early studies showed 6-fluoromevalonate (6-F-Mev) (Fig. 3-30) to be a fairly potent inhibitor of the incorporation of labeled acetate and mevalonate into cholesterol by rat liver homogenates (Singer *et al.*, 1959; Tschesche and Machleidt, 1960). This hypocholesterolemic effect led Quisted *et al.* (1981) to explore the possibility that 6-F-Mev, as well as the corresponding difluoro and trifluoro derivatives, could be used for insect control by inhibiting juvenile hormone biosynthesis at an early stage. The anti-juvenile hormone activity shown by 6-F-Mev, and to a lesser extent by difluoro-Mev, was manifested only in lepidoptera. Although activity was significant, with 50% effective dose (ED_{50}) values ranging from 0.7 to >200 mg per gram, it was deemed insufficient to merit commercial development.

Two recent studies have explored the mechanism of the inhibitory action of 6-F-Mev on cholesterol and juvenile hormone biosynthesis. Fluorinated analogues of DMAPP and IPP had been shown to be poor

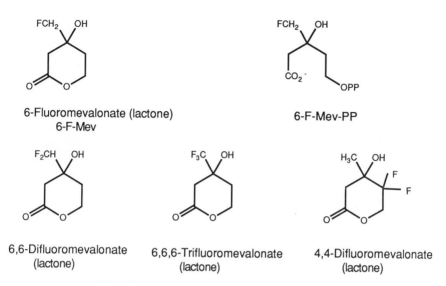

6-Fluoromevalonate (lactone)
6-F-Mev

6-F-Mev-PP

6,6-Difluoromevalonate
(lactone)

6,6,6-Trifluoromevalonate
(lactone)

4,4-Difluoromevalonate
(lactone)

Figure 3-30. Fluorinated analogues of mevalonic acid.

substrates and, in some cases, potent inhibitors of the enzymes processing biosynthetic intermediates formed from Mev (see below, Section 3.4.2). The demonstration by Nave *et al.* (1985) that 6-F-Mev blocks the incorporation of Mev, but not of IPP, into lipids by a rat liver multienzyme system indicates that inhibition occurs before conversion to IPP (see Fig. 3-29). Accumulation of Mev-pyrophosphate (Mev-PP) and, to a lesser extent, Mev-phosphate implicated the decarboxylation of Mev-PP as the site of inhibition. During decarboxylation, the hydroxyl group on C-3 is lost. Inhibition of $C-O$ bond cleavage by the presence of the very electronegative fluorine atom at C-6 was suggested as a reason for the inability of the decarboxylase to catalyze the conversion of 6-F-Mev-PP to IPP. Although reaction does not occur, binding of 6-F-Mev-PP to the enzyme is some two orders of magnitude more efficient ($K_i = 37$ nM) than binding of Mev-PP ($K_m = 10\ \mu M$), making 6-F-Mev-PP the most potent known inhibitor of avian liver Mev-PP decarboxylase. Nave *et al.* suggest that 6-F-Mev-PP resembles a transition state analogue. Despite this potency, the conversion of 6-F-Mev to 6-F-Mev-PP is inefficient, and the latter is prone to deactivation by phosphatases, casting doubt on *in vivo* efficacy (Nave *et al.*, 1985).

Reardon and Abeles (1987) carried out a similar study using 6-F-Mev, as well as on 6,6-difluoro-Mev, 6,6,6-trifluoro-Mev, and 4,4-difluoro-Mev (Fig. 3-30). Again, the most effective inhibitor of cholesterol biosynthesis was 6-F-Mev, with the site of action identified as an inhibition of the conversion of the Mev-PP to IPP. They also noted the remarkably low K_i (10 nM) of 6-F-Mev for the decarboxylase enzyme and postulated an inhibition of phosphorylation of the C-3 hydroxyl group by fluorine. Accumulation of an enzyme–6-F-Mev–ATP complex with a low K_{diss} could account for the low K_i.

The potent inhibition of a key enzyme in the early stages of the biosynthesis of cholesterol (and other products of terpene biosynthesis) shown by 6-F-Mev suggests that, with proper manipulation of analogue structures, an effective *in vivo* agent may be developed. At the least, these studies demonstrate again the profound effects produced by the presence of fluorine on enzyme substrates.

3.4.2 Fluorinated Analogues of IPP and DMAPP as Mechanistic Probes and Inhibitors of Terpene Biosynthesis

Reardon and Abeles (1986) prepared a series of substrate analogues to study the mechanism of isopentenyl pyrophosphate isomerase, the enzyme that catalyzes the interconversion of IPP and DMAPP by an antarafacial [1,3] transposition of hydrogen. Protonation of the double bond to give a tertiary carbonium ion or removal of a proton to produce an allylic carbanion have been considered as possibilities for the initial step. Evidence supports the participation of two bases, one functioning as its conjugate acid, to facilitate the ionization. (Z)-3-(Trifluoromethyl)-2-butenyl pyrophosphate (Fig. 3-31) reacts at a rate 1.8×10^{-6} times the rate for IPP. This was attributed to a destabilizing effect of the electronegative

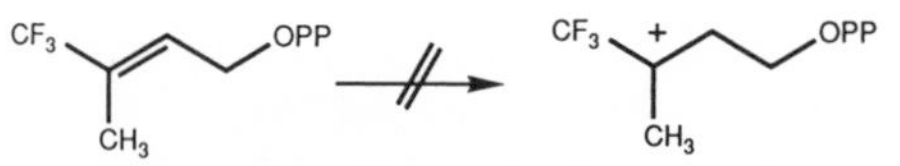

(Z)-3-(Trifluoromethyl)-2-butenyl pyrophosphate

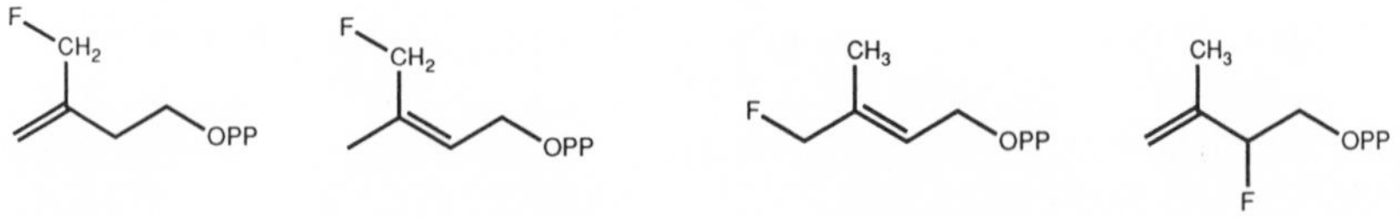

4-F-IPP   (Z)-4-F-DMAPP   (E)-4-F-DMAPP   2-F-IPP

Figure 3-31. (Z)-3-(Trifluoromethyl)-2-butenyl pyrophosphate and fluorinated analogues of DMAPP and IPP as probes for the mechanism of the isomerase reaction.

trifluoromethyl group on an intermediate carbonium ion, giving support to the cationic mechanism (Fig. 3-31).

The fluoromethyl analogues of IPP and DMAPP (Fig. 3-31) were included in a series of compounds prepared by Poulter and co-workers to study the isomerase reaction. 2-F-IPP and (Z)- and (E)-4-F-DMAPP were not substrates for the reaction, but instead were found to be active-site-directed inhibitors. Covalent attachment of these analogues to the enzyme presumably results from S_N2 or S_N2' attack of the enzyme-bound nucleophile responsible for the antarafacial deprotonation–protonation of the isomerase reaction, while lack of reaction again can be attributed to a retardation of ionization due to the presence of fluorine. Adding support to the cationic mechanism is the fact that 2-(dimethylamino)ethyl diphosphate functioned as a very potent transition state analogue (Muehlbacher and Poulter, 1988, 1985).

Poulter and co-workers have also made extensive use of fluorinated substrate analogues in studying the mechanism of the prenyl transfer reaction, the next step in terpene biosynthesis (Poulter et al., 1979; Poulter and Rilling, 1978). An S_N2 displacement or ionization prior to condensation

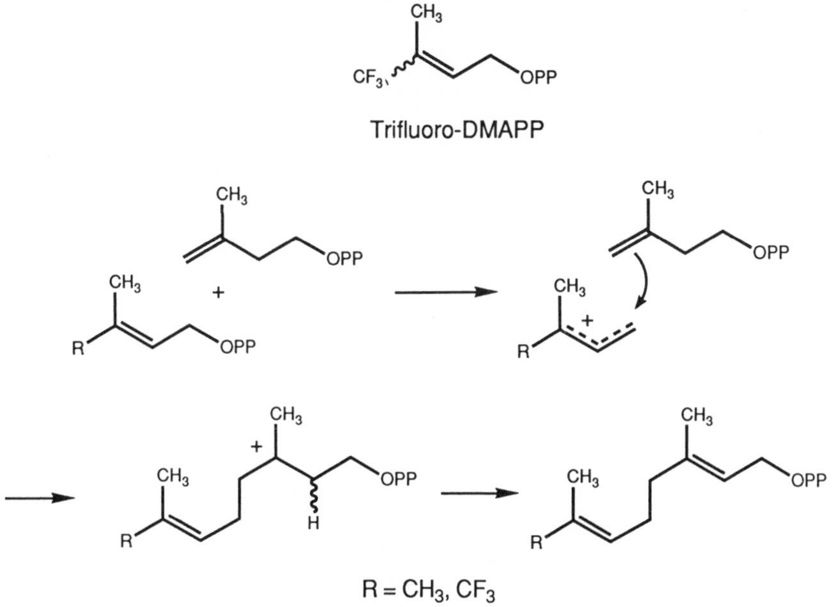

Trifluoro-DMAPP

R = CH₃, CF₃

Figure 3-32. A greatly decreased rate of reaction of trifluoromethyl analogues of DMAPP provides evidence that the prenyl transfer reaction occurs by initial ionization (an S_N1 mechanism) (Poulter and Rilling, 1978).

(an S_N1 mechanism) had both been considered. A rate depression of 3×10^7 was found for the trifluoromethyl analogues (Fig. 3-32) of DMAPP, even though the binding constants were only 20–60 times greater than the K_m of the natural substrate. The trifluoromethyl group has been shown to slightly accelerate an S_N2 reaction but to strongly retard an S_N1 process. Thus, this provides strong evidence that the prenyl transfer reaction proceeds by initial ionization of the allylic pyrophosphate, prior to nucleophilic attack of IPP (Fig. 3-32) (Poulter and Rilling, 1978; Poulter and Satterwhite, 1977).

During the condensation of IPP with the carbonium ion generated enzymatically from DMAPP, a positive charge develops at the 3 position of IPP. While removal of a proton from C-2 would stabilize this intermediate, participation of a nucleophile at this stage has been considered as another possibility (path A, Fig. 3-33). The presence of such a nucleophile was probed using 2-fluoro-IPP and 2,2-difluoro-IPP. Previously established stereospecific enzymatic removal of the *pro-R* proton from IPP implies that (R)-fluoro-IPP should trap an added X group if a nucleophilic mechanism were involved, since removal of a proton required for the E2 elimination of this nucleophile to give product would be blocked by the presence of fluorine. With 2-fluoro-IPP as a substrate analogue, 2-fluorofarnesyl pyrophosphate was formed, and no evidence for nucleophilic addition was found (Fig. 3-33). 2,2-Difluoro-IPP failed to condense with DMAPP, showing that proton abstraction from C-2 is a requirement for reaction (Poulter *et al.*, 1979).

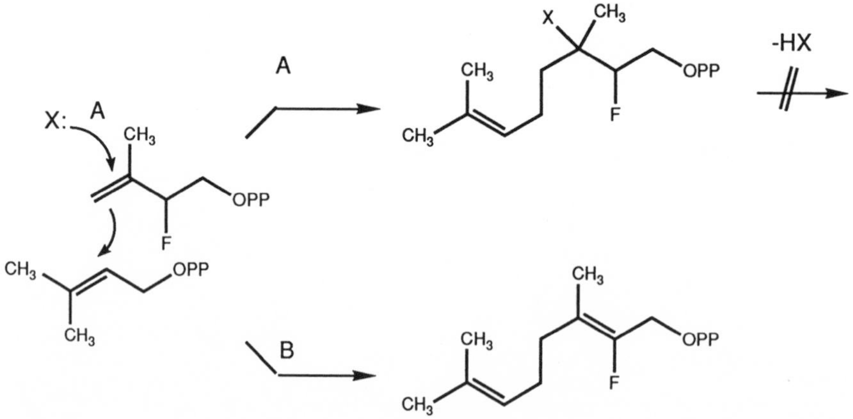

Figure 3-33. The formation of 2-fluorofarnesyl pyrophosphate from 2-fluoro-IPP and the failure to detect any substrate–nucleophile complex show that addition of a nucleophile (X:) is not involved in the prenyl transfer reaction (path A) (Poulter *et al.*, 1979).

3.5 Halogenated Terpenoids—Summary

Halogenated steroids, vitamin D_3 analogues, retinoids, and other isoprenoids have been discussed together because of the biochemical relationships between these classes of compounds. A review of this length clearly cannot cover every significant development in such a large field, and this chapter again should be considered a selected sampling of research. In each group, halogen substitution has been carried out as a strategy for developing more potent and/or more selectively acting analogues, and major successes have been realized. The impressive list of medicinally important steroid analogues that contain halogen is an obvious example of this success. Halogen substitution has also been used to probe biochemical mechanisms of ligand–receptor interactions as well as of the actions of enzymes involved in the biosynthesis and metabolism of these compounds. The use of radiohalogenation to produce imaging agents for a number of applications promises to be an area of particularly active future research.

References

Asato, A. E., and Liu, R. S. H., 1986. The preparation of vicinal difluoroolefinic carbonyl compounds and their application to the synthesis of difluororetinal analogues, *Tetrahedron Lett.* 27:3337–3340.

Asato, A. E., Matsumoto, H., Denny, M., and Liu, R. S. H., 1978. Fluorinated rhodopsin analogues from 10-fluoro- and 14-fluororetinal, *J. Am. Chem. Soc.* 100:5957–5960.

Beierwaltes, W. H., Wieland, D. M., Yu, T., Swanson, D. P., and Mosley, S. T., 1978. Adrenal imaging agents: Rationale, synthesis, formulation and metabolism, *Semin. Nucl. Med.* VIII:5–21.

Benisek, W. F., Ogez, J. R., and Smith, S. B., 1982. Design of site-specific pharmacologic reagents. Illustration of some alternative approaches by reagents directed towards steroid-hormone-specific targets, in *Modification of Proteins. Food, Nutritional, and Pharmacological Aspects* (R. E. Fenney and J. R. Whitaker, eds.), American Chemical Society, Washington, D.C., pp. 267–323.

Brandes, S. J., and Katzenellenbogen, J. A., 1987. Fluorinated androgens and progestins: Molecular probes for androgen and progesterone receptors with potential use in positron emission tomography, *Mol. Pharmacol.* 32:391–403.

Brandes, S. J., and Katzenellenbogen, J. A., 1988. Fundamental considerations in the design of fluorine-18 labeled progestins and androgens as imaging agents for receptor-positive tumors of the breast and prostate, *Nucl. Med. Biol.* 15:53–57.

Chen, P. S., Jr., and Borrevang, P., 1970. Fluorine-containing steroids, in *Handbook of Experimental Pharmacology*, Vol. XX/2, *Pharmacology of Fluorides* (O. Eichler, A. Farah, H. Herken, and A. D. Welch, eds.), Springer-Verlag, Berlin, pp. 193–252.

Cherbas, P., Cherbas, L., Lee, S.-S., and Nakanishi, K., 1988. 26-[^{125}I]Iodoponasterone A is a potent ecdysone and a sensitive radioligand for ecdysone receptors, *Proc. Natl. Acad. Sci. USA* 85:2096–2100.

Chin, C.-C., Murdock, G. L., and Warren, J. C., 1982. Identification of two histidyl residues in the active site of human placental estradiol 17β-dehydrogenase, *Biochemistry* 21:3322–3326.

Counsell, R. E., Klausmeier, W. H., Weinhold, P. A., and Skinner, R. W. S., 1981. Radiolabeled androgens and their analogues, in *Radiopharmaceuticals: Structure–Activity Relationships* (R. P. Spencer, ed.), Grune and Stratton, New York, pp. 425–448.

Crescitelli, F., 1988. The gecko visual pigment: The chromophore dark exchange reaction, *Exp. Eye Res.* 46:239–248.

Crescitelli, F., and Liu, R. S. H., 1988. The spectral properties and photosensitivities of analogue photopigments regenerated with 10- and 14-substituted retinal analogues, *Proc. Roy. Soc. London* 233:55–76.

Filler, R., 1979. Fluorine-containing drugs, in *Organofluorine Chemicals and Their Industrial Applications* (E. Banks, ed.), Horwood, Chichester, England, pp. 123–153.

Filler, R., and Naqvi, S. M., 1982. Fluorine in biomedicinal chemistry. An overview of recent advances and selected topics, in *Biomedicinal Aspects of Fluorine Chemistry* (R. Filler and Y. Kobayashi, eds.), Kodansha Ltd., Tokyo; Elsevier Biomedical Press, Amsterdam, pp. 1–32.

Fried, J., 1957. Structure–activity relationships in the field of halogenated steroids, *Cancer* 10:752–756.

Fried, J., and Borman, A., 1958. Synthetic derivatives of cortical hormones, *Vit. Horm.* 16:303–374.

Fried, J., and Sabo, E. F., 1953. Synthesis of 17α-hydroxycorticosterone and its 9α-halo derivatives from 11-epi-17α-hydroxycorticosterone, *J. Am. Chem. Soc.* 75:2273–2274.

Fried, J., and Sabo, E. F., 1954. 9α-Fluoro derivatives of cortisone and hydrocortisone, *J. Am. Chem. Soc.* 76:1455–1456.

Gronemeyer, H., and Govindan, M. V., 1986. Affinity labelling of steroid receptors, *Mol. Cell. Endocrin.* 46:1–19.

Hanzawa, Y., Yamada, A., and Kobayashi, Y., 1985. Preparation of 19,19,19-trifluororetinal (9-trifluoromethylretinal), *Tetrahedron Lett.* 26:2881–2884.

Haynes, R. C., Jr., and Murad, F., 1985. Adrenocorticotropic hormone; Adrenocortical steroids and their synthetic analogues; Inhibitors of adrenocortical steroid biosynthesis, in *The Pharmacological Basis of Therapeutics* (A. G. Gilman, L. S. Goodman, T. W. Rall, and F. Murad, eds.), Macmillan, New York, pp. 1459–1489.

Hochberg, R. B., 1979. Iodine-125-labeled estradiol: A gamma-emitting analogue of estradiol that binds to the estrogen receptor, *Science* 205:1138–1140.

Hochberg, R. B., and Rosner, W., 1980. Interaction of 16α-[125I]iodo-estradiol with estrogen receptor and other steroid-binding proteins, *Proc. Natl. Acad. Sci. USA* 77:328–332.

Hochberg, R. B., MacLusky, N. J., Chambers, J., Eisenfeld, A. J., Naftolin, F., and Schwartz, P. E., 1985a. Concentration of [125I]iodoestradiol in human ovarian tumors *in vivo* and correlation with estrogen receptor content, *Steroids* 46:775–788.

Hochberg, R. B., Hoyte, R. M., and Rosner, W., 1985b. E-17α-(2-[16α-125I]Iodovinyl)-19-nortestosterone: The synthesis of a gamma-emitting ligand for the progesterone receptor, *Endocrinology* 117:2550–2552.

Hoyte, R. M., Rosner, W., and Hochberg, R. B., 1982. Synthesis of 16α-[125I]iodo-5α-dihydrotestosterone and evaluation of its affinity for the androgen receptor, *J. Steroid Biochem.* 16:621–628.

Hoyte, R. M., Rosner, W., Johnson, I. S., Zielinski, J., and Hochberg, R. B., 1985. Synthesis and evaluation of potential radioligands for the progesterone receptor, *J. Med. Chem.* 28:1695–1699.

Kamata, S., Naga, N., Mitsugi, T., Kondo, E., Nagata, W., Nakamura, M., Miyata, K., Odaguchi, K., Shimizu, T., Kawabata, T., Suzuki, T., Ishibashi, M., and Yamada, F., 1985. Aldosterone antagonists 1. Synthesis and biological activities of 11β,18-epoxypregnane derivatives, *J. Med. Chem.* 28:428:433.

Katzenellenbogen, J. A., 1977. Affinity labeling as a technique in determining hormone mechanisms, *Vit. Horm.* 41:1–84.
Katzenellenbogen, J. A., and Katzenellenbogen, B. S., 1984. Affinity labeling of receptors for steroid and thyroid hormones, in *Vitamins and Hormones. Advances in Research and Applications* (G. D. Aurbach and D. B. McCormick, eds.), Academic Press, Orlando, Florida. pp. 213–274.
Katzenellenbogen, J. A., Hsiung, H. M., Carlson, K. E., McGuire, W. L., Kraay, R. J., and Katzenellenbogen, B. S., 1975. Iodohexestrols. II. Characterization of the binding and estrogenic activity of iodinated hexestrol derivatives, in vitro and in vivo, *Biochemistry* 14:1742–1750.
Katzenellenbogen, J. A., MeElvany, K. D., Senderoff, S. G., Carlson, K. E., Landvatter, S. W., Welch, M. J., and the Los Alamos Medical Radioisotope Group, 1982. 16α-[^{77}Br]Bromo-11β-methoxyestradiol-17β: A gamma-emitting estrogen imaging agent with high uptake and retention in target organs, *J. Nucl. Med.* 23:411–419.
Kiesewetter, D. O., Kilbourn, M. R., Landvatter, S. W., Heiman, D. F., Katzenellenbogen, J. A., and Welch, M. J., 1984. Preparation of four fluorine-18-labeled estrogens and their selective uptakes in target tissues by immature rats, *J. Nucl. Med.* 25:1212–1221.
Kobayashi, Y., and Taguchi, T., 1982. Fluorinated vitamin D$_3$ analogues. Synthesis and biological activities, in *Biomedicinal Aspects of Fluorine Chemistry* (R. Filler and Y. Kobayashi, eds.), Kodansha Ltd., Tokyo; Elsevier Biomedical Press, Amsterdam, pp. 33–53.
Kobayashi, Y., Nakazawa, M., Kumadaki, I., Taguchi, T., Ohshima, E., Ikekawa, N., Tanaka, Y., and DeLuca, H. F., 1986. Studies on organic fluorine compounds. L. Synthesis and biological activity of 2α-fluorovitamin D$_3$, *Chem. Pharm. Bull.* 34:1568–1572.
Koeffler, P., Amatruda, T., Ikekawa, N., Kobayashi, Y., and DeLuca, H. F., 1984. Induction of macrophage differentiation of human normal and leukemic myeloid stem cells by 1,25-dihydroxyvitamin D$_3$ and its fluorinated analogues, *Cancer Res.* 44:5624–5628.
Kollman, P. A., Giannini, D. D., Duax, W. L., Rothenberg, S., and Wolff, M. E., 1973. Quantitation of long-range effects in steroids by molecular orbital calculations, *J. Am. Chem. Soc.* 95:2869–2873.
Kowalsky, R. J., and Perry, J. R., 1987. *Radiopharmaceuticals in Nuclear Medicine Practice*, Appleton and Lange, Norwalk, Connecticut.
Labrie, F., Dupont, A., Cusan, L., Giguere, M., Bergeron, N., Borsanyi, J. P., Lacourciere, Y., Blanger, A., Emond, J., Monfetti, G., Boucher, H., and Lachance, R., 1988. Combination therapy with flutamide and castration (LHRH agonist or orchiectomy) in previously untreated patients with clinical stage D2 prostate cancer, *J. Steroid Biochem.* 30:107–117.
Lamb, D. J., Bullock, D. W., Hoyte, R. M., and Hochberg, R. B., 1988. Δ^9-[16α-^{125}I]Iodo-19-nortestosterone: A gamma-emitting photoaffinity label for the progesterone receptor, *Endocrinology* 122:1923–1932.
Liu, R. S. H., and Browne, D. T., 1986. A bioorganic view of the chemistry of vision: H.T.-*n* and B.P.-*m,n* mechanisms for the reactions of confined, anchored polyenes, *Acc. Chem. Res.* 19:42–48.
Liu, R. S. H., Matsumoto, H., Asato, A. E., Denny, M., Shichida, Y., Yoshizawa, T., and Dahlquist, F. W., 1981. Synthesis and properties of 12-fluororetinal and 12-fluororhodopsin. A model system for ^{19}F NMR studies of visual pigments, *J. Am. Chem. Soc.* 103:7195–7201.
Liu, R. S. H., Crescitelli, F., Denny, M., Matsumoto, H., and Asato, A. E., 1986. Photosensitivity of 10-substituted visual pigment analogues: Detection of a specific secondary opsin–retinal interaction, *Biochemistry* 25:7026–7030.
Lovey, A. J., and Pawson, B. A., 1982. Fluorinated retinoic acids and their analogues, 3. Synthesis and biological activity of aromatic 6-fluoro analogues, *J. Med. Chem.* 25:71–75.

Marcotte, P. A., and Robinson, C. H., 1982. Inhibition and inactivation of estrogen synthetase (aromatase) by fluorinated substrate analogues, *Biochemistry* 21:2773–2778.

Mazaitis, J. K., Gibson, R. E., Komai, G., Eckelman, W. C., Francis, B., and Reba, R., 1980. Radioiodinated estrogen derivatives, *J. Nucl. Med.* 21:142–146.

McElvany, K. D., Carlson, K. E., Welch, M. J., Senderoff, S. G., Katzenellenbogen, J. A., and the Los Alamos Medical Radioisotope Group, 1982. In vivo comparison of 16α-[^{77}Br]bromoestradiol-17β and 16α-[^{125}I]iodoestradiol-17β, *J. Nucl. Med.* 23:420–424.

Mead, D., Loh, R., Asato, A. E., and Liu, R. S. H., 1985. Fluorinated retinoids via crossed aldol condensation of 1,1,1-trifluoroacetone, *Tetrahedron Lett.* 26:2873–2876.

Mintun, M. A., Welch, M. J., Siegel, B. A., Mathias, C. J., Brodack, J. W., McGuire, A. H., and Katzenellenbogen, J. A., 1988. Breast cancer: Pet imaging of estrogen receptors, *Radiology* 169:45–48.

Muehlbacher, M., and Poulter, C. D., 1985. Isopentenyl diphosphate:demethylallyl diphosphate isomerase. Irreversible inhibition of the enzyme by active-site-directed covalent attachment, *J. Am. Chem. Soc.* 107:8307–8308.

Muehlbacher, M., and Poulter, C. D., 1988. Isopentenyl-diphosphate isomerase: Inactivation of the enzyme with active-site-directed irreversible inhibitors and transition-state analogues, *Biochemistry* 27:7315–7328.

Nakada, M., Tanaka, Y., DeLuca, H. F., Kobayashi, Y., and Ikekawa, N., 1985. Biological activities and binding properties of 23,23-difluoro-25-hydroxyvitamin D_3 and its 1α-hydroxy derivative, *Arch. Biochem. Biophys.* 241:173–178.

Nave, J.-F., d'Orchymont, J., Ducep, J.-B., Piriou, F., and Jung, M. J., 1985. Mechanism of the inhibition of cholesterol biosynthesis by 6-fluoromevalonate, *Biochem. J.* 227:247–254.

Nozaki, T., 1983. Other cyclotron radionuclides, in *Radionuclides Production*, Vol. II (F. Helus and L. G. Colombetti, eds.), CRC Press, Boca Raton, Florida, pp. 104–124.

Ohshima, E., Sai, H., Takatsuto, S., Ikekawa, N., Kobayashi, Y., Tanaka, Y., and DeLuca, H. F., 1984. Synthesis and biological activity of 1α-fluoro-25-hydroxyvitamin D_3, *Chem. Pharm. Bull.* 32:3525–3531.

Okamoto, S., Tanaka, Y., DeLuca, H. F., Kobayashi, Y., and Ikekawa, N., 1983. Biological activity of 24,24-difluoro-1,25-dihydrovitamin D_3, *Am. J. Physiol.* 244:E159–E163.

Onoda, M., Haniu, M., Yanagibashi, K., Sweet, F., Shively, J. E., and Hall, P. F., 1987. Affinity alkylation of the active site of C_{21} steroid side-chain cleavage cytochrome P-450 from neonatal porcine testis: A unique cysteine residue alkylated by 17-(bromoacetoxy)progesterone, *Biochemistry* 26:657–662.

Oshida, J.-I., Morisaki, M., and Ikekawa, N., 1980. Synthesis of 2β-fluoro-1α-hydroxyvitamin D_3, *Tetrahedron Lett.* 21:1755–1756.

Paaren, H. E., Fivizzani, M. A., Schnoes, H. K., and DeLuca, H. F., 1981. 1α,25-Difluorovitamin D_3: An inert vitamin D analogue, *Arch. Biochem. Biophys.* 209:579–583.

Pomper, M. G., Katzenellenbogen, J. A., Welch, M. J., Brodack, J. W., and Mathias, C. J., 1988. 21-[^{18}F]Fluoro-16α-ethyl-19-norprogesterone: Synthesis and target tissue selective uptake of a progestin receptor based radiotracer for positron emission tomography, *J. Med. Chem.* 31:1360–1363.

Poulter, C. D., and Rilling, H., 1978. The prenyl transfer reaction. Enzymatic and mechanistic studies of the 1'-4 coupling reaction in the terpene biosynthetic pathway, *Acc. Chem. Res.* 11:307–313.

Poulter, C. D., and Satterwhite, D. M., 1977. Mechanism of the prenyl-transfer reaction. Studies with (*E*)- and (*Z*)-3-trifluoromethyl-2-buten-1-yl pyrophosphate, *Biochemistry* 16:5470–5478.

Poulter, C. D., Mash, E. A., Argyle, J. C., Muscio, O. J., and Rilling, H., 1979. Farnesyl pyrophosphate synthase. Mechanistic studies of the 1'-4 coupling reaction in the terpene biosynthetic pathway, *J. Am. Chem. Soc.* 101:6761–6763.

Prestwich, G. D., 1986. Fluorinated sterols, hormones and pheromones: Enzyme-targeted disruptants in insects, *Pestic. Sci.* 37:430–440.

Prestwich, G. D., Yamaoka, R., Phirwa, S., and DePalma, A., 1984. Isolation of 2-fluorocitrate produced by *in vivo* dealkylation of 20-fluorostigmasterol in an insect, *J. Biol. Chem.* 259:11022–11026.

Quisted, G. B., Cerf, D. C., Schooley, D. A., and Staal, G. B., 1981. Fluoromevalonate acts as an inhibitor of insect juvenile hormone biosynthesis, *Nature* 289:176–177.

Reardon, J. E., and Abeles, R. H., 1986. Mechanism of action of isopentenyl pyrophosphate isomerase: Evidence for a carbonium ion intermediate, *Biochemistry* 25:5609–5616.

Reardon, J. E., and Abeles, R. H., 1987. Inhibition of cholesterol biosynthesis by fluorinated mevalonate analogues, *Biochemistry* 26:4717–4722.

Ringold, H. J., Lawrence, H., Jr., and Graves, J. M. H., 1964. The influence of unsaturation and of fluorine substitution on ketone–alcohol equilibrium constants. A measure of α,β-unsaturated ketone resonance energy and of halogen destabilization, *J. Am. Chem. Soc.* 86:4510–4512.

Robinson, S. P., and Jordon, V. C., 1988. Metabolism of steroid-modifying anticancer agents, *Pharmacol. Ther.* 36:41–103.

Shichida, Y., Ono, T., Yoshizawa, T., Matsumoto, H., Asato, A., Zingoni, J. P., and Liu, R. S. H., 1987. Electrostatic interaction between retinylidene chromophore and opsin in rhodopsin studied by fluorinated rhodopsin analogues, *Biochemistry* 26:4422–4428.

Shiuey, S.-J., Partridge, J. J., and Uskokovic, M. R., 1988. Triply convergent synthesis of $1\alpha,25$-dihydroxy-24(R)-fluorocholecalciferol, *J. Org. Chem.* 53:1040–1046.

Singer, F. M., Januszka, J. P., and Borman, A., 1959. New inhibitors of *in vitro* conversion of acetate and mevalonate to cholesterol, *Proc. Soc. Exp. Biol.* 102:370–373.

Stern, P. H., Tanaka, Y., DeLuca, H., Ikekawa, N., and Kobayashi, Y., 1981. Bone resorptive activity of side-chain fluoro derivatives of 25-hydroxy- and $1\alpha,25$-dihydroxy D_3 in culture, *Mol. Pharmacol.* 20:460–462.

Stern, P. H., Mavreas, T., Tanaka, Y., DeLuca, H. F., Ikekawa, N., and Kobayashi, Y., 1984. Fluoride substitution of vitamin D analogues at C-26 and C-27: Enhancement of activity of 25-hydroxyvitamin D but not of 1,25-dihydroxyvitamin D on bone and intestine *in vitro*, *J. Pharmacol. Exp. Ther.* 229:9–13.

Swain, S. M., Steinberg, S. M., Bagley, C., and Lippman, M. E., 1988. Tamoxifen and fluoxymesterone versus tamoxifen and danazol in metastatic breast cancer—a randomized study, *Breast Cancer Res. Treat.* 12:51–57.

Thrall, J. H., Freitas, J. E., and Beierwaltes, W. H., 1978. Adrenal scintigraphy, *Semin. Nucl. Med.* VIII:23–41.

Tschesche, R., and Machleidt, H., 1960. Synthesen von substituierten β-Hydroxy-β-methylglutarsäuren und Mevalonsäuren, *Justus Liebigs Ann. Chem.* 631:61–76.

Weeks, C. M., Daux, W. L., and Wolff, M. E., 1973. A comparison of the molecular structures of six corticosteroids, *J. Am. Chem. Soc.* 95:2865–2868.

Welch, J. T., 1987. Advances in the preparation of biologically active organofluorine compounds, *Tetrahedron* 43:3123–3197.

Wilhelm, F., Dauben, W. G., Kohler, B., Roesle, A., and Norman, A. W., 1984. 6-Fluorovitamin D_3: A new antagonist of the biological actions of vitamin D_3 and its metabolites which interacts with the intestinal receptor for $1\alpha,25(OH)_2$-vitamin D_3, *Arch. Biochem. Biophys.* 233:127–132.

Wolff, M. E., and Hansch, C., 1973. Correlation of physicochemical parameters and biological activities of steroids. 9α-Substituted corticol derivatives, *Experientia* 29:1111–1113.

Wong, T. C., Rutar, V., and Wang, J.-S., 1984. Study of ^1H chemical shifts and couplings with ^{19}F in 9α-fluorocortisol. Application of a novel $^1H–^{13}C$ chemical shift correlation technique with homonuclear decoupling, *J. Am. Chem. Soc.* 106:7046–7051.

Halogenated Analogues of Products of the Arachidonic Acid Cascade: Prostaglandins, Thromboxanes, and Leukotrienes

4.1 Introduction

The eicosanoids are a group of C-20 unsaturated acids that have important roles in the regulation of such physiological functions as fertility control and induction of labor, regulation of blood platelet aggregation, gastric acid secretion, stimulation of smooth muscle activity, and inflammation response. Since the initial isolation and identification of prostaglandins from sheep seminal vesicle glands and from human semen ($PGF_{2\alpha}$ and PGE, respectively) in the 1950s by Bergström and Sjövall [Bergström *et al.*, 1963; reviewed by von Euler and Eliasson (1963)], a diverse series of acids have been identified. These include the primary prostaglandins and prostacyclins, thromboxanes, and unsaturated eicosanoic acids having perhydroxyl and hydroxyl substitution (leukotrienes). These compounds, collectively named eicosanoids, are derived enzymatically from C-20 acids, most notably arachidonic acid, as shown in Fig. 4-1 (the "arachidonic acid cascade"). A recent review of the biosynthesis, nomenclature, and physiological role of eicosanoids is provided by Moore (1985).

4.2 Halogenated Analogues of Prostaglandins

In the area of drug development, the success realized in producing analogues having enhanced and more selective biological activities by selective halogenation of steroid hormones is ample demonstration of the value of this approach. Accordingly, the recognition of the diverse and important biological functions of prostaglandins has prompted extensive synthetic efforts, which have resulted in a host of halogenated analogues with the aim of achieving similar beneficial alterations in biological activities. There

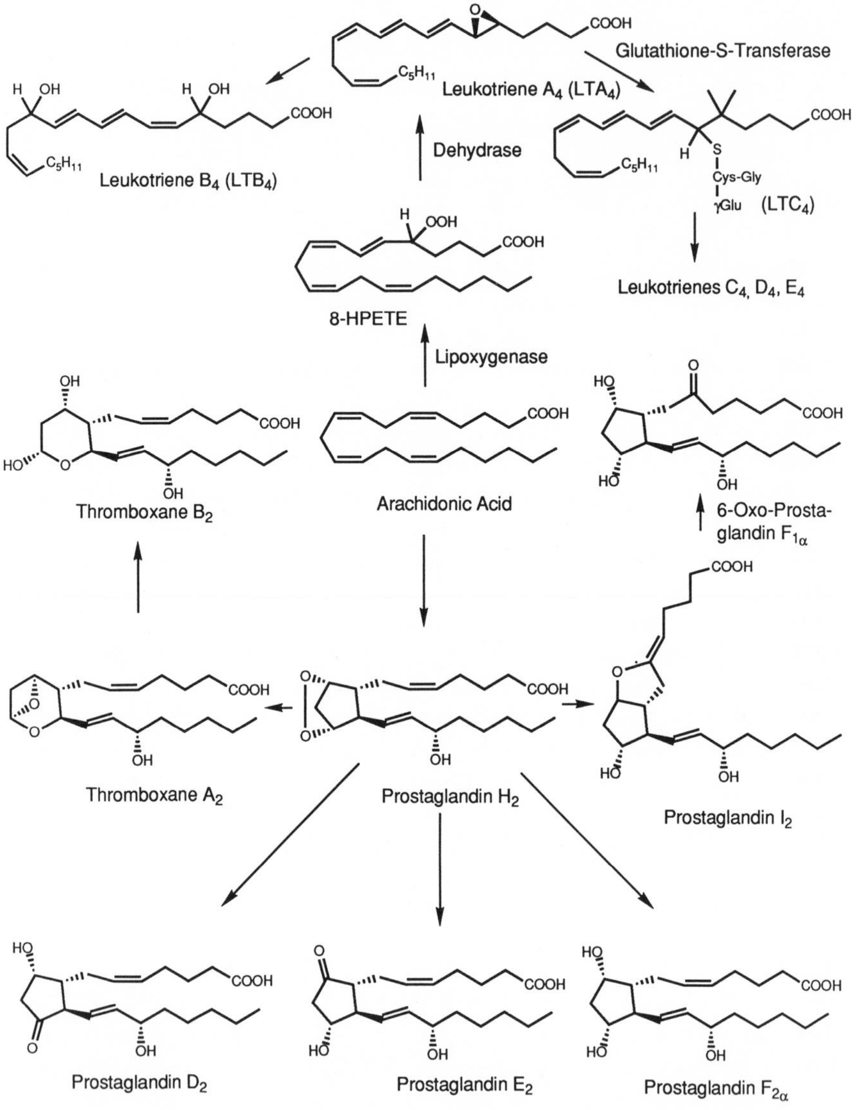

Figure 4-1. The "arachidonic acid cascade."

have been notable successes, although in some cases the results have not been as impressive as hoped. Selective fluorination has been especially stressed, and this aspect has been reviewed recently by Welch (1987), Barnette (1984), and Filler and Naqvi (1982). A brief overview will be presented here. To facilitate this discussion, the structures and nomenclature of several prostaglandins are given in Fig. 4-2.

4.2.1 Halogenated Primary Prostaglandins

4.2.1.1 Fluorination to Block Metabolism

There have been several rational approaches to the improvement of prostaglandin activity by selective introduction of a carbon–fluorine bond. Included have been several attempts to prolong activity by incorporation of fluorine at a site where it could impede metabolism. Thus, 2,2-difluoro analogues of $PGF_{2\alpha}$ and PGE_2 (Fig. 4-3) were found to have longer duration of action, presumably as a result of the blocking of β-oxidation, a major pathway of prostaglandin metabolism [Axen, 1977; reviewed by Barnette (1984)].

In addition to having smooth-muscle stimulatory activity, $PGF_{2\alpha}$ has the most potent luteolytic activity of the naturally occurring prostaglandins. Much effort has been directed toward producing $PGF_{2\alpha}$ analogues with increased luteolytic activity and decreased smooth-muscle stimulatory properties, one goal being the development of postcoital contraceptive agents. For example, (+)-12-fluoro-$PGF_{2\alpha}$ (Fig. 4-4) shows a 10-fold increase in antifertility activity, with a considerable decrease in effect on smooth muscle (Greico *et al.*, 1980a). Increased luteolytic activity was attributed to the fact that (+)-12-fluoro-$PGF_{2\alpha}$ is not a substrate for placental 15-hydroxyprostaglandin dehydrogenase. Barnette (1984) has suggested, on the basis of studies carried out with fluorinated steroids (Ringold *et al.*, 1964), that a fluorine-induced increase in electrophilicity of the enone may shift the enol/enone equilibrium in the enzymatic redox process toward the reduced form. Although 14-fluoro-$PGF_{2\alpha}$ analogues (Fig. 4-4) also were expected to show resistance to oxidation by 15-hydroxyprostaglandin dehydrogenase, antifertility activity of these analogues was comparable to or only slightly increased relative to that of $PGF_{2\alpha}$. Fluorination again resulted in a decrease in smooth-muscle stimulation effects (Greico *et al.*, 1980b). (+)-13-Fluoro-$PGF_{2\alpha}$ methyl ester (Fig. 4-4) was 2.5 times more effective than natural $PGF_{2\alpha}$ in an antifertility assay. Unfortunately, there was no decrease in smooth-muscle stimulatory properties (Greico *et al.*, 1985). Other fluorinated derivatives designed to impede or block inactivation by oxidation of the 15-hydroxyl group include 16-fluoro- and 16,16-difluoro-$PGF_{2\alpha}$ and -PGE_2 analogues (Fig. 4-4). Both difluoro analogues displayed significantly greater antifertility activity compared to that of the natural relatives. Difluoro-$PGF_{2\alpha}$ had smooth muscle activity comparable to that of $PGF_{2\alpha}$ (Magerlein and Miller, 1975).

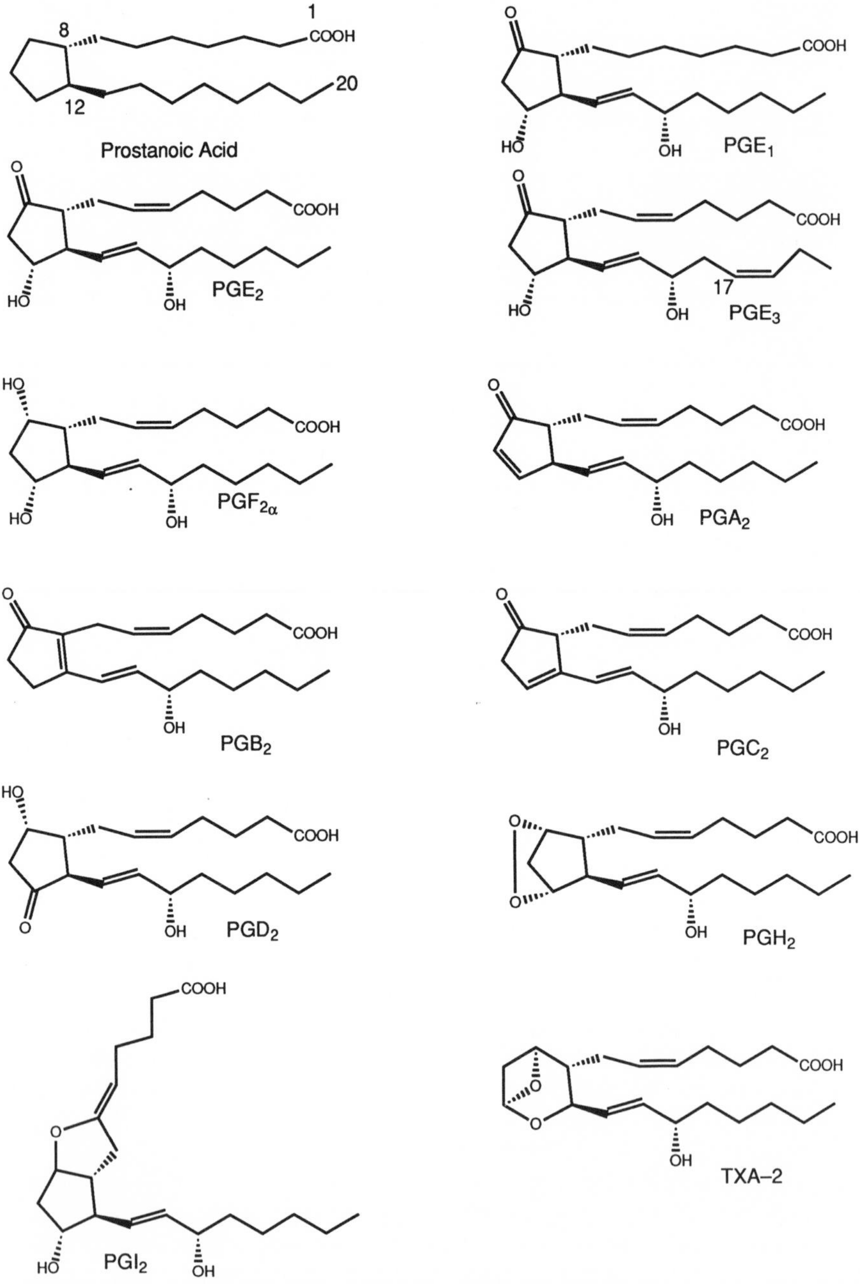

Figure 4-2. Examples of common prostaglandins.

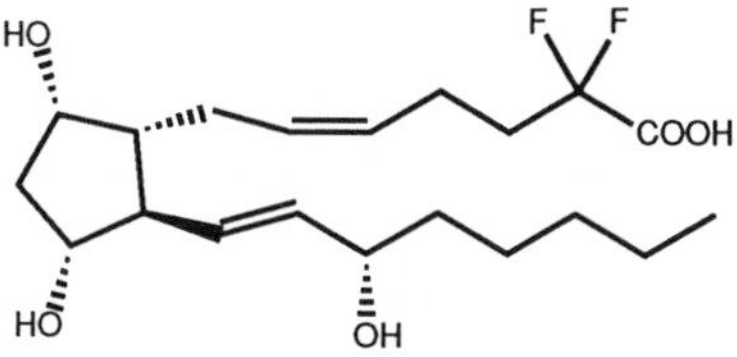

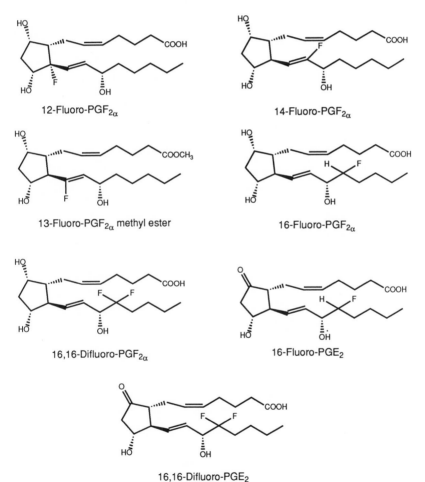

2,2-Difluoro-PGF$_{2\alpha}$ 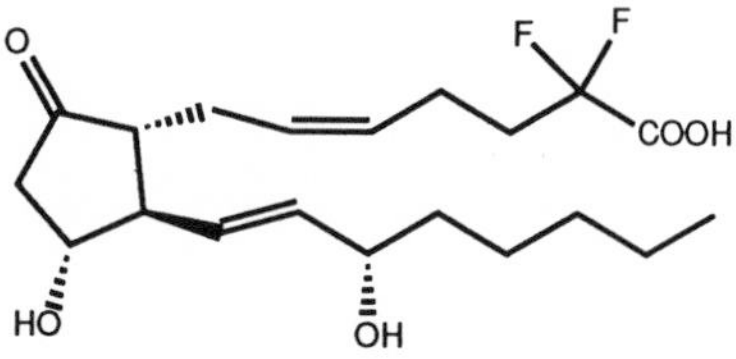   2,2-Difluoro-PGE$_2$

Figure 4-3. Prostaglandins fluorinated at the 2 position as a strategy to block β-oxidation.

12-Fluoro-PGF$_{2\alpha}$ 14-Fluoro-PGF$_{2\alpha}$

13-Fluoro-PGF$_{2\alpha}$ methyl ester   16-Fluoro-PGF$_{2\alpha}$

16,16-Difluoro-PGF$_{2\alpha}$ 16-Fluoro-PGE$_2$

16,16-Difluoro-PGE$_2$

Figure 4-4. Fluorinated PGF$_{2\alpha}$ and PGE$_2$ analogues developed as potential antifertility drugs.

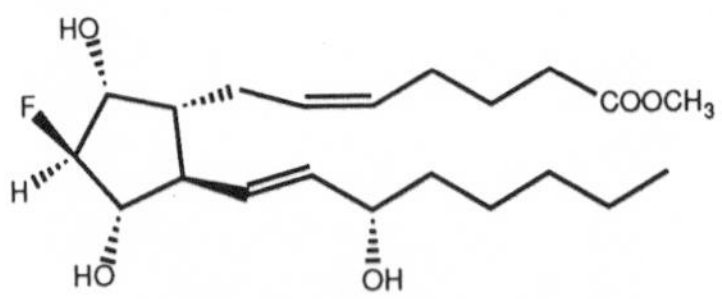

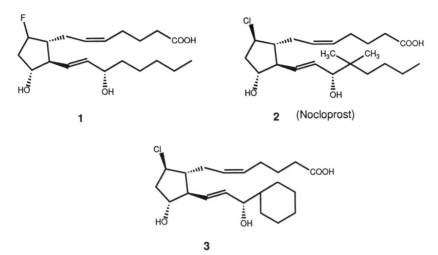

10β-Fluoro-PGF$_{2\alpha}$ methyl ester 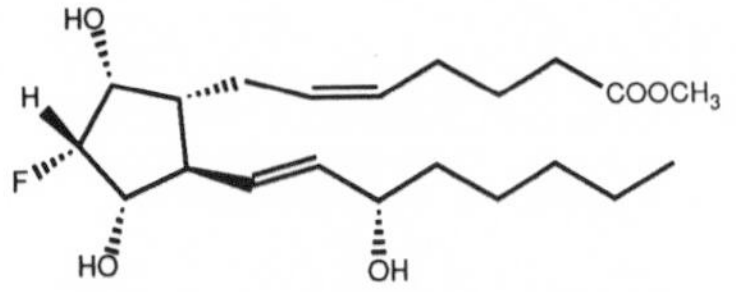   10α-Fluoro-PGF$_{2\alpha}$ methyl ester

Figure 4-5. 10-Fluorinated PGF$_{2\alpha}$ analogues that possess the β-fluorohydrin structural feature.

4.2.1.2 Incorporation of the trans-β-Fluorohydrin Moiety to Increase Activity

10β-Fluoro-PGF$_{2\alpha}$ (Fig. 4-5) incorporates the *trans*-β-fluorohydrin structural feature present in 9α-fluorocorticosteroids. Methyl esters of both 10α- and 10β-fluoro-PGF$_{2\alpha}$ have been prepared, but no biological data have been reported (Greico *et al.*, 1979a, b).

4.2.1.3 Replacement of Hydroxyl with Halogen

A series of PGF$_{2\alpha}$ analogues have been prepared wherein chlorine or fluorine replaces a hydroxyl group at C-9 and/or C-11. When substituted at C-9 or C-11, halogen can mimic the carbonyl group of the PGE or PGD series, respectively. For example, the 9-deoxy-9β-fluoro-11α-hydroxy analogue (1; Fig. 4-6) was three to four times more potent than PGE$_2$ as a bronchodilator (Arroniz *et al.*, 1978).

Figure 4-6. PGF$_{2\alpha}$ analogues that have halogen as a replacement for hydroxyl at C-9.

PGE_2 both inhibits gastric acid secretion and is cytoprotective (inhibits ulcer formation and promotes healing of ulcers) [for a discussion of this and other pharmacological issues, see Moore (1985)]. For this reason, chemically stable PGE_2 analogues produced by the replacement of the 9-keto function with a 9-halo substituent have received recent attention with respect to the development of new antiulcer agents. For example, one such PGE_2 analogue which has promise as an antiulcer drug is 9-deoxy-9β-chloro-16,16-dimethyl-$PGF_{2\alpha}$ (nocloprost) (**2**; Fig. 4-6), which is comparable to 16,16-dimethyl-PGE_2 in preventing gastrointestinal erosion, more potent in preventing gastric acid secretion, and less potent in causing diarrhea (Loge and Radüchel, 1984).

PGD_2 inhibits aggregation of platelets in several species and has several effects on vascular action depending on the system being studied (Thierauch *et al.*, 1988, and references therein). Based on the effectiveness of halogen at C-9 in mimicking the carbonyl group of prostaglandins of the E series, a series of analogues containing an 11α- or 9β-chloro or -fluoro substituent were prepared in an effort to develop stable and potent PGD_2 analogues. In this study, the 9-deoxy-9β-chloro-15-cyclohexyl analogue (**3**; Fig. 4-6) was found to have the highest PGE_2 mimetic activity and, moreover, was orally active (Thierauch *et al.*, 1988).

4.2.2 Halogenated Marine Prostaglandins

4.2.2.1 Punaglandins

Marine prostaglandins, which are produced by a variety of marine organisms and were first discovered in 1969 (Weinheimer and Spraggins, 1969), feature unusual C-12 oxygen functionality, presumably a reflection of a distinctive biosynthetic pathway. The clavulones (e.g., clavulone II; Fig. 4-7) were isolated from the Japanese coral *Clavularia viridis* in 1982. These prostanoids, which feature an unusual C-7–C-8 double bond, attracted much interest because of marked antitumor activity. In 1985, prostaglandins isolated from the coral *Telesto riisei* were shown to have, in addition to a C-12 hydroxyl group and the exocyclic double bond at C-8, a unique C-10 chlorine (Baker *et al.*, 1985). The demonstration that these eicosanoids, named punaglandins after *puna* (Hawaiian for coral), were potent inhibitors of L1210 leukemia cell proliferation, more potent than the clavulones, intensified interest in these compounds as potential chemotherapeutic agents (Fukushima and Kato, 1985). The structure of punaglandin-4 (**4**; Fig. 4-7), which differs from the originally tentatively

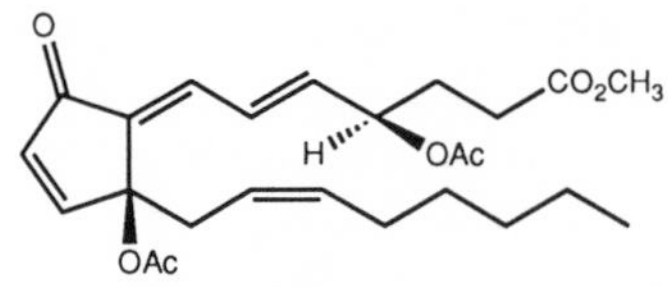

Clavulone II

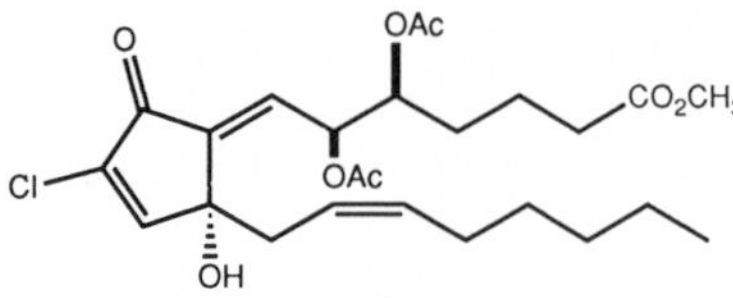

4   (7E)-Punaglandin-4

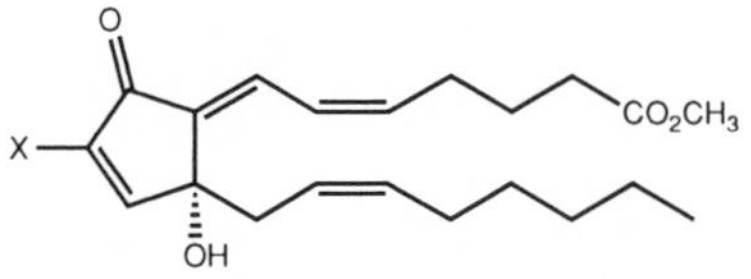

5   X = Cl   Chlorovulone I   6
     X = Br   Bromovulone I
     X = I   Iodovulone I

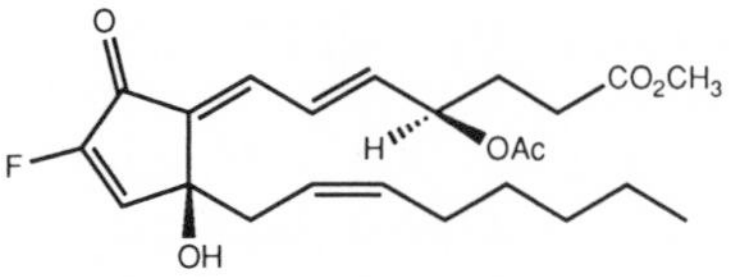

Figure 4-7. Examples of marine prostaglandins.

assigned structures with respect to stereochemical configuration at C-12 (Baker *et al.*, 1985), is based on total synthesis (Suzuki *et al.*, 1988, and references therein).

4.2.2.2 Halogenated Clavulones

Further research on marine prostanoids from *Clavularia viridis* resulted in the isolation of clavulones having halogen situated at C-10. Thus, chloro- (Iguchi *et al.*, 1985) and iodo- and bromovulones (Iguchi *et al.*, 1986) (e.g., **5**; Fig. 4-7) were characterized and found to have increased antitumor activity relative to that of the nonhalogenated relatives.

4.2.2.3 Structural Requirements for Antitumor Activity of Coral Prostanoids

A comparative study of antitumor and cytotoxic properties of prostanoids gave a rank order of activity of chlorovulone I > bromovulone I = iodovulone I > clavulone I or II > prostaglandin. Chlorovulone I was about 13 times more potent than clavulone as an antiproliferative agent. Synthetic 10-fluoro-12-*O*-desacetylclavulone II (**6**; Fig. 4-7) was comparable in activity to chlorovulone I (Honda *et al.*, 1987). Results from investigation of several analogues related to the halogenated clavulones suggest that the activity of these compounds is resident in the cross-conjugated dienone moiety and that reactivity of this functionality is increased by introduction of halogen at C-10. Thus, the absence of unsaturation at C-10 greatly diminishes activity. The presence of the hydroxyl group at C-12 enhances activity, but the stereochemistry at this position apparently is not critical (Honda *et al.*, 1987, 1988).

4.2.3 Halogenated Prostacyclins

In addition to PGF, PGD, and PGE, the endoperoxide produced by prostaglandin synthase is converted enzymatically to two other biologically important and potent products, prostacyclin (PGI_2) and thromboxane A_2 (TXA_2) (Fig. 4-8). The very unstable TXA_2 ($t_{1/2} = 30$ s at 37°C) strongly contracts the aorta and induces platelet aggregation. In contrast, PGI_2, which has a half-life in blood at 37°C of only two to three minutes, is the most potent endogenous inhibitor of platelet aggregation known. The opposing actions of TXA_2 and PGI_2 are thought to play an important role in the regulation of vasculature activity [for a more complete discussion, see Moncada and Vane (1978)].

Analogues of PGI_2 having similar antiaggregating and vasodilating properties, but possessing longer biological half-lives, would have many important clinical applications. For this reason, many analogues of PGI_2 have been prepared with the aim of increasing biostability and potency, as well as achieving selectivity of action. A concise description of strategies used is provided in the review by Barnette (1984). A more thorough discussion of the use of the selective introduction of fluorine into PGI_2 analogues is given in the same review. A summary of the important points will be given here, along with a survey of recent developments.

Chemical reactivity of PGI_2 clearly is resident in the enol ether functionality. Barnette (1984) has divided fluorinated PGI_2 analogues into three classes: those prepared from available fluorinated prostaglandins, and which retain the reactive enol ether, those that have fluorine situated strategically to stabilize the enol ether, and fluorinated analogues of hydrolytically stable PGI_2 derivatives.

4.2.3.1 Fluorinated prostacyclins Having a Labile Enol Ether—Introduction of Fluorine to Block Oxidation of 15-OH

Attempts to increase potency and selectivity of PGI_2 based on favorable fluorine substitution patterns observed with $PGF_{2\alpha}$ and other

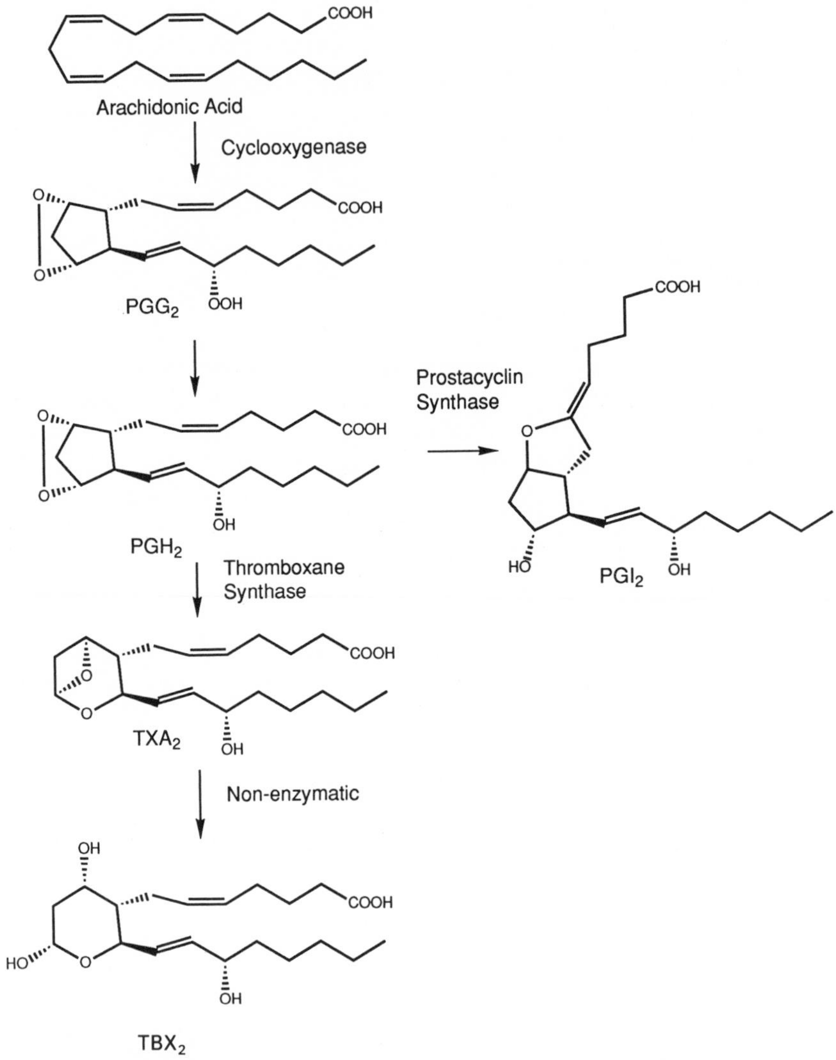

Figure 4-8. Biosyntheses of PGI_2 and TXA_2 from arachidonic acid.

primary prostaglandins have met with little success. 12-Fluoro-PGI_2 and 14-fluoro-PGI_2 (Fig. 4-9) had little, if any, increased activity over PGI_2, in contrast to results seen in the $PGF_{2\alpha}$ series [reviewed by Barnette (1984)]. A preliminary report suggested that 16,16-difluoro- and 2,2,16,16-tetrafluoro-PGI_2 (Fig. 4-9) had increased duration of activity [cited by Barnette (1984)].

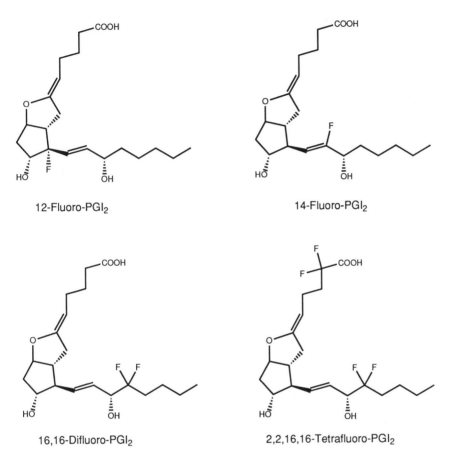

12-Fluoro-PGI₂

14-Fluoro-PGI₂

16,16-Difluoro-PGI₂

2,2,16,16-Tetrafluoro-PGI₂

Figure 4-9. Examples of fluorinated analogues of PGI_2 designed to retard oxidation of the 15-OH group.

4.2.3.2 Introduction of Fluorine to Increase Chemical Stability

Hydrolytic deactivation of PGI_2 is initiated by protonation of the enol ether. Several attempts have been made to impede this process by decreasing the electron density of the enol ether by placing fluorine in close proximity to this group. A notable example of this is seen in the synthesis of 10,10-difluoro-13-dehydro-PGI_2 (Fig. 4-10) (Fried et al., 1980). Both hydrolytic instability and oxidative deactivation by 15-hydroxyprostaglandin dehydrogenase were expected to decrease. This analogue was three- to fourfold more potent than PGI_2 with respect to in vivo relaxation of arteries and was 70% as active as PGI_2 in inhibition of human platelets

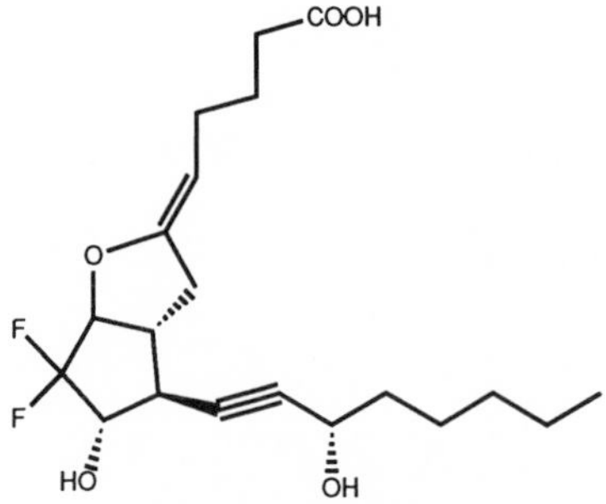

10,10-Difluoro-13-dehydro-PGI₂

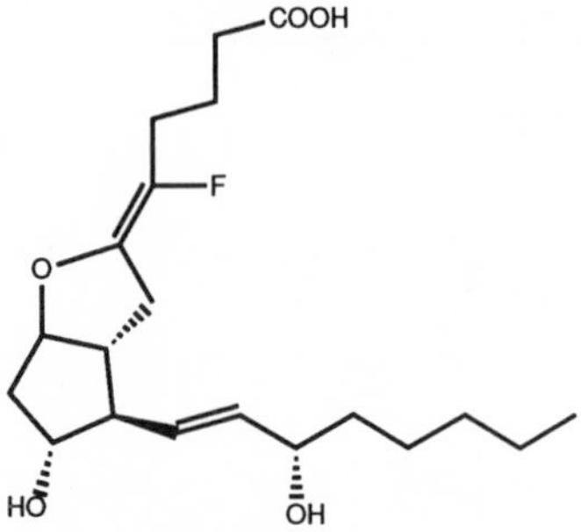

5-Fluoro-PGI₂

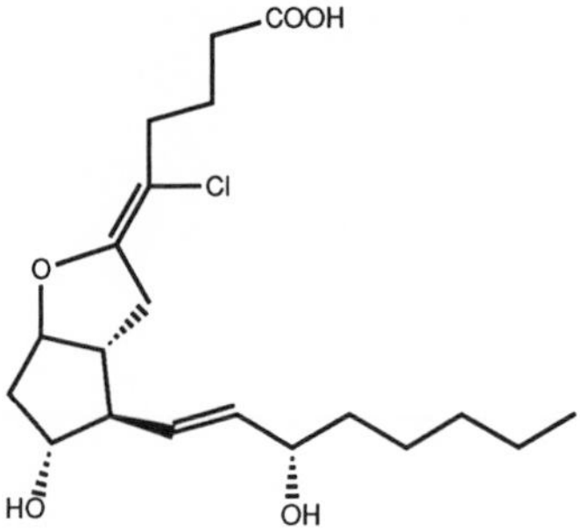

5-Chloro-PGI₂

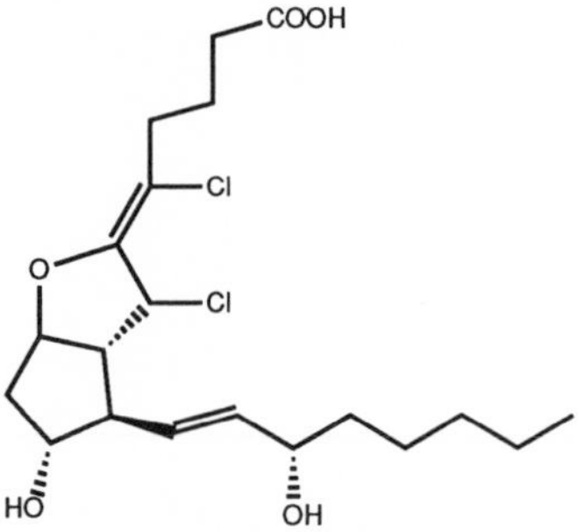

5,7-Dichloro-PGI₂

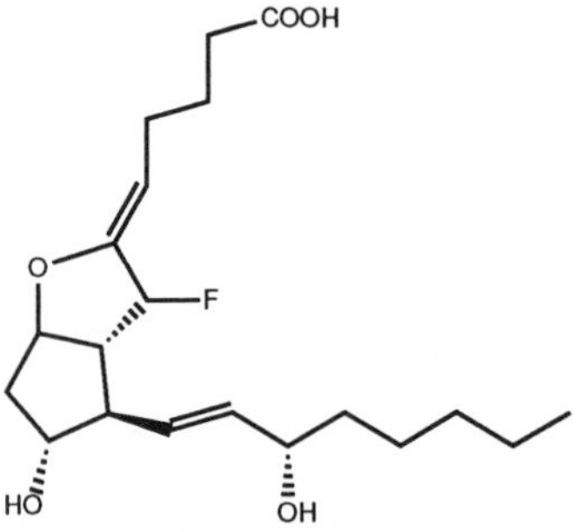

7-Fluoro-PGI₂

Figure 4-10. Examples of analogues of PGI₂ wherein halogen has been introduced to decrease the hydrolytic lability of the enol ether.

in vitro. As hoped, *in vitro* studies showed that the difluoro analogue had a biological half-life of about 24 h, compared to 10–15 min for PGI₂ under the same conditions. Unfortunately, the duration of action *in vivo* was unchanged, suggesting that rapid removal of the drug from its site of action or other pharmacokinetic factors may be as important as metabolic oxidation and chemical instability as deactivation processes (Hatano *et al.*, 1980).

5-Fluoro-PGI$_2$ (Fig. 4-10) has been reported in the patent literature (Nysted and Pappo, 1982). This analogue would be expected to have greater chemical stability as a result of decreased electron density of the double bond. 5-Chloro- and 5,7-dichloro-PGI$_2$ (Fig. 4-10) were prepared by direct chlorination of PGI$_2$ with *N*-chlorosuccinimide. 5-Chloro-PGI$_2$ had a half-life of 1.5 h in pH 4.7 buffer, compared to a half-life of 22 s for PGI$_2$ in pH 5.98 buffer. An inhibition of platelet aggregation was retained after 2 h in pH 7.4 buffer (Bannai *et al.*, 1983a).

Included in a series of chemically stable prostacyclin analogues synthesized by Bannai *et al.* (1983b) was 7-fluoro-PGI$_2$ (Fig. 4-10), which had a half-life of more than one month in pH 7.4 buffer. This analogue is a potent inhibitor of platelet aggregation, with an initial activity some 10-fold less than that of PGI$_2$. Unlike PGI$_2$, which rapidly lost activity, the activity of 7-fluoro-PGI$_2$ was unchanged after 4 h. Advantage was taken of this stability to study the relationship between PGI$_2$ receptor binding and cAMP formation (Mizuno *et al.*, 1983).

4.2.3.3 Fluorinated Analogues of Chemically Stable Prostacyclins

Data from a large number of PGI$_2$ analogues show that the 5,6 double bond is important for activity. Fluorinated analogues of hydrolytically stable dihydro- and Δ^4-iso-PGI$_2$ (Fig. 4-11) are consistent with these data and show little biological activity (Barnette, 1984).

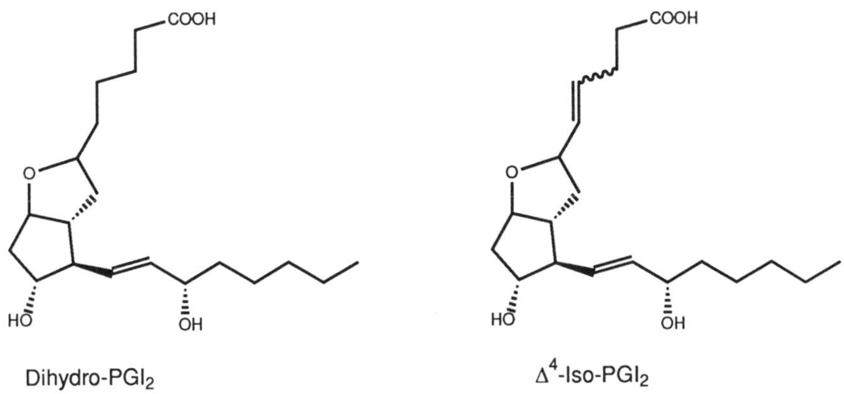

Dihydro-PGI$_2$ Δ^4-Iso-PGI$_2$

Figure 4-11. Examples of chemically stable PGI$_2$ analogues, fluorinated derivatives of which show little biological activity.

4.3 Fluorinated Thromboxanes

The bicyclic oxetane-oxane structure incorporated into thromboxane (TXA_2) (Fig. 4-12) renders this compound extremely sensitive to hydrolysis. In fact, unlike PGI_2, TXA_2 is not yet accessible by synthesis, although several more stable analogues have been prepared wherein one or both of the oxygen atoms in the bicyclic ring have been replaced with carbon or sulfur. Fried *et al.* (1984) initiated an alternative approach, relying once again on the electron-withdrawing effects of fluorine to destabilize the transition state for hydrolysis. Two model compounds (Fig. 4-12) were prepared. In reactivity studies, the fluorinated analogue was shown to be hydrolyzed at a rate 10^8 slower than that for TXA_2, and some 100 times slower than that for common, unstrained acetals. The conclusion was reached that the 7,7-difluoro-2,6-dioxa[3.1.1]bicyclo-heptane system is sufficiently stable to be compatible with many synthetic operations. 10,10-Difluoro-TXA_2 has been synthesized recently and, with respect to stimulation of platelet aggregation, was found to be four to five times more potent than TXA_2 (Morinelli *et al.*, 1989).

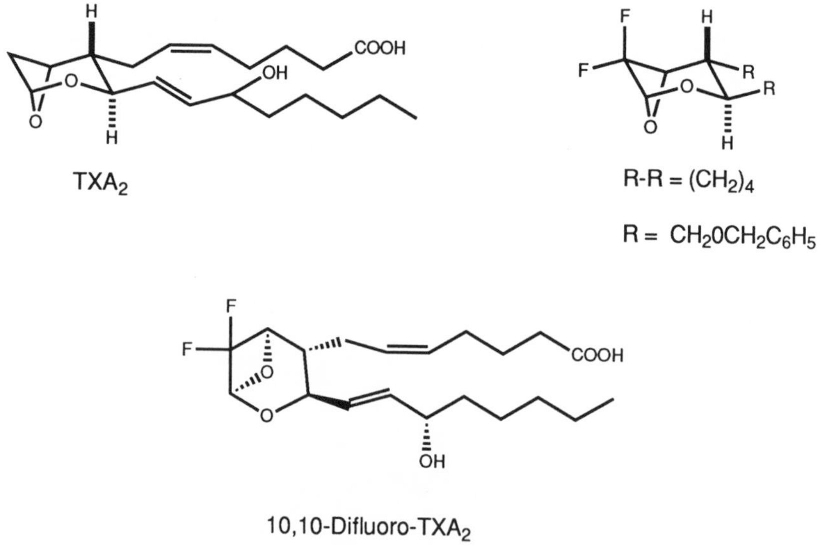

TXA$_2$

R-R = (CH$_2$)$_4$

R = CH$_2$OCH$_2$C$_6$H$_5$

10,10-Difluoro-TXA$_2$

Figure 4-12. TXA$_2$, fluorinated model bicyclic oxetane–oxane structures, and 10,10-difluoro-TXA$_2$.

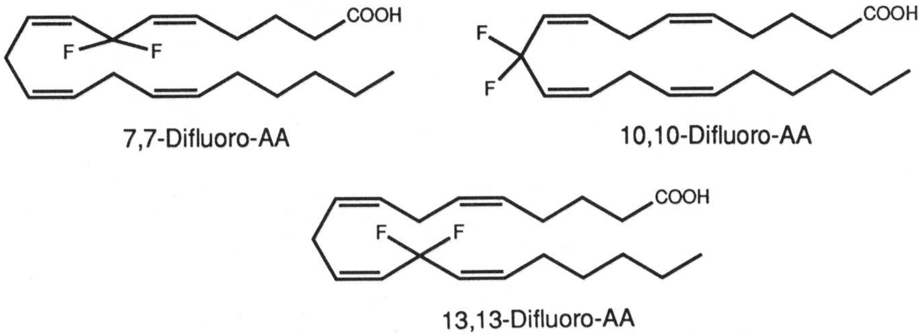

7,7-Difluoro-AA   10,10-Difluoro-AA

13,13-Difluoro-AA

Figure 4-13. Fluorinated arachidonic acid analogues as potential biochemical precursors of fluorinated PGI_2 and TXA_2 analogues.

4.4 Fluorinated Analogues of Arachidonic Acid

Fried and co-workers recently initiated a fundamentally different approach to prostacyclins and thromboxanes incorporating geminal difluoro substitution. Operation of the enzymatic processes which produce PGI_2 and TXA_2 on fluorinated analogues of arachidonic acid (AA) might be expected to give the corresponding fluorinated PGI_2 and TXA_2 analogues. 7,7-, 10,10-, and 13,13-Difluoroarachidonic acids (Fig. 4-13)

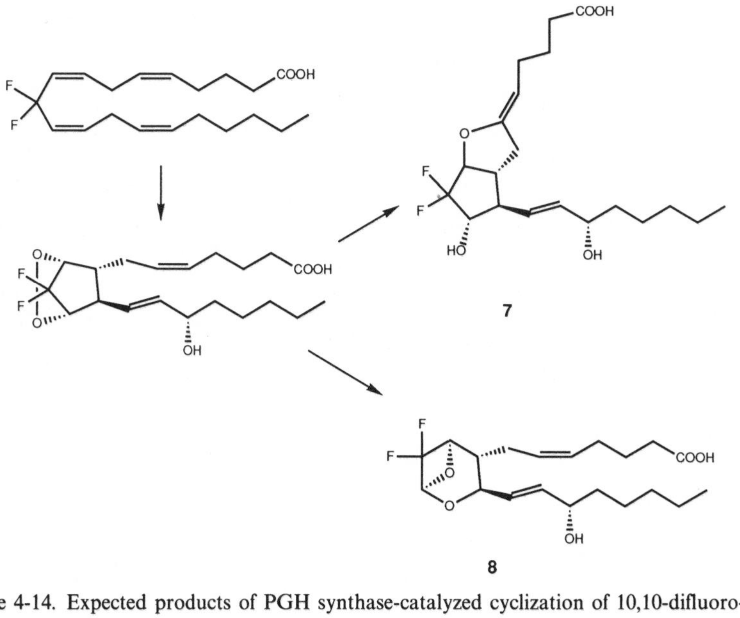

7

8

Figure 4-14. Expected products of PGH synthase-catalyzed cyclization of 10,10-difluoro-AA.

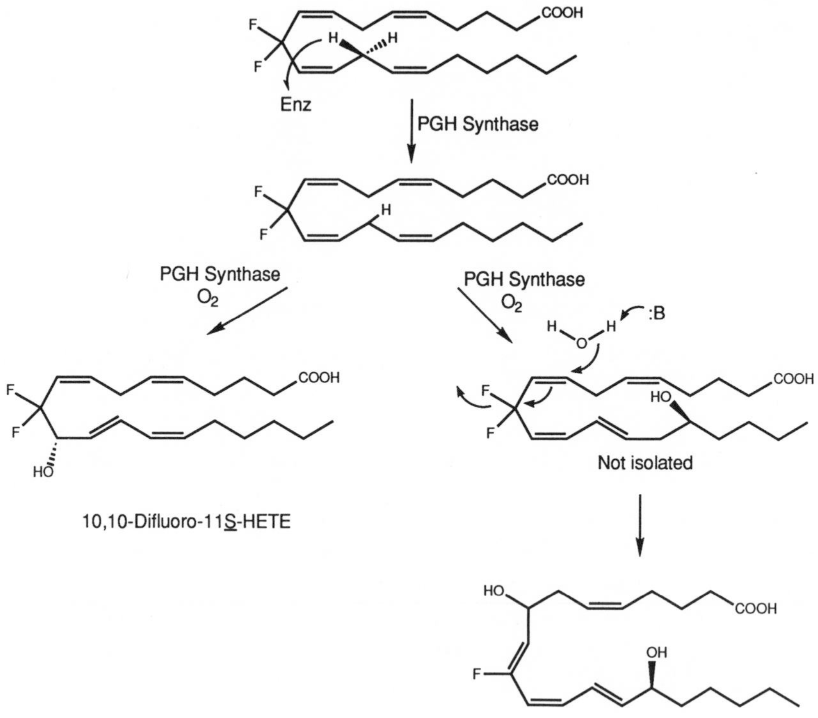

Figure 4-15. Products produced by the action of PGH synthase on 10,10-difluoro-AA.

were prepared to examine this possibility (Kwok *et al.*, 1987a). For example, 10,10-difluoro-AA was expected to produce chemically stabilized 10,10-difluoro-PGI$_2$ and 10,10-difluoro-TXA$_2$ (**7** and **8**; Fig. 4-14). Whereas 10,10-difluoro-AA was a substrate for both prostaglandin endoperoxide (PGH) synthase and soybean lipoxidase, no PGH synthase-catalyzed cyclization to prostaglandins was observed. Instead, fluorinated analogues of hydroxyeicosenoic acids (HETE) were formed, including 10,10-difluoro-11(*S*)-HETE and 10-fluoro-8,15-diHETE (Fig. 4-15). The mechanism

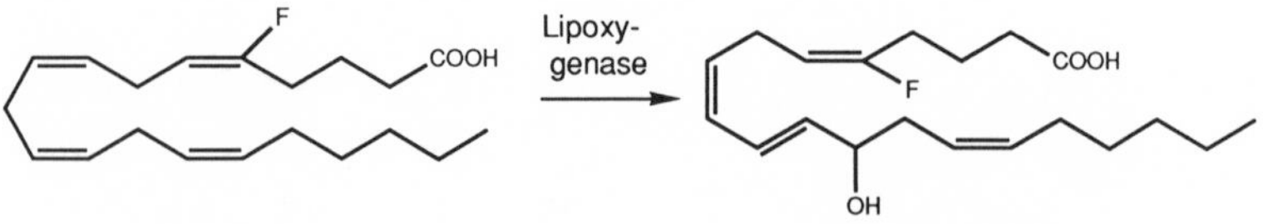

Figure 4-16. Biosynthesis of 5-fluoro-12-HETE from 5-fluoro-AA.

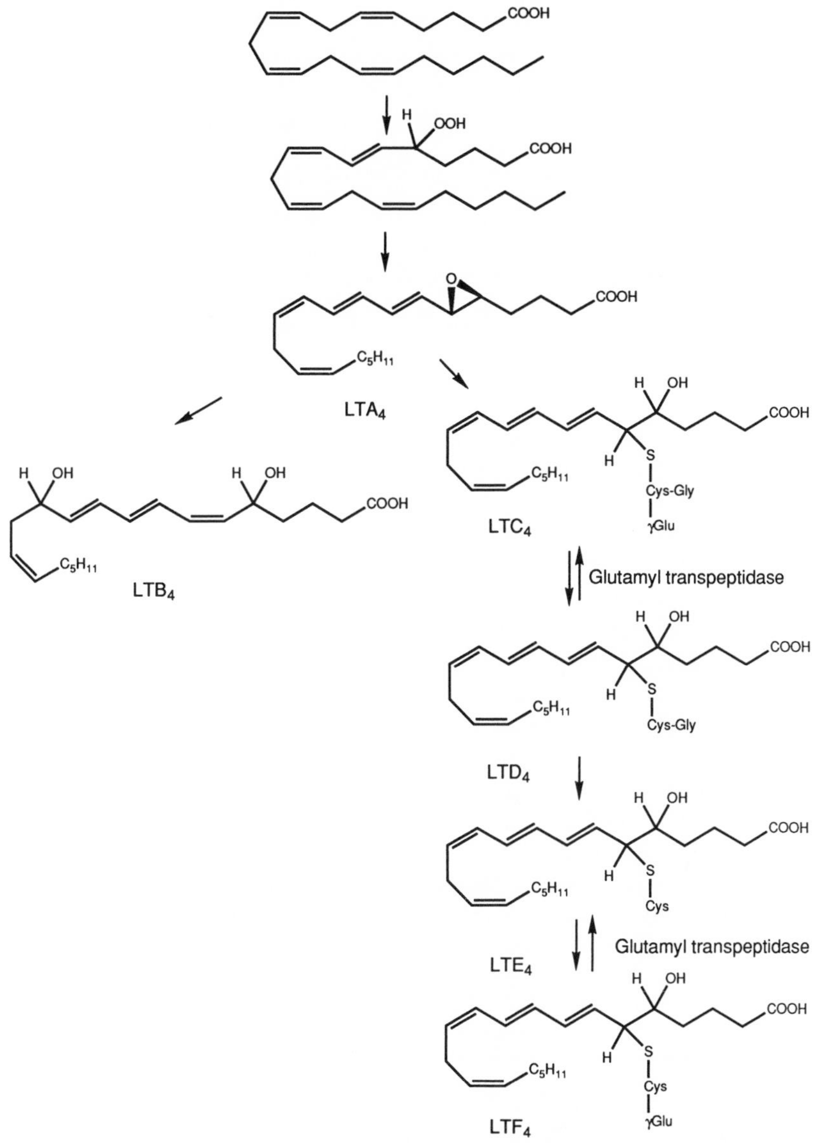

Figure 4-17. Biosynthesis of leukotrienes.

shown in Fig. 4-15 is consistent with the formation of these products (Kwok *et al.*, 1987b).

While the aforementioned study failed to produce cyclooxygenase-type products, the lipoxygenase products themselves are topics of much current research. Kobayashi and co-workers prepared 5-fluoroarachidonic acid (Fig. 4-16) and used it as a substrate for human platelet lipoxygenase (Taguchi *et al.*, 1987). As with arachidonic acid itself, the main product was hydroxylated at C-12, thus giving an enzymatic synthesis of 5-fluoro-12-HETE (Fig. 4-16). Although 12-HETE is the major product derived from arachidonic acid in platelets, its physiological role is uncertain. Selective fluorination of precursor arachidonic acid is proposed as a strategy to study the functional importance and metabolism of 12-HETE (Taguchi *et al.*, 1987).

The leukotrienes are formed from arachidonic acid through an unstable epoxide, leukotriene A_4 (LTA_4), which is in turn transformed either enzymatically or nonenzymatically to a series of hydroxy derivatives (leukotrienes B_4, C_4, D_4, E_4, and F_4) (Fig. 4-17). These eicosanoids are mediators of hypersensitivity reactions and inflammation and, accordingly, have attracted recent intense scrutiny [reviewed by Samuelsson (1983)]. For example, leukotriene B_4 (LTB_4) stimulates the functions of neutrophils, including chemotaxis, aggregation, and degranulation, and thus is an important mediator of various inflammatory responses. The study of LTB_4 action has been hampered by its rapid metabolism through ω-oxidation, preventing its detection under certain biological conditions. Two groups recently reported the enzymatic synthesis of 20,20,20-trifluoro-LTB_4, a metabolically stable analogue of LTB_4, from 20,20,20-trifluoroarachidonic acid (Fig. 4-18) (Tanaka *et al.*, 1988; Tsai *et al.*, 1989). Trifluoro-LTB_4 stimulated chemotaxis in human neutrophils to the same extent as did LTB_4 and, unlike LTB_4, was detected readily when precursor arachidonic acid was incubated with neutrophils stimulated with immunoglobulin G (Tanaka *et al.*, 1988). In contrast, trifluoro-LTB_4 acted

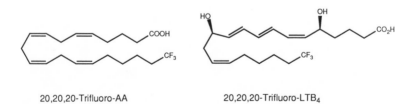

20,20,20-Trifluoro-AA   20,20,20-Trifluoro-LTB$_4$

Figure 4-18. 20,20,20-Trifluoro-LTB$_4$, an analogue that has increased stability over LTB$_4$, and 20,20,20-trifluoro-AA, its precursor in an enzymatic synthesis.

as an antagonist to LTB$_4$-induced degranulation (Tsai *et al.*, 1989). These studies indicate that this analogue should be particularly useful in the study of LTB$_4$ biochemistry.

4.5 An Iodinated Inhibitor of Lipoxygenase

In strategies to modulate the functions of leukotrienes, the mammalian lipoxygenases involved in the biosynthesis of leukotrienes represent an alternative target for drug design. Lipoxygenases catalyze the conversion of *cis,cis*-1,4-dienes to diene hydroperoxides. Lipoxygenase I from soybeans, frequently used in biochemical studies as a model for the mammalian enzyme, is specific for fatty acids in which the diene unit begins on the sixth carbon from the methyl terminus (the $\omega6$ carbon). To study the mechanism of $C-H$ bond cleavage, analogues were synthesized wherein the $\omega6-\omega7$ double bond was replaced by halogen. The observation that 12-iodo-*cis*-9-octadecenoic acid (Fig. 4-19) functioned as an irreversible inhibitor was unexpected, and the mechanism for this inactivation is not obvious. A possibility consistent with available evidence involves the oxidation of the iodide to the iodoso compound, followed by rapid loss of IO$^-$ to form a carbocation (Rotenberg *et al.*, 1988) (see Chapter 9 in Vol. 9A of this series).

4.6 Halogenated Eicosanoids—Summary

As with other classes of compounds, halogenation can have significant effects on the biological activity of eicosanoids. In the discussion above, examples of synthetic analogues have been chosen to illustrate strategies employed to increase activity and selectivity of eicosanoids by selective halogenation. Other examples can be found in the reviews cited. It is important to stress that halogenation is one of many approaches to this end and that other structural manipulations—frequently in combination

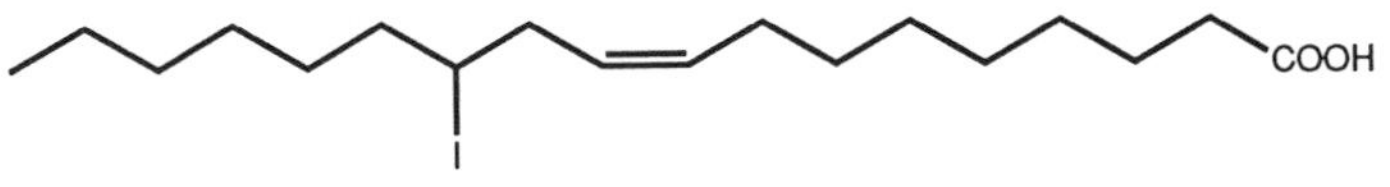

12-Iodo-<u>cis</u>-9-octadecenoic acid

Figure 4-19. An irreversible inhibitor of lipoxygenase I.

with halogenation—have produced important advances in this area. Enzymatic syntheses—exemplified by the synthesis of fluorinated HETE analogues and trifluoro-LTB$_4$—represent an intriguing approach to analogues that are otherwise difficult to obtain. The halogenated prostanoids isolated from marine organisms have provided interesting lead compounds for new chemotherapeutic agents, and new synthetic compounds emanating from these leads undoubtedly will appear soon.

References

Arroniz, C. E., Gallina, J., Martinez, E., Muchowski, J. M., Velarde, E., and Rooks, W. H., 1978. Synthesis of ring halogenated prostagladins (1), *Prostaglandins* 16:47–65.

Axen, U. F., 1977. 2,2-Difluoro-16-phenoxy-PGF$_{2\alpha}$ analogues, U.S. Patent 4,001,300.

Baker, B. J., Okuda, R. K., Yu, P. T. K., and Scheuer, P. J., 1985. Punaglandins: Halogenated antitumor eicosanoids from the octacoral *Telesto riisei*, *J. Am. Chem. Soc.* 107:2976–2977.

Bannai, K., Toru, T., Oba, T., Tanaka, T., Okamura, N., Watanabe, K., Hazato, A., and Kurozumi, S., 1983a. Halogenation of PGI$_2$-enol ether with *N*-halosuccinimide: Synthesis of new stable PGI$_2$ analogues, 5-chloro- and 5,7-dichloro-PGI$_2$, *Tetrahedron Lett.* 23:3707–3710.

Bannai, K., Toru, T., Oba, T., Tanaka, T., Okamura, N., Watanabe, K., Hazato, A., and Kurozumi, S., 1983b. Synthesis of chemically stable prostacyclin analogues, *Tetrahedron* 39:3807–3819.

Barnette, W. E., 1984. The synthesis and biology of fluorinated prostacyclins, *Crit. Rev. Biochem.* 15:201–235.

Bergström, S., Ryhage, R., Samuelsson, B., and Sjövall, J., 1963. Prostaglandins and related factors, *J. Biol. Chem.* 238:3555–3564.

Filler, R., and Naqvi, S. M., 1982. Fluorine in biomedicinal chemistry. An overview of recent advances and selected topics, in *Biomedicinal Aspects of Fluorine Chemistry* (R. Filler and Y. Kobayashi, eds.), Kodansha Ltd., Tokyo; Elsevier Biomedical Press, Amsterdam, pp. 1–32.

Fried, J., Mitra, D. K., Nagarajan, M., and Mehrotra, M. M., 1980. 10,10-Difluoro-13-dehydroprostacyclin: A chemically and metabolically stabilized potent prostaglandin, *J. Med. Chem.* 23:234–237.

Fried, J., Hallinan, E. A., and Szwedo, M. J., Jr., 1984. Synthesis and properties of 7,7-difluoro derivatives of the 2,6-dioxa[3.1.1]bicycloheptane ring system present in thromboxane A$_2$, *J. Am. Chem. Soc.* 106:3871–3892.

Fukushima, M., and Kato, T., 1985. Antitumor marine eicosanoids: Clavulones and punaglandins, *Adv. Prostaglandin Thromboxane Leukotriene Res.* 15:415–418.

Greico, P. A., Williams, E., and Sugahara, T., 1979a. Ring-fluorinated prostaglandins: Total synthesis of ($\pm$)-10α-fluoro prostaglandin F$_{2\alpha}$ methyl ester, *J. Org. Chem.* 44:2194–2199.

Greico, P. A., Sugahara, T., Yokoyama, Y., and Williams, E., 1979b. Fluoroprostaglandins: Synthesis of ($\pm$)-10β-fluoro prostaglandin F$_{2\alpha}$ methyl ester, *J. Org. Chem.* 44:2189–2194.

Greico, P. A., Owens, W., Wang, C.-L. J., Williams, E., and Schillinger, W. J., 1980a. Fluoroprostaglandins: Synthesis and biological evaluation of the methyl esters of (+)-12-fluoro-, (−)-*ent*-12-fluoro-, (+)-15-*epi*-fluoro-, and (−)-*ent*-15-*epi*-fluoroprostaglandin F$_{2\alpha}$, *J. Med. Chem.* 23:1072–1077.

Greico, P. A., Schillinger, W. J., and Yokoyama, Y., 1980b. C(14)-Fluorinated prostaglandins: Synthesis and biological evaluation of the methyl esters of (+)-14-fluoro-, (+)-15-*epi*-14-fluoro-, (+)-13(*E*)-14-fluoro-, and (+)-13(*E*)-15-*epi*-14-fluoro-prostaglandin F$_{2\alpha}$, *J. Med. Chem.* 23:1077–1083.

Greico, P. A., Takigawa, T., and Vedananda, T. R., 1985. Fluoroprostaglandins: Total synthesis of (+)-13-fluoroprostaglandin F$_{2\alpha}$ methyl ester, *J. Org. Chem.* 50:3111–3115.

Hatano, Y., Kohli, J. D., Goldberg, L. I., Fried, J., and Mehrotra, M. M., 1980. Vascular relaxing activity and stability of 10,10-difluoro-13,14-dehydroprostacyclin, *Proc. Natl. Acad. Sci. USA* 77:6846–6850.

Honda, A., Mori, Y., Iguchi, K., and Yamada, Y., 1987. Antiproliferative and cytotoxic effects of newly discovered halogenated coral prostanoids from the Japanese stolonifer *Clavularia viridis* on human myeloid leukemia cells in culture, *Mol. Pharmacol.* 32:530–535.

Honda, A., Mori, Y., Iguchi, K., and Yamada, Y., 1988. Structure requirements for antiproliferative and cytotoxic activities of marine coral prostanoids from the Japanese stolonifer *Clavularia viridis* against human myeloid leukemia cells in culture, *Prostaglandins* 36:621–630.

Iguchi, K., Kaneta, S., Mori, K., Yamada, Y., Honda, A., and Mori, Y., 1985. Chlorovulones, new halogenated marine prostanoids with an antitumor activity from the stolonifer *Clavularia viridis* Quoy and Gaimard, *Tetrahedron Lett.* 26:5787–5790.

Iguchi, K., Kaneta, S., Mori, K., Yamada, Y., Honda, A., and Mori, Y., 1986. Bromovulone I and iodovulone I, unprecedented brominated and iodinated marine prostanoids with antitumor activity isolated from the Japanese stolonifer *Clavularia viridis* Quoy and Gaimard, *J. Chem. Soc.* 1986:981–982.

Kwok, P.-Y., Muellner, F. W., Chen, C.-K., and Fried, J., 1987a. Total synthesis of 7,7-, 10,10-, and 13,13-difluoroarachidonic acids, *J. Am. Chem. Soc.* 109:3684–3692.

Kwok, P.-Y., Muellner, F. W., and Fried, J., 1987b. Enzymatic conversions of 10,10-difluoroarachidonic acid with PGH synthase and soybean lipoxygenase, *J. Am. Chem. Soc.* 109:3692–3698.

Loge, O., and Radüchel, B., 1984. Gastrointestinal properties of nocloprost, a stable prostaglandin E$_2$ analogue, *Naunyn-Schmiedeberg's Arch. Pharmacol., Suppl.* 325:R33.

Magerlein, B. J., and Miller, W. L., 1975. 16-Fluoroprostaglandins, *Prostaglandins* 9:527–530.

Mizuno, Y., Ichikawa, A., and Tomita, K., 1983. Effect of 7-fluoro prostacyclin, a stable prostacyclin analogue, on cAMP accumulation and prostaglandin binding in mastocytoma P-815 cells, *Prostaglandins* 26:785–795.

Moncada, S., and Vane, J. R., 1978. Prostacyclin formation and effect, in *Chemistry, Biochemistry, and Pharmacological Activity of Prostanoids* (S. M. Roberts and F. Scheinmann, eds.), Pergamon Press, Oxford, pp. 258–273.

Moore, P. K., 1985. *Prostanoids; Pharmacological, Physiological and Clinical Relevance*, Cambridge University Press, London.

Morinelli, T. A., Okwu, A. K., Mais, D. E., Halushka, P. V., John, V., Chen, C.-K., and Fried, J., 1989. Difluorothromboxane A$_2$ and stereoisomers: Stable derivatives of thromboxane A$_2$ with differential effects on platelets and blood vessels, *Proc. Natl. Acad. Sci. USA* 86:5600–5604.

Nysted, L., and Pappo, R., 1982. 5-Fluoro-PGI$_2$ compounds, Eur. Pat. Appl. EP 62,303 [*CA* 98:71787w (1983)].

Ringold, H. J., Lawrence, H., Jr., and Graves, J. M. H., 1964. The influence of unsaturation and of fluorine substitution on ketone–alcohol equilibrium constants. A measure of α,β-unsaturated ketone resonance energy and of halogen destabilization, *J. Am. Chem. Soc.* 86:4510–4512.

Rotenberg, S. A., Grandizio, A. M., Selzer, A. T., and Clapp, C. H., 1988. Inactivation of soybean lipoxygenase 1 by 12-iodo-*cis*-octadecenoic acid, *Biochemistry* 27:8813–8818.

Samuelsson, B., 1983. Leukotrienes: Mediators of immediate hypersensitivity reactions and inflammation, *Science* 220:568–575.

Suzuki, M., Morita, Y., Yanagisawa, A., Baker, B. J., Scheuer, P. J., and Noyori, R., 1988. Synthesis and structural revision of (7*E*)- and (7*Z*)-punaglandin 4, *J. Org. Chem.* 53:286–295.

Taguchi, T., Takigawa, T., Igarashi, A., Kobayashi, Y., Tanaka, Y., Jubiz, W., and Briggs, R. G., 1987. Synthesis of 5-fluoroarachidonic acid and its transformation to 5-fluoro-12-hydroxyeicosatetraenoic acid, *Chem. Pharm. Bull.* 35:1666–1669.

Tanaka, Y., Klauck, T. M., Jubiz, W., Taguchi, T., Hanzawa, Y., Igarashi, A., Inazawa, K., Kobayashi, Y., and Briggs, R. G., 1988. Biosynthesis of 20,20,20-trifluoroleukotriene B_4 from 20,20,20-trifluoroarachidonic acid: A metabolically stable analogue of leukotriene B_4 and its application to a study of stimulation of leukotriene B_4 synthesis by immunoglobulin G, *Arch. Biochem. Biophys.* 263:178–190.

Thierauch, K.-H., Stürzebecher, C.-St., Schillinger, E., Rehwinkel, H., Radüchel, B., Skuballa, W., and Vorbrüggen, H., 1988. Stable 9β- or 11α-halogen-15-cyclohexyl-prostaglandins with high affinity to the PGD_2-receptor, *Prostaglandins* 35:855–868.

Tsai, B. S., Keith, R. H., Villani-Price, D., Haack, R. A., Bauer, R. F., Leonard, R., Abe, Y., and Nicolaou, K. C., 1989. Differential effects of 20-trifluoromethyl leukotriene B_4 on human neutrophil functions, *Prostaglandins* 37:287–302.

von Euler, U. S., and Eliasson, R., 1967. *Prostaglandins* (Monographs in Medicinal Chemistry), Academic Press, New York.

Weinheimer, A. J., and Spraggins, R. L., 1969. The occurrence of two new prostaglandin derivatives (15-*epi*-PGA$_2$ and its acetate, methyl ester) in the gorgonian *Plexaura homomalla*. Chemistry of coelenterates. XV, *Tetrahedron Lett.* 1969:5185–5188.

Welch, J. T., 1987. Advances in the preparation of biologically active organofluorine compounds, *Tetrahedron* 43:3123–3197.

Biochemistry of Halogenated Nucleosides and Nucleotides

5.1 Biochemistry of Fluorinated Pyrimidines and Their Nucleosides

5.1.1 5-Fluorouracil—A Substrate for Lethal Synthesis

The classic biochemical studies of Peters and co-workers, demonstrating the biosynthesis of the extremely toxic fluorocitrate from fluoroacetate, provided the initial example of the profound biological consequences of lethal synthesis (Chapter 1). The syntheses of fluorinated steroids by Fried and co-workers, based in part on the precedent provided by the altered biological activity of fluorocitrate, was a second development of historical proportions (Chapter 3). Of comparable scientific impact was the synthesis of 5-fluorouracil (fl^{5}ura), 5-fluoroorotic acid (fl^{5}oro), and 5-fluorocytosine (fl^{5}cyt) (Fig. 5-1) by Heidelberger et $al.$ (1957).

Several principles were combined in the rational design of these pyrimidine analogues. The recognition that certain tumors have an unusually high uracil (ura) demand for DNA synthesis suggested that an effective antimetabolite of ura might be selectively toxic to these cells. The substitution of fluorine for hydrogen was considered a good approach to such an antimetabolite, based on the profound biochemical consequences of such substitution seen previously, for example, with fluorocitrate. Before incorporation of ura into DNA, thymidylate synthase-catalyzed methylation of deoxyuridine monophosphate (dUMP) at C-5 to produce thymidine monophosphate (dTMP) is required. If fl^{5}ura served as a substrate for ura anabolic enzymes to produce fluorodeoxyuridine monophosphate (fl^{5}dUMP), blockade of C-5 methylation was expected to result. Additional toxicity was expected to arise from biosynthesis of fluorouridine triphosphate (fl^{5}UTP) and incorporation of this analogue into RNA (Heidelberger et $al.$, 1957).

Many of the above expectations were realized, and fl^{5}ura was found to

fl⁵ura   fl⁵oro   fl⁵cyt

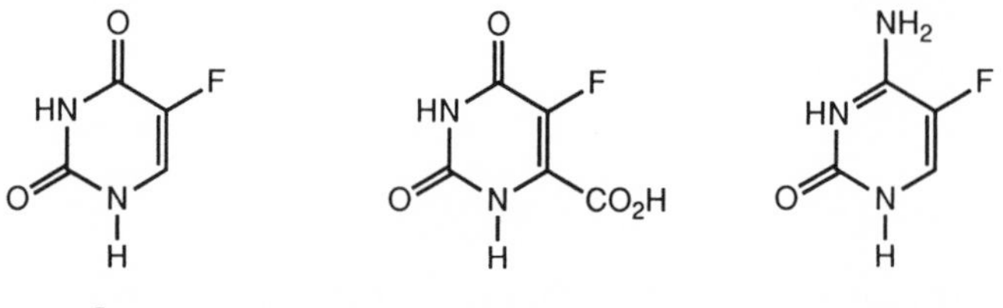

Figure 5-1. Early examples of fluorinated pyrimidines: 5-fluorouracil, 5-fluoroorotic acid, and 5-fluorocytidine (Heidelberger *et al.*, 1957).

be, and remains, an important anticancer drug, mainly for the treatment of solid tumors of the breast, ovary, and gastrointestinal tract. Although the clinical response rate has been somewhat disappointing, the impact of this research has been considerably broader than the development of an effective chemotherapeutic agent. Thus, attempts to improve on the effectiveness of fl⁵ura and to capitalize otherwise on this lead have spurred synthetic efforts to produce additional substituted nucleosides in a search for analogues with comparable or improved activity. In addition, extensive studies on the mechanism of action of fl⁵ura have revealed many details of the anabolism and catabolism of fl⁵ura (and ura) and have provided valuable pharmacokinetic data which, in turn, have aided in the development of combination chemotherapeutic schemes involving fl⁵ura and related nucleosides. Adding interest to these mechanistic studies is the fact that, despite three decades of study, the mechanism of action of fl⁵ura remains controversial.

In this section the development and mechanisms of action of fl⁵ura and other fluorinated pyrimidines will be considered. Efforts in several scientific disciplines have generated an enormous volume of literature. Fortunately, several excellent reviews are available, including the recent ones by Douglas (1987), Heidelberger *et al.* (1983), Heidelberger (1975a), Santi *et al.* (1982, 1976), and Myers (1981).

The inhibition by fl⁵dUMP of thymidylate synthase and the incorporation of fl⁵UTP into RNA have been demonstrated unequivocally and studied extensively for years as probable mechanisms of cytotoxicity, and, more recently, the incorporation of fl⁵dUTP into DNA has received attention as possibly contributing significantly to cytotoxicity of fl⁵ura. As noted, the relative importance of these mechanisms as determinants of antitumor effectiveness remains a topic of debate, the resolution of which is made difficult by complex pharmacokinetic relationships which cloud interpretation of experimental results (Myers, 1981). It appears increasingly likely that more than one mechanism can be operative, with the dominant mode of action being dependent on several factors, including the type of cell being studied.

5.1.2 Anabolic and Catabolic Reactions of Fl^{5}ura and Related Pyrimidines

The relationship between activities of enzymes involved in pyrimidine processing and fluoropyrimidine cytotoxicity has been studied extensively. Considering first the "lethal synthesis" of fl^{5}dUMP, the inhibitor of thymidylate synthase, three routes are available for the activation of fl^{5}ura to fl^{5}dUMP (Fig. 5-2). Thus, fl^{5}ura is a substrate for orotate (pyrimidine) phosphoribosyltransferase (OPRT), which converts fl^{5}ura to fluorouridine 5'-monophosphate (fl^{5}UMP), using 5-phosphoribose-1-pyrophosphate (PRPP) as a cosubstrate. PRPP is also the first intermediate in *de novo* purine biosynthesis. A second route to fl^{5}UMP is provided by the action of uridine-cytidine kinase on fluorouridine (fl^{5}U), available from fl^{5}ura and ribose-1-phosphate by the action of uridine-deoxyuridine phosphorylase. Phosphorylation of fl^{5}UMP to fl^{5}UDP, followed by reduction with ribonucleotide reductase, produces fl^{5}dUDP, which in turn is dephosphorylated to fl^{5}dUMP, the active antimetabolite. A third route to fl^{5}dUMP is provided by uridine-deoxyuridine phosphorylase-catalyzed synthesis of fluorodeoxyuridine (fl^{5}dU) from fl^{5}ura and deoxyribose-1-phosphate. Phosphorylation catalyzed by thymidine kinase produces fl^{5}dUMP (Myers, 1981).

Further phosphorylation of fl^{5}UDP provides fl^{5}UTP, which can serve as a substrate for RNA polymerase and become incorporated into RNA (Section 5.1.4). Formation of fl^{5}dUTP, which is a substrate for DNA polymerase, also occurs. Rapid hydrolysis by dUTPase and removal by uracil-DNA glycosylase of fl^{5}dUTP incorporated into DNA has complicated the detection of this incorporation (Myers, 1981) (Section 5.1.5).

Catabolism of fl^{5}ura proceeds normally through α-fluoro-β-ureido propionate to α-fluoro-β-alanine (Myers, 1981). This initial reduction is catalyzed by pyrimidine reductase or dihydrouracil dehydrogenase.

Strategies to modulate these enzymatic pathways have been directed toward determining the relative importance of potential mechanisms of toxicity and achieving more selective toxicity of tumor cells. This latter goal could be achieved either by increasing toxicity to tumor cells or by decreasing toxicity to normal cells. For example, addition of thymidine, an addition that can either protect against or enhance the cytotoxicity of fl^{5}ura, depending on the cell type, reverses the effects of inhibition of thymidylate synthase. Residual toxicity, referred to as thymidine-nonreversible toxicity,

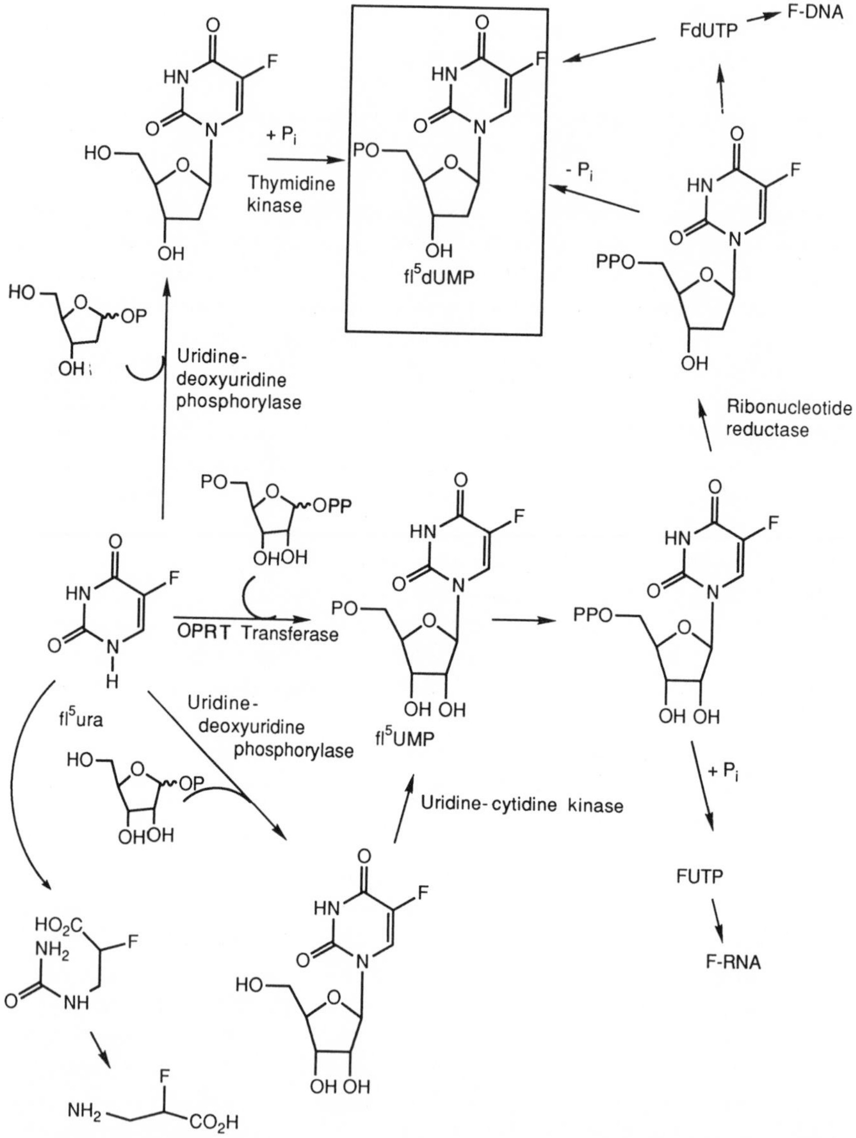

Figure 5-2. Three available routes for the activation of fl⁵ura to fl⁵dUMP (see text for details).

presumably occurs through a different mechanism, such as effects on RNA structure and function. In another approach, blockade of the thymidylate synthase pathway by drugs such as methotrexate channels the fluorinated metabolites into polynucleotide synthesis, where correlations with toxicity are expected to reflect effects of modified biosynthesis of RNA (or DNA). The design of prodrugs for fl⁵ura of fl⁵ura metabolites is based, at least in part, on the potential for achieving selectivity, taking advantage of the fact that certain tumor cells may have higher levels of the enzyme activities required for formation of fl⁵ura from the prodrug (Santi et al., 1982). The discussion below will include examples of these strategies.

5.1.3 5-Fluorouracil and the Thymidylate Synthesis Cycle

Thymidylate synthase catalyzes the conversion of dUMP to dTMP, an essential early step in the biosynthesis of DNA. The cofactor 5,10-methylene tetrahydrofolate (CH_2FAH_4) serves as the methyl group donor and is converted to 7,8-dihydrofolate (FAH_2) during the reaction, through a sequence of transformations shown in Fig. 5-3. Regeneration of CH_2FAH_4 by the action of dihydrofolate reductase and serine hydroxymethyltransferase completes the thymidylate synthesis cycle. Early studies showed that fl⁵dUMP, derived from fl⁵ura, is a potent inhibitor of thymidylate synthase (Cohen et al., 1958), but the nature—competitive or noncompetitive—of this inhibition remained a matter of considerable controversy for several years. Studies initiated by the groups of Santi (Santi and McHenry, 1972) and Heidelberger (Langenbach et al., 1972) using radiolabeled fl⁵dUMP provided the necessary breakthrough which has led to detailed knowledge of the interactions of fl⁵dUMP with thymidylate synthase, providing at the same time a sensitive assay for the enzyme [reviewed by Santi et al. (1982)].

The formation of a stable, covalent, ternary complex between fl⁵dUMP, thymidylate synthase, and CH_2FAH_4 is the critical step in the inhibition mechanism. As with the natural substrate, conjugate addition of a cysteinyl SH group to the 6 position of fl⁵dUMP is followed by a Mannich-type addition of the C-5 carbanion of the pyrimidine ring to CH_2FAH_4. However, deprotonation at C-5 is impossible, due to presence of fluorine; thus, elimination of FAH_4 is blocked, and the ternary complex accumulates (Fig. 5-4).

The importance of minimizing levels of the natural substrate, dUMP, and of maximizing levels of reduced folate cofactor has been shown by a more detailed analysis of the mechanism of fl⁵dUMP inhibition of thymidylate synthase (Heidelberger et al., 1983; Myers, 1981). The isola-

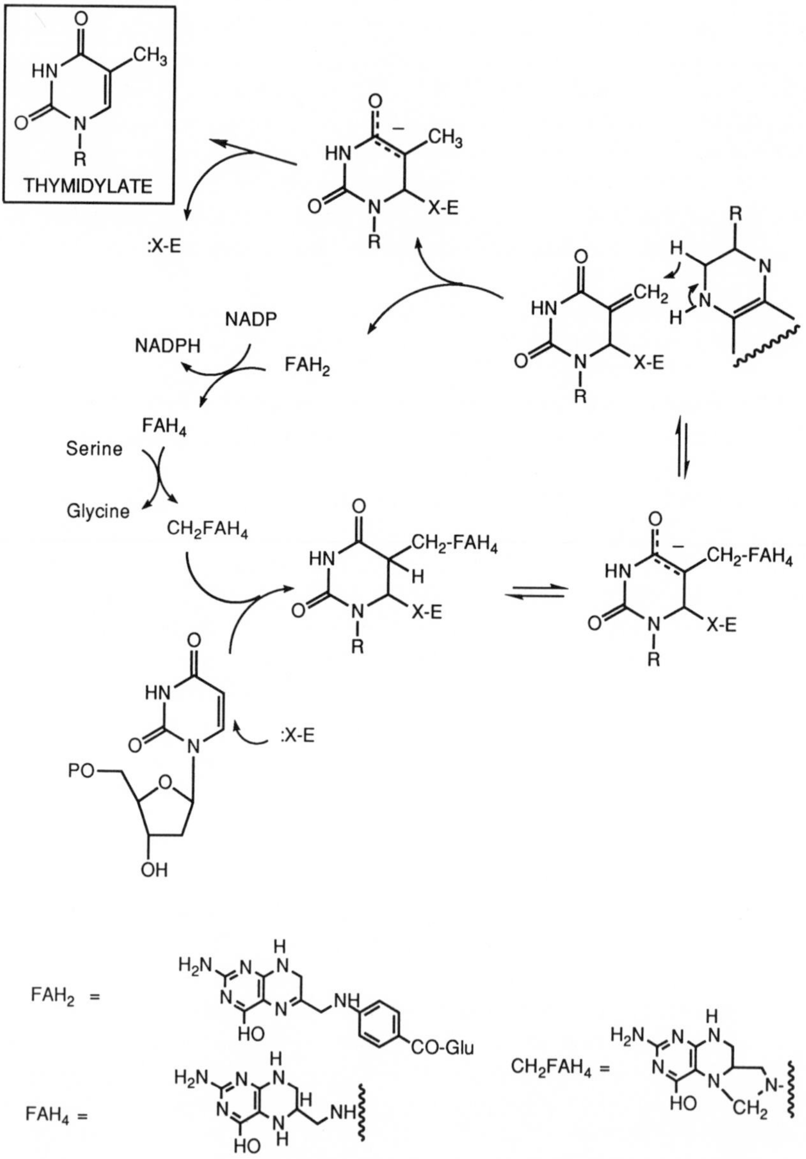

Figure 5-3. The thymidylate synthase cycle.

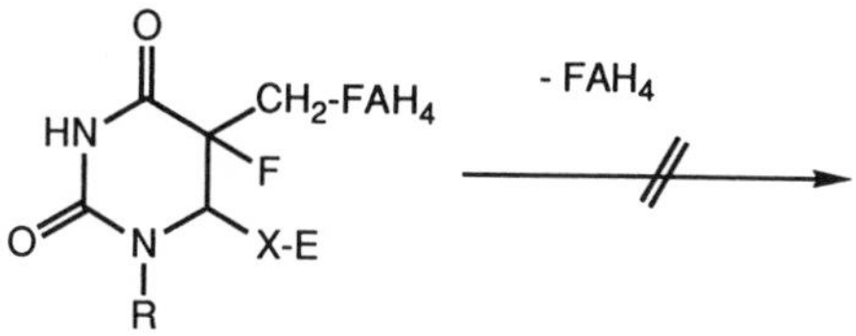

Figure 5-4. The stable covalent ternary complex between thymidylate synthase, CH_2THF_4, and fl^5dUMP—the presence of fluorine at C-5 blocks deprotonation and elimination of FAH_4.

tion of stable covalent structures following denaturation of the ternary complex was initially difficult to reconcile with the fact that inhibition can be reversed, for example, by the presence of increasing concentrations of dUMP. Subsequent investigations of kinetic isotope effects have shown that the covalent bond is formed after a rate-determining step, assumed to be a conformational change in the enzyme from an "open" to a "closed" state. A model consistent with these and other kinetic studies involves an ordered sequence for dissociation (and formation) of the ternary complex. In this model, in the formation of the ternary complex, fl^5dUMP or dUMP binding always precedes binding of CH_2FAH_4. Indeed, binding studies have shown that the cofactor does not bind to thymidylate synthase in the absence of dUMP or fl^5dUMP. Likewise, after denaturation of the complex, the enzymatic mechanism for dissociation is no longer available, and the covalent adduct can be isolated. Analysis of the rate associated with this model reveals that the effects of high levels of dUMP will combine with those of suboptimal levels of cofactor in a multiplicative manner. For this reason, accumulation of dUMP resulting from inhibition of thymidylate synthase decreases the effectiveness of that inhibition (Heidelberger et al., 1983). Thus, the presence of high levels of fl^5dUMP is a necessary, but not sufficient, condition for inhibition of thymidylate synthase.

5.1.4 Incorporation of 5-Fluorouracil into RNA and Antitumor Activity

The expectation, based on the preferential incorporation of ura into tumors, that fl^5ura would be incorporated into RNA of tumor cells (Heidelberger et al., 1957) was borne out shortly after its synthesis (Chaudhuri et al., 1958), and an early study produced a correlation between this incorporation and antitumor activity (Kessel et al., 1966). In contrast to the readily demonstrable inhibition of thymidylate synthase by fl^5dUMP—the blockade of dTMP biosynthesis—extensive efforts to define the biological consequences of incorporation of fl^5UTP into RNA have produced scientific controversy (Myers, 1981), along with evidence that multiple biochemical lesions may result.

5.1.4.1 Incorporation of Fl⁵ura into Ribosomal RNA (rRNA)

The nucleoli of cells contain the biochemical machinery for the biosynthesis of rRNA, mediated by RNA polymerase I. A 45S precursor rRNA is synthesized first and is subsequently processed into the 28S, 18S, and 5.8S molecules present in the mature ribosomes. The 40S and 60S ribosomal subunits are next assembled from the mature rRNAs, a 5S rRNA synthesized elsewhere in the nucleus by RNA polymerase III, and proteins. The nearly completed 40S and 60S subunits are then transported to the cytoplasm, where they are transformed to functional ribosomes (Darnell *et al.*, 1986).

Fl⁵ura, fl⁵U, and fl⁵oro inhibit the synthesis of mature rRNA (Heidelberger, 1983). The synthesis of the 45S precursor rRNA appears to be only slightly affected. However, the processing of this to 18S and 28S species is markedly inhibited, as a result of incorporation of fl⁵UTP into the 45S molecule (Wilkinson *et al.*, 1975). Moreover, the inhibition of rRNA maturation caused by fl⁵ura was shown in an early study to correlate well with the sensitivity of the cells to the drug (Wilkinson and Crumley, 1977). Methylation of 45S rRNA (2'-O-methylation and base methylation)—an important part of the processing—has been found to be fl⁵ura sensitive (Cory *et al.*, 1979; Carrico and Glazer, 1979). The importance of inhibition of rRNA maturation with respect to cytotoxicity remains unclear. Generally, this inhibition has been demonstrated by radiolabel pulsing experiments, and the duration of the effect has been questioned. In particular, no gross changes in mature rRNA have been observed (Armstrong *et al.*, 1986a).

5.1.4.2 Effect of Fl⁵ura on Transcription, Processing, and Translation of Messenger RNA (mRNA)

Fl⁵ura is readily incorporated into a major class of primary nuclear RNA transcripts, the heterogeneous nuclear RNA, or hnRNA. Heterogeneous in size, as the name suggests, these are synthesized outside the nucleolus, and thus are not precursors of rRNA, but include precursors of mRNA. A distinctive feature of mRNA, and of hnRNA precursors of mRNA, is the presence of a 3'-terminal sequence of adenylate residues [poly(A)]. Using L1210 cells, Glazer and Peale (1979) confirmed earlier work which showed that fl⁵ura readily replaces ura in hnRNA but has little effect on transcription. In a human colon carcinoma cell line, HT-29, cell lethality correlated with incorporation of fl⁵ura and fl⁵U into nRNA (Glazer and Lloyd, 1982). However, in the same cell line, the fl⁵ura-modified mRNA which formed showed no major differences in translational activity, suggesting that modified mRNA is not associated with the cytotoxicity of the drugs (Glazer and Hartman, 1983).

Despite repeated failures to demonstrate aberrant translation of mRNA due to incorporation of fl⁵ura, such a lesion has remained a candidate for contributing to toxicity. Recently, detection of miscoding of fl⁵ura-containing mRNA has been achieved using a gene-amplified KB cell line which overproduces dihydrofolate reductase (DHFR) (Dolnick and Pink, 1985). Dolnick and Pink suggested that the biosynthesis of fraudulent enzyme may account for the increased protein synthesis observed in the presence of fl⁵ura. Thus, by a feedback mechanism, the cell will correct for lower enzyme activity by synthesizing more enzyme. They further hypothesized that the metabolic demands so induced may cause the cell to die of exhaustion (Dolnick and Pink, 1985, 1983).

In the same cell line, Will and Dolnick (1986) found evidence that incorporation of fl⁵ura may inhibit the maturation of DHFR precursor mRNA into mature mRNA. In a similar study using C_3-L5178Y murine leukemia cells, another DHFR gene-amplified cell line, Armstrong *et al.* (1986a) found that fl⁵ura blocked the transport of nascent DHFR-mRNA into the cytoplasm, as a result of a defect either in processing or in nuclear–cytoplasmic transport.

5.1.4.3 *Incorporation of Fl⁵ura into Small-Molecular-Weight Nuclear RNA (snRNA)*

From the above data, it is increasingly evident that one effect of fl⁵ura is to alter mRNA metabolism at post-transcriptional sites. In parallel with the mechanism proposed for a similar effect on rRNA processing, this could be caused by fl⁵ura-induced changes in secondary structure of the precursor mRNA. However, as pointed out by Armstrong *et al.* (1986b) and by Will and Dolnick (1986), this inhibition may be secondary to fl⁵ura effects on still another RNA fraction, the small-molecular-weight nuclear RNA (snRNA). The snRNA fractions, some of which are uracil-rich, are proposed to have several important biological functions, including roles in the processing of rRNA, as well as in splicing, processing, and polyadenylation of mRNA. In cultured sarcoma-180 murine tumor cells, Armstrong *et al.* (1986b) found that the presence of fl⁵U induced changes in secondary structures in two snRNA fractions and reduced the rate of turnover in a third, presumably as a result of analogue incorporation. This raises the possibility that altered biochemical properties of the snRNA fractions may be reflected in changes in larger-molecular-weight RNA fractions and may contribute to cytotoxicity of the fluorinated pyrimidines.

5.1.4.4 Effect of Incorporation of Fluorinated Pyrimidines on Transfer RNA (tRNA)

Although fl^{5}ura is incorporated readily into tRNA, early studies suggested that this incorporation had little effect on the ability of the fl^{5}ura-containing tRNAs to accept amino acids. However, a quantitative study of bacterial tRNAs has shown that rates of acceptance of Glu, Asp, GluNH$_2$, and His are moderately retarded, while Lys is accepted by tRNA^{Lys} at a rate only 3–7% of normal. Contributing to this may be the fact that the substitution of fl^{5}ura for ura in tRNA causes a depletion of minor bases that are produced by enzymatic processing of ura subsequent to incorporation [reviewed by Heidelberger et al. (1983)].

5.1.4.5 Effect on RNA of Long-Term Treatment with Fl^{5}ura

Spears et al. (1985) have developed a procedure for determination of the incorporation of nonradiolabeled fl^{5}ura into RNA which is applicable to tissues in vivo. Using this procedure, the time course (up to 96 h following bolus injection) of in vivo incorporation of fl^{5}ura into murine L1210 lymphocytic leukemia cells was followed. Spears et al. suggested that the observed increasingly high percentage of total fl^{5}ura metabolites present as fluorine-containing RNA (F-RNA) (91.2% at 96 h) may indicate that fl^{5}ura-containing RNA serves as a storage depot for fl^{5}dUMP. This would explain the persistence of fl^{5}dUMP long after treatment that is sometimes observed. In addition, based on protein content, or on DNA content, there was a dramatic 40% drop in the amount of total cellular RNA present, presumably a reflection of the importance of RNA effects in cellular toxicity. In contrast, relative to protein content, there was only a 10% drop in DNA levels after 96 h. The potential sensivity of the assay developed (detection of F-RNA in 1 part per 30,000 normal bases in a 100-mg sample), together with the use of unlabeled drug, suggests important clinical applications.

5.1.5 Influence of Fl^{5}ura and Its Nucleosides on DNA Biosynthesis

5.1.5.1 Incorporation of dUTP and Fl^{5}dUTP and Repair Mechanisms

Under normal conditions, the incorporation of dUPT into DNA is minimized by the removal of dUTP by dUTPase. What little uracil is incorporated into DNA is rapidly removed by DNA repair mechanisms. Thus, uracil-DNA glycosylase removes uracil, leaving an apyrimidinyl site which is subject to cleavage by endo-nucleases. Following cleavage, enzymatic reinsertion of the proper base completes the repair. This same repair mechanism serves to remove uracil formed by spontaneous deamination of cytosine residues. Assuming comparable behavior of fl^5dUTP, its incorporation into DNA could be expected to play a minor role in the biological effects of fl^5ura and its nucleosides. Indeed, failures in early attempts to detect such incorporation seemed to confirm this belief. On the other hand, if thymidine levels are depleted, by such agents as methotrexate, dUTP can compete with dTTP and become incorporated into DNA (Goulian et al., 1980). Furthermore, fl^5dUMP inhibition of thymidylate synthetase can be expected to also cause a depletion of thymidine, a depletion that also could lead to greater incorporation of dUTP, or of fl^5dUTP.

Caradonna and Cheng (1980) have confirmed the efficient hydrolysis of fl^5UTP by dUTPase, as well as the ability of uracil-DNA glycosylase to excise a fluorouracil moiety. From these results, they proposed an association between cellular toxicity and incorporation of fl^5ura or ura into DNA. As part of the normal repair mechanism, incorporated dUTP and fl^5dUTP would be removed by uracil-DNA glycosylase. However, because of the depletion of thymidine in the normal nucleoside pool (caused by the inhibition of thymidylate synthase by fl^5dUMP), reincorporation of dUMP (or fl^5dUMP) would be more likely to occur, eventually leading to degradation of DNA and toxicity.

Using tritiated fl^5dU of high specific activity, Kufe et al. (1981) provided the first demonstration of incorporation of fl^5ura into eucaryotic DNA (L1210 cells). In a similar study, Ingraham et al. (1982) showed that fl^5dU does cause a marked lowering of dTTP levels, which would favor utilization of dUTP and fl^5dUTP as suggested by Caradonna and Cheng. In addition, depleted dTTP levels activated thymidine biosynthetic enzymes, which further increased the levels of dUMP, increasing the competitive utilization of fl^5dUTP and dUTP as substrates for DNA synthesis. A contribution to cell toxicity was suggested to result from repetitive insertion/removal of ura and participation of fl^5ura in this process. On the other hand, Danenberg et al. (1981) concluded that the level of incorporation of fl^5ura into L1210 cell DNA was too low (1 molecule per 1.6×10^8 nucleotides) to contribute to toxicity.

Research by several groups on a variety of normal and malignant cells has added evidence for the involvement of DNA repair mechanisms in fl^5ura toxicity. Cheng and Nakayama (1983) found that fl^5U was incorporated into mainly smaller DNA fragments of HeLa cells. A decrease in the dTTP pool, a general inhibition of DNA synthesis, formation of fragmented DNA, and damage to preexisting DNA all indicated an effect on DNA repair mechanisms. Similar observations were made by Scheutz *et al.* (1986) with murine bone marrow cells. In human adenocarcinoma cells, Lonn and Lonn (1986) found evidence for two mechanisms for the induction of DNA lesions by fl⁵ura, along with a correlation between the number of lesions and cell death. Incorporation of fl⁵ura residues is one mechanism. If this was blocked by inhibition of DNA polymerase α, lesions still were produced, a result which again implicates DNA repair mechanisms. Lonn and Lonn stressed that a properly functioning repair process is required even under normal circumstances, due to natural alterations of several hundred pyrimidine residues in DNA daily by such natural processes as deamination of cyt to ura residues. In contrast, Parker *et al.* (1987) found no evidence for incorporation of fl⁵dUMP or fl⁵ura into DNA of murine T-lymphoma cells, and a thymidine-sensitive DNA fragmentation appeared not to be associated with cell toxicity.

5.1.5.2 Structural Consequences of Incorporation of Fl⁵dUMP Residues into DNA

Accumulated evidence, examples of which are presented above, suggests that incorporation of fl⁵dUMP into DNA produces strand breaks and that this may contribute to cytotoxicity. The effects of incorporation of fl⁵dUMP on the physicochemical properties of model oligonucleotides have been the subject of recent NMR studies. (Kremer *et al.*, 1987; Sowers *et al.*, 1987). Sowers *et al.* attributed decreased stability of a normal Watson–Crick helix to decreased stacking of the fl⁵dUMP residue, rather than to instability caused by the increased acidity of N-3 of the fluoropyrimidine, as proposed by Kremer *et al.*

5.1.6 Formation and Biochemical Properties of Fl⁵U-Diphosphohexoses

Nucleoside diphosphate esters of sugars are requisite precursors for the incorporation of sugars into macromolecules (see Chapter 6, Section 6.3.1). For example, UDP-glucose is the glycosyl donor in glycogen biosynthesis. Fl⁵UDP-hexoses have been detected in tumor and liver cells, and treatment with fl⁵ura or fl⁵U leads to the synthesis of fl⁵UDP-*N*-acetylhexosamines in bacteria or mammalian tumors (Weckbecker and Keppler, 1984, and references therein). Weckbecker and Keppler (1984) prepared several fl⁵UDP-sugars to assess their behavior *in vitro* as substrates for the enzymes of UDP-sugar metabolism. Enzyme kinetic parameters were found to be comparable to those of UDP-sugars.

5.1.7 Relative Importance of Cytotoxic Mechanisms to Chemotherapy and the Development of Drug Resistance

The accumulated data from studies in cell cultures, and in animal and human tumors, provide strong evidence that fl⁵ura can exert toxic effects through enzymatic processing to three active nucleotide derivatives. To summarize, fl⁵dUMP inhibits thymidylate synthase by forming a ternary complex with the enzyme and the reduced folate cofactor; fl⁵UTP is incorporated efficiently into RNA, where processing and maturation of several forms of RNA are affected; and fl⁵dUTP can affect DNA biosynthesis through incorporation or by interfering with repair mechanisms. The mechanism which predominates for a given cell depends upon such factors as the biochemical characteristics of the cell and the experimental conditions (whether *in vivo* or *in vitro*, for example). The apparent site of action can be modulated through the coadministration of a second drug, and the sensitivity of a cell to such modulation has been used extensively as a method to determine the biochemical lesion leading to cytotoxicity (Heidelberger *et al.*, 1983). In addition, synergistic toxicity which can result from the combined effect of fl⁵ura (or related fluoropyrimidines) and modulators of pyrimidine metabolism has been the basis for approaches to improved therapeutic schemes.

The existence of thymidine-reversible and thymidine-nonreversible components of the toxicity of fl⁵ura was considered in early work to correspond to thymidylate synthase and RNA effects, respectively. In murine mammary and colon carcinomas, increased antitumor activity of fl⁵ura elicited by coadministration of thymidine (thy) was attributed to the more effective utilization of fl⁵UTP in RNA synthesis, and increased incorporation was observed. While the fl⁵ura/thy combination was promising in initial animal studies, it proved quite toxic in human trials (Benz and Cadman, 1981), demonstrating that extrapolation of data from model cell lines and tumors to human tumors *in vivo* remains a serious problem (Clark *et al.*, 1987). For a more thorough discussion and original references to earlier papers exploring these complex issues, the reviews by Heidelberger *et al.* (1983) and Myers (1981) are recommended.

Several recent studies have reinforced a belief that the inhibition of thymidylate synthase (the thymidine-reversible component) by fl⁵dUMP is the important mechanism of action in most tumors. For example, Houghton *et al.* (1986) have shown that during treatment of mice bearing human adenocarcinoma xenografts, the tumors intrinsically resistant to fl⁵ura have higher levels of thymidylate synthase and more rapidly recover thymidylate synthase activity after treatment. An important factor in the development of resistance to fl⁵ura treatment is thought to be a

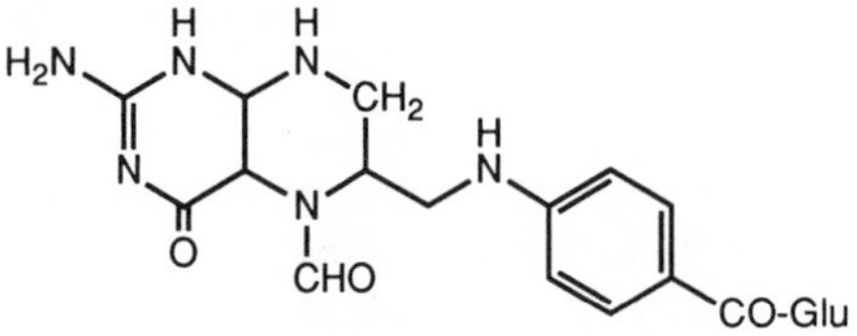

Leucovorin

Figure 5-5. Leucovorin, an intermediate product of the metabolism of folic acid, used as an antidote to folic acid antagonists.

gradual buildup of dUMP. Thymidylate synthase that dissociates from the fl^{5}dUMP ternary complex will be protected from re-formation of the complex due to association with a large excess of dUMP, and the effects of the drug will diminish. Thus, blocking this dissociation by excess folate should increase the effectiveness of the drug, as discussed in Section 5.1.3. Indeed, while the fl^{5}ura/thy combination has produced little improvements in therapeutic response, the addition of leucovorin (Fig. 5-5) to increase folate levels has shown significant promise (Clark *et al.*, 1987, and references therein). Included in recent studies are reports that the combination of fl^{5}ura and an exogeneous folate source is effective against both murine and human leukemia cells, known to be unresponsive to fl^{5}ura as a single agent (Keyomarsi and Moran, 1986; Mini *et al.*, 1987). On the other hand, Clark *et al.* (1987) have found evidence for thymidylate synthase gene amplification accompanying the onset of resistance in a colon tumor that originally responded to fl^{5}ura/leucovorin treatment, showing that other mechanisms of resistance are available [additional examples of combination chemotherapeutic schemes are reviewed by Santi *et al.* (1982)].

5.1.8 Prodrugs of Fl^{5}ura and of Its Metabolites

Santi *et al.* (1982) and Filler and Naqvi (1982) have provided effective discussions of the use of prodrugs to improve the efficacy of fl^{5}ura. Examples of prodrugs recently developed are shown in Fig. 5-6. In general, the strategy involves targeting quantitative differences between enzymes in tumor and normal tissue required to activate a given prodrug. For example, 5'-deoxy-5-fluorouridine requires uridine phosphorylase for activation to fl^{5}U, and this enzyme has elevated levels in some tumors. The related 5'-deoxy-4',5-difluorouridine is a better substrate for this enzyme (see Section 5.4.3).

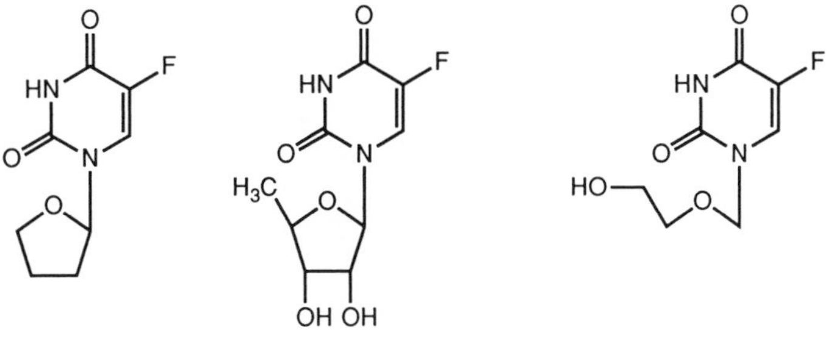

| Ftorafur | 5'-Deoxy-5-fluorouridine | 5-Fluoro-1-(hydroxy-ethoxymethyl)uracil |

Figure 5-6. Examples of prodrugs that release fl^{5}ura.

5.1.9 5-Fluorodeoxycytidine (Fl^{5}dC)

Although the potent antitumor activity of fl^{5}dC has been known for over 25 years, this compound had received relatively little attention because the site and mechanism of action were thought to be similar to those for fl^{5}ura and fl^{5}dU (Boothman et al., 1987a, and references therein). In addition, fl^{5}dC was known to be more toxic than fl^{5}ura and fl^{5}dU, and also less effective. Inhibition of thymidylate synthase by fl^{5}dU derived from fl^{5}dC apparently represents a major part of the cytotoxic activity of this compound, although, here again, multiple mechanisms have been implicated (Kaysen et al., 1986). Fl^{5}dUMP is formed by phosphorylation of fl^{5}dC by deoxycytidine kinase (dC kinase), producing fluorodeoxycytidine monophosphate (fl^{5}dCMP), followed by deamination of fl^{5}dCMP by deoxycytidylate (dCMP) deaminase. An alternate route to fl^{5}dUMP involves initial deamination of fl^{5}dC to fl^{5}dU by cytidine-deoxycytidine deaminase (Fig. 5-7) (Newman and Santi, 1982).

The recognition that levels of dCMP deaminase are elevated in many human malignant tumors has led to a strategy designed to render fl^{5}dC more selective toward tumor cells in its cytotoxic activity. The coadministration of an optimal dose of tetrahydrouridine, a potent inhibitor of C-dC deaminase, was expected to channel the metabolism of fl^{5}dC exclusively through the dC kinase–dCMP deaminase pathway. Because of the elevated dCMP deaminase activity in the tumor cell, the desired effect of this altered metabolism, initially demonstrated in cell culture (Boothman et al., 1985), would be a tumor-selective production of fl^{5}dUMP. The cytotoxicity was expected to be increased further by the known resistance of fl^{5}dC to phosphorolytic enzymes and by the bypassing of fl^{5}dU as an

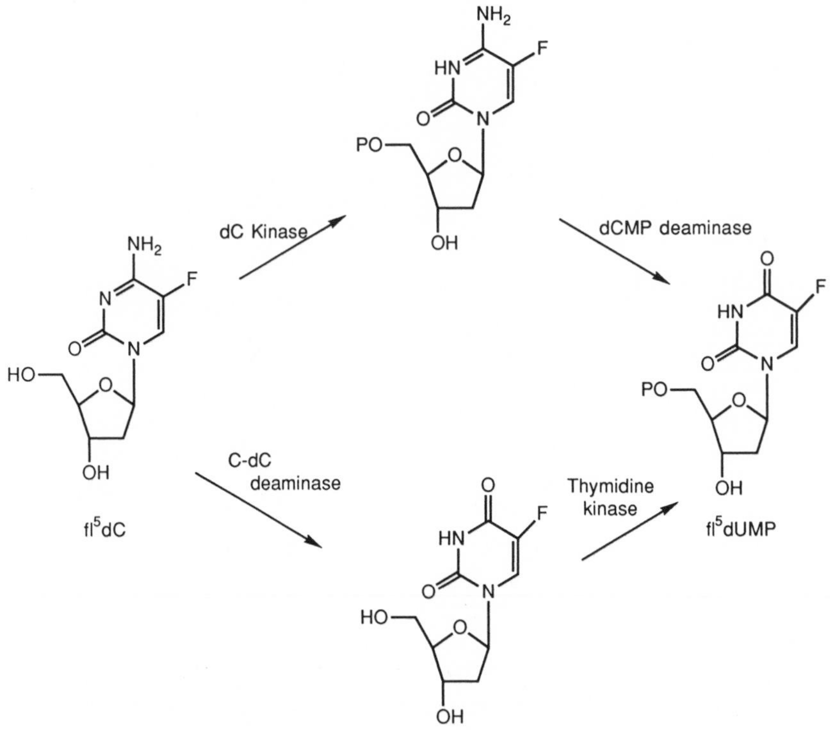

Figure 5-7. Two available routes for the conversion of fl⁵dC to fl⁵dUMP.

intermediate. Fl⁵dU is normally converted efficiently to fl⁵ura by the high levels of thymidine phosphorylase found in normal and tumor cells. Bypassing fl⁵dU and fl⁵ura formation also would serve to diminish RNA-directed toxicity, thus targeting the drug to thymidylate synthase inhibition. Finally, the presence of tetrahydrouridine and the low levels of dC kinase should prevent the formation of fl⁵dUMP in normal cells (Boothman *et al.*, 1987b).

The potential for this strategy was demonstrated in mice bearing ascitic mammary adenocarcinoma-755. Overall protection of normal tissue was achieved using tetrahydrouridine, as shown by the presence of drastically lower levels of fl⁵dUMP and its metabolites, including fl⁵UMP-containing RNA, compared to tumor levels. However, in the tumor, selective channeling of the drug through the dC kinase–dCMP deaminase pathway was not achieved due to an abnormally high overall level of dC metabolism, including C-dC deaminase activity. Tumor cells are killed by effective metabolism of fl⁵dC by both pathways as a result of inhibition of thymidylate synthase, incorporation of fl^{5}dUMP into DNA, and incorporation of fl^{5}UMP into RNA. Fl^{5}dC also is incorporated selectively into tumor DNA as a result of elevated dC kinase levels. Normal cells are protected from these toxic effects by inhibition of their normal level of dC deaminase activity (Boothman et al., 1987c). The combination of fl^{5}dC and tetrahydrouracil was also very effective in the treatment of mice bearing Lewis lung carcinoma, a tumor considered refractory to fluoropyrimidine treatment (Mekras et al., 1984). Much higher levels of fl^{5}dUMP (between 45- and >5400-fold) were formed in the tumor compared to normal tissues, as were RNA- and DNA-level metabolites, when tetrahydrouridine was coadministered (Boothman et al., 1987b).

The presence of fl^{5}cyt in DNA has been shown to affect postreplicative modifications. Osterman et al. (1988) have shown that the inhibition of DNA methylation by fl^{5}dC involves incorporation of fl^{5}cyt residues into DNA and formation of a covalent complex with DNA-cytosine methyl transferase. This inhibition by fl^{5}dC (and by 5-azacytidine) had been shown previously to affect gene expression and cell differentiation.

5.1.10 1-β-D-Arabinofuranosyl-5-fluorouracil (Ara-fl^{5}U)

1-β-D-Arabinofuranosyluracilmonophosphate (ara-UMP) is methylated by thymidylate synthase, albeit poorly. Interest in the development of ara-fl^{5}U (Fig. 5-8) as a potential inhibitor of thymidylate synthase was stimulated by the knowledge that arabinofuranosyl nucleosides are metabolically more stable than the corresponding ribosyl derivatives because they serve as poor substrates for glycosidases (Yung et al., 1961).

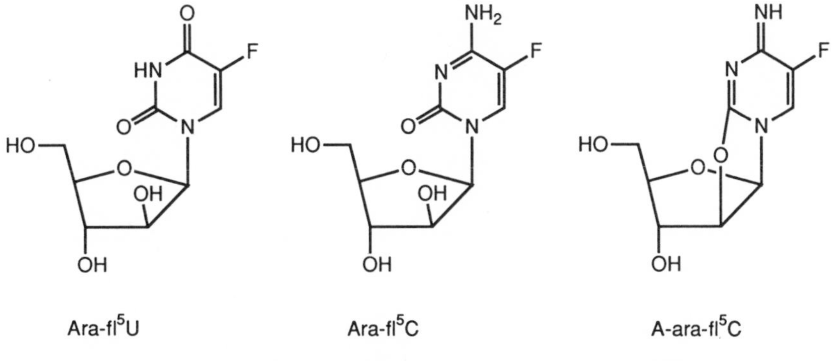

Ara-fl^{5}U   Ara-fl^{5}C   A-ara-fl^{5}C

Figure 5-8. Examples of fluorinated arabinofuranosyl nucleosides. A-ara-fl^{5}C functions as a prodrug both for ara-fl^{5}C (by slow hydrolysis of the anhydro linkage) and for ara-fl^{5}U (by enzymatic deamination of ara-fl^{5}C) (Fox et al., 1972).

In transplanted mouse leukemia B82 cells, ara-fl^{5}U had a therapeutic index comparable to that of fl^{5}dU but was fourfold less potent. Ara-fl^{5}U was considerably more potent than fl^{5}U. Despite the rather stringent specificity of thymidylate synthase for the 5′-phospho-2′-deoxyribosyl moiety, ara-fl^{5}UMP behaves similarly to fl^{5}dUMP as a mechanism-based inhibitor of thymidylate synthase (Nakayama *et al.*, 1981). The antiviral activity of ara-fl^{5}U will be discussed in Section 5.4.1.5.

5.1.11 1-β-D-Arabinofuranosyl-5-fluorocytosine (Ara-fl^{5}C) and 2,2′-Anhydro-1-β-D-arabinofuranosyl-5-fluorocytosine (A-ara-fl^{5}C)

1-β-D-Arabinofuranosyl-cytosine (ara-C) has proven activity against lymphomas and leukemias and is a very important drug for the treatment of human acute myeloblastic leukemia. Its effectiveness, however, is limited by the rapid development of resistance and by a short biological half-life, due in part to enzymatic deamination to inactive ara-U. The effectiveness

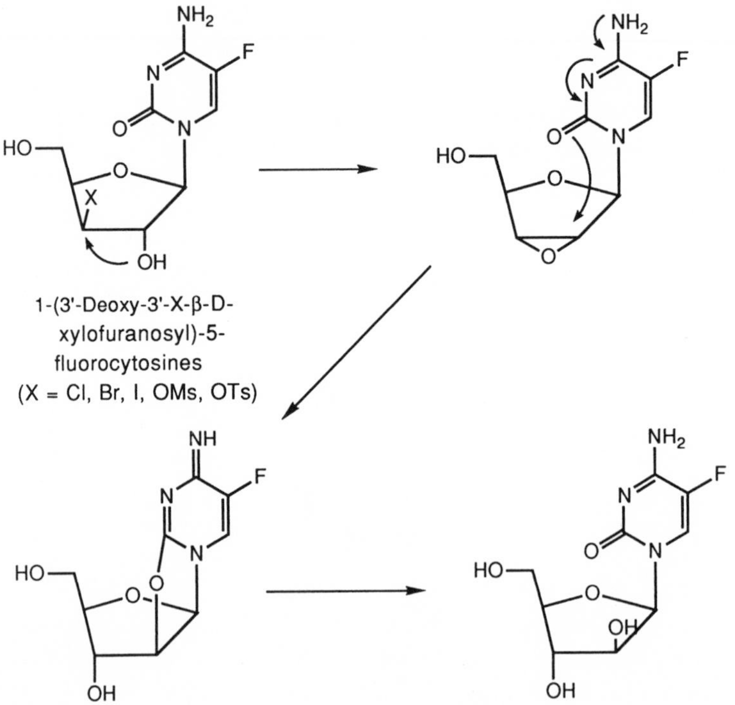

1-(3′-Deoxy-3′-X-β-D-
xylofuranosyl)-5-
fluorocytosines
(X = Cl, Br, I, OMs, OTs)

Figure 5-9. "Double-barreled" masked precursors of anticancer nucleosides (Watanabe *et al.*, 1980).

of ara-C is improved by the fact that certain tumor cells lack the ability to deaminate ara-C, while both ara-C and ara-fl^{5}C (Fig. 5-8) are rapidly deaminated by human kidney and mouse liver enzymes. Ara-fl^{5}C is active against ara-C-sensitive cells, and the effects of ara-C and of ara-fl^{5}C are reversed by dC, but not by dT, indicating that cytotoxicity is mediated by triphosphates at the DNA level and is unrelated to thymidylate synthase activity (Dollinger *et al.*, 1967). This observation that ara-fl^{5}C and ara-fl^{5}U have different mechanisms of toxicity prompted the development of "double-barreled" masked precursors to both analogues. In order to develop drugs with prolonged biological half-life, Fox *et al.* (1972) prepared the 2,2'-anhydro derivative of ara-fl^{5}C (A-ara-fl^{5}C; Fig. 5-8), based on the expectation that this analogue, under physiological conditions, would hydrolyze slowly to ara-fl^{5}C. In addition, since ara-fl^{5}C was expected to undergo enzymatic deamination, A-ara-fl^{5}C would serve also as a slow releaser of ara-fl^{5}U. This drug was shown to be effective against ara-C-sensitive cells (as an ara-C analogue) and ara-C-resistant cells (acting as a 5-fluoropyrimidine nucleoside) and to have prolonged *in vivo* activity. This strategy has been extended to the development of masked precursors of the "double-barreled" masked precursor of ara-fl^{5}C and ara-fl^{5}U (Fig. 5-9) (Watanabe *et al.*, 1980, and references therein).

5.1.12 Biochemistry of 5-Trifluoromethylpyrimidines

5.1.12.1 5-Trifluoromethyluracil and Its Nucleosides

Using similar mechanistic considerations to those which prompted the synthesis of fl^{5}ura, Heidelberger *et al.* (1962) prepared 5-trifluoromethyl-2'-deoxyuridine (CF$_3$dU; Fig. 5-10). Behaving similarly to fl^{5}dU, CF$_3$dU is

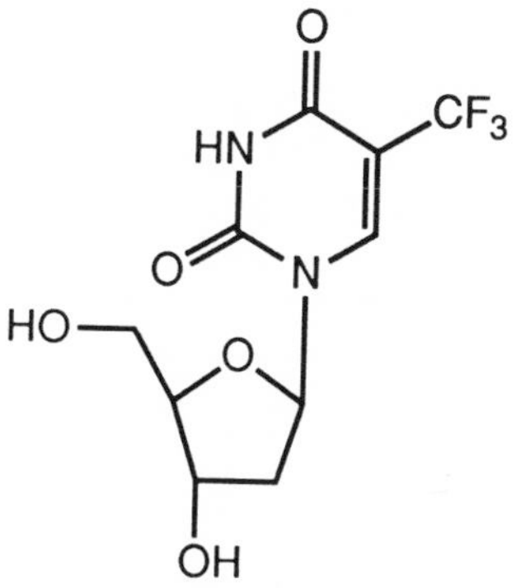

Figure 5-10. Structure of CF$_3$dU.

converted *in vivo* by thymidine kinase to trifluoromethyl-2'-deoxyuridine monophosphate (CF$_3$dUMP), a potent inhibitor of thymidine synthase. The mechanism of inhibition of thymidine synthase by CF$_3$dUMP has been studied extensively by Danenberg (1977) and by Santi *et al.* (1976) and has been found to differ significantly from that of fl^{5}dUMP inhibition. Despite the strength of the C−F bond, CF$_3$ura and its nucleosides acylate amines in aqueous media and, in the absence of amines, are hydrolyzed readily to the 5-carboxylate. These observations provided the basis for a mechanistic proposal for the inactivation of thymidylate synthase. The labilization of the CF$_3$ group can be explained by transient formation of a C-5 carbanion, either by ionization of N-1 (in CF$_3$ura) or by addition of a nucleophile to C-6 of the nucleosides. Displacement of fluoride will yield the reactive exocyclic difluoromethylene group. Reaction of a nucleophile with this moiety, or with an acyl fluoride derived therefrom, results in amide or carboxylate formation (Fig. 5-11) (Santi *et al.*, 1976). Addition of an enzyme-bound nucleophile, such as an SH group, to C-6 of CF$_3$dUMP would initiate a similar process, leading to acylation of thymidylate synthase and inactivation. However, there has been no direct demonstration of irreversible covalent labeling of an enzyme active site with radiolabeled CF$_3$dUMP. Moreover, short incubation periods produced isolable, but noncovalent, enzyme-CF$_3$dUMP complexes. Thymidylate synthase inhibition by CF$_3$dUMP also differs from inhibition by fl^{5}dUMP in that there appears to be no requirement for folate cofactor (Danenberg, 1977).

5.1.12.2 Metabolism and Activation of CF$_3$dUMP

Heidelberger and King (1979) have reviewed the biochemistry and pharmacology of CF$_3$dU. In contrast to metabolism of fl^{5}ura and fl^{5}dU, metabolism of CF$_3$ura or CF$_3$dU does not lead to degradation of the pyrimidine ring, and the sole catabolites recovered are CF$_3$ura and 5-carboxyuracil. The nucleoside linkage in CF$_3$dU is cleaved readily by thymidine phosphorylase. Phosphorylation of CF$_3$dU produces CF$_3$dUMP, CF$_3$dUDP, and CF$_3$dUTP, and the latter is incorporated into DNA. However, the low levels of incorporation found in cellular and tissue DNA suggested that cytotoxicity and tumor inhibition were caused primarily by thymidylate synthase inhibition (Heidelberger and King, 1979).

5.1.12.3 Antitumor Properties of CF$_3$dU

Despite early evidence of greater efficacy relative to that of fl^{5}dU in certain tumors, chemotherapeutic studies with CF$_3$dU have been limited.

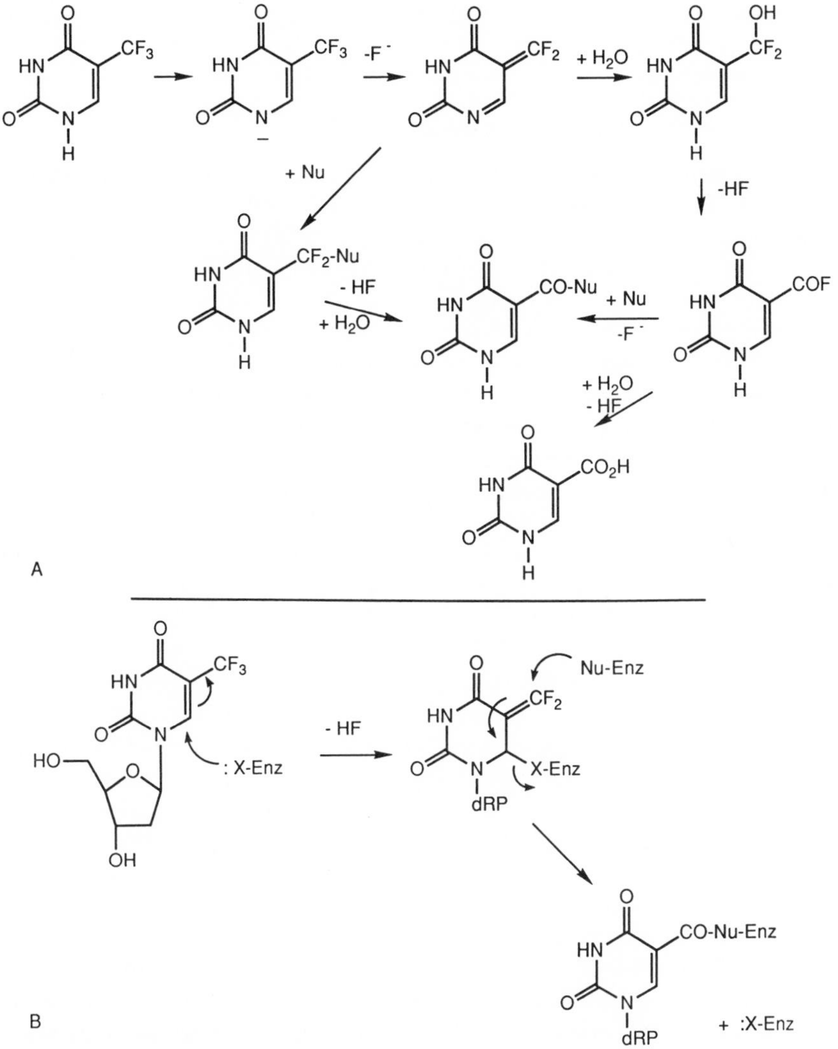

Figure 5-11. Mechanism for the formation of reactive intermediates from 5-CF$_3$ura (A) and 5-CF$_3$dU (B) (see text for details).

Major problems were associated with toxicity and rapid catabolism by thymidine phosphorylase to the inactive pyrimidine base. Interest was diminished further by the belief that the results of CF$_3$dU treatment probably would mimic the effects of fl^{5}dU, since the primary target for both compounds was felt to be thymidylate synthase. Thus, Heidelberger (1975b) reported that CF$_3$dU had been abandoned as a cancer chemotherapeutic agent (see Section 5.1.13 for a prodrug approach to CF_3dU as an antitumor agent).

5.1.12.4 Antiviral Properties of CF_3dU

The demonstration in 1962 that 5-iodo-2'-deoxyuridine (io^5dU) is clinically effective against herpes simplex viral (HSV) infection of the eye (see Section 5.2.2) prompted evaluation of CF_3dU as an antiviral agent. Subsequent research showed that CF_3dU is more effective than io^5dU, as well as 5-bromo-2'-deoxyuridine (br^5dU) and ara-C. The potent inhibition of thymidylate synthase by CF_3dUMP and the selective incorporation of CF_3dUTP into viral DNA are important factors in this antiviral efficacy. The decreased formation of dTMP as a consequence of the former activity increases the relative phosphorylation of CF_3dUMP by thymidine kinase. Virus-selective toxicity results from the markedly higher thymidine kinase activity in virus-infected cells, as well as from the fact that CF_3dU is a better substrate for viral DNA polymerase than it is for DNA polymerase in uninfected cells (Prusoff et al., 1985). Incorporation of CF_3ura residues into viral DNA adversely affects the structure and function of the transcribed RNA [reviewed by Heidelberger and King (1979) and by Heidelberger (1975b)]. In contrast to the limited development of this analogue as an antitumor drug, CF_3dU is now licensed by the U.S. Food and Drug Administration for clinical use as an antiviral agent, most notably for the topical treatment of herpetic keratitis. (This application now largely has been replaced by the use of other agents, such as acyclovir and derivatives.)

CF_3dU has been shown to have potent activity against a laboratory strain of the potentially lethal human cytomegalovirus (HCMV) (Wingard et al., 1981). Because this strain of herpes virus does not induce thymidylate kinase in infected cells, finding selective agents for treatment of HCMV has been very difficult. Spector et al. (1983) have evaluated the in vitro effectiveness of CF_3dU in combination with other antiviral agents in six different patient strains of HCMV and found additive or synergistic effects, with no increased toxicity to cultured human embryonic lung cells.

5.1.13 5-Trifluoromethyldeoxycytidine (CF_3dC)—A Prodrug for CF_3dUMP

In their discussion of CF_3dU as a potential chemotherapeutic agent, Mekras et al. (1985) emphasized the potential differences in metabolic behavior between CF_3dUMP and fl^5dUMP. Thus, the effects of the former will be exclusively DNA directed, since thymidine and its analogues are not taken up into RNA. On the other hand, stable incorporation of CF_3dUTP

into DNA is catalyzed readily by DNA polymerases, causing the formation of low-molecular-weight DNA. CF_3dUTP inhibits DNA polymerase competitively with respect to dUTP and noncompetitively with respect to incorporation of other nucleotides. While both fl^5dUMP and CF_3dUMP inhibit thymidylate synthase, the absence of a folate requirement for thymidylate synthase inhibition for the latter is an additional potentially exploitable difference. Finally, since the pyrimidine ring remains intact during metabolism of CF_3dUMP, there is no potential for the formation of fluoroacetate, implicated as a possible source of neurotoxicity associated with fl^5ura and fl^5dU therapy.

In a study of the clinical problems associated with the use of CF_3dU in chemotherapy, the suggestion was made by Heidelberger and co-workers that a biological precursor of CF_3dUMP resistant to thymidine phosphorylase could be an effective agent (Dexter *et al.*, 1972). Using a

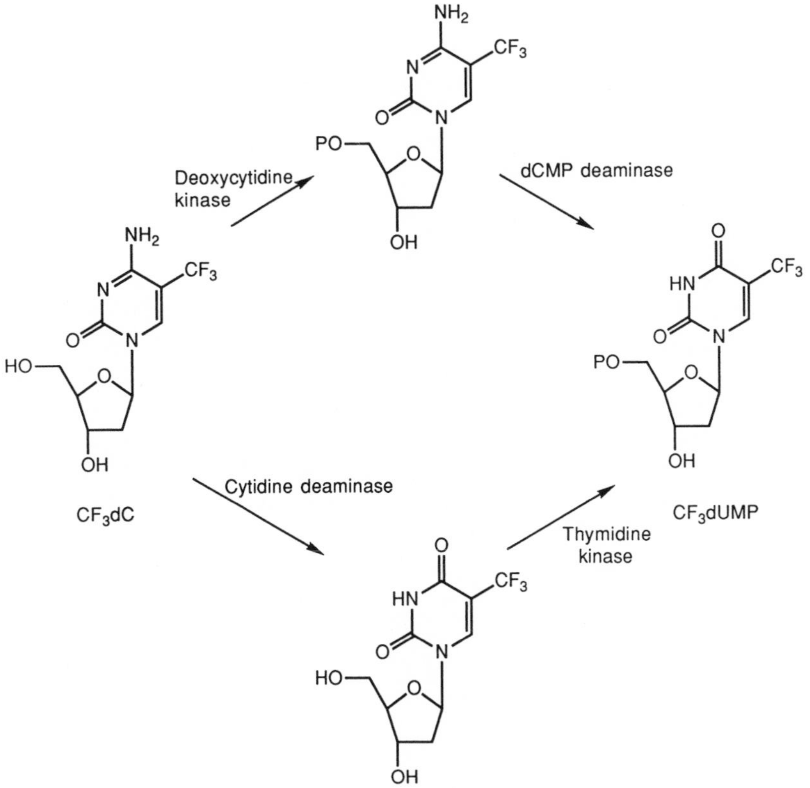

Figure 5-12. Mechanism of activation of CF_3dC to CF_3dUMP.

strategy similar to that applied to fl^{5}dC described above (Section 5.1.9), Mekras *et al.* (1985) have examined the use of CF$_3$dC, an analogue resistant to thymidine phosphorylase cleavage, as a prodrug for CF$_3$dUMP. As a consequence of elevated tumor levels of cytidine deaminase, moderate concentrations of tetrahydrouridine inhibited systemic cytidine deaminase to a much greater extent than tumor cytidine deaminase (Fig. 5-12). The capability of tumor cells to deaminate CF$_3$dC in the presence of tetrahydrouridine appears to cause selective delivery of CF$_3$dU to these cells. Benefits to chemotherapeutic efficacy include the circumvention of systemic catabolism by thymidine phosphorylase, exclusive DNA-, rather than RNA-, directed toxicity, and tumor-selective cytotoxicity by selective deaminase inhibition in normal tissue. A far superior efficacy of the CF$_3$dC–tetrahydrouridine combination compared to CF$_3$dC alone was demonstrated in two murine tumor models, mammary adenocarcinoma and Lewis lung carcinoma. In the latter model, this combination also was shown to be superior both to fl^{5}dU and to fl^{5}ura (Mekras *et al.*, 1985).

5.2 Biochemistry of Chloro-, Bromo-, and Iodopyrimidines and Their Nucleosides

5.2.1 5-Chloro- (Cl^{5}dU), 5-Bromo- (Br^{5}dU), and 5-Iododeoxyuridine (Io^{5}dU): Introduction

The similar size of fluorine and hydrogen causes fl^{5}ura to mimic the action and properties of ura in many biochemical processes. In contrast, the van der Waals radii of bromine (1.95 Å) and iodine (2.15 Å) are comparable to that of the methyl group (2.0 Å). For this reason, br^{5}dU and io^{5}dU (Fig. 5-13) closely resemble dT, and the metabolism and other enzymatic processing of these analogues reflect this fact [reviewed by Prusoff and Goz (1975) and Goz (1978)]. Whereas a major part of the biochemistry of fl^{5}ura involves its incorporation into various RNA fractions, br^{5}dU and io^{5}dU and their phosphorylated derivatives readily serve as substrates for the enzymes involved with DNA synthesis and are incorporated into DNA, but not into RNA, of mammalian, plant, and bacterial cells, as well as into phage and animal viruses. The biochemistry of cl^{5}dU has received somewhat less attention than that of br^{5}dU and io^{5}dU.

Io^{5}dU was developed initially as an anticancer drug (Prusoff, 1959), while br^{5}dU was synthesized as a potential inhibitor of DNA biosynthesis (Beltz and Visser, 1955). Visser *et al.* (1960) also reported the synthesis of

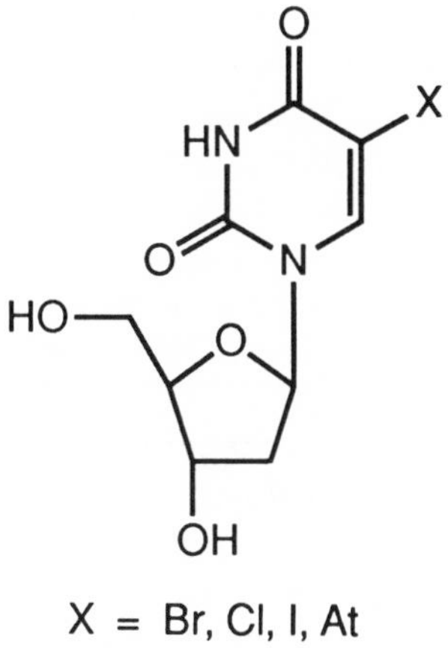

X = Br, Cl, I, At

Figure 5-13. Cl^{5}dU, bf^{5}dU, io^{5}dU, and at^{5}dU.

cl^{5}dU (Fig. 5-13). A rare reference to the biochemistry of astatine (the fifth halogen) is found in the report of the preparation and biodistribution studies of 5-astatodeoxyuridine (^{211}At) (Rossler et al., 1977). In addition to extensive studies of br^{5}dU and io^{5}dU as potential antitumor and antiviral agents, these analogues have become important biochemical tools in other areas such as the study of cellular mechanisms of mutagenesis and of cellular differentiation and gene expression. A precise and efficient procedure for monitoring cell growth kinetics has relied on flow cytometry combined with a radioimmunological assay for br^{5}dU in DNA, and there is current renewed interest in these analogues as clinical radiosensitizers.

5.2.2 Antiviral Properties of Br^{5}dU and Io^{5}dU

Because viral infections exploit host cell biochemical machinery, the design of specific and effective antiviral agents has been especially challenging, while the search for such agents has taken on even more urgency with the onset of the AIDS epidemic. Included in approaches to viral-specific agents has been the development of a large number of pyrimidine and purine nucleoside analogues, including halogenated analogues. Several extensive reviews are available which describe in detail these and other approaches to antiviral agents (e.g., Harnden, 1985; De Clercq, 1984, 1985a).

Io^{5}dU was shown early to be an effective agent against herpes simplex virus type 1 (HSV-1), and in 1962 io^{5}dU became the first antiviral agent to be licensed by the U.S. Food and Drug Administration [reviewed by Prusoff *et al.* (1985)]. However, *in vivo* toxicity has restricted its use to topical treatment of herpetic keratitis. The cytotoxicities of io^{5}dU and br^{5}dU are produced by their incorporation into DNA, an incorporation that leads to physicochemical alterations in structure and function [reviewed by Prusoff *et al.* (1984)]. Since DNA is responsible for the initiation of cellular functions, the biological consequences of this are myriad [for reviews, see Prusoff *et al.* (1979), Goz (1978), and Prusoff and Goz (1975)].

Br^{5}dU and io^{5}dU must be phosphorylated by thymidine kinase to be active. The selectivity of these analogues is based in large part on greater thymidine kinase activity in virus-infected cells relative to that in uninfected cells. Br^{5}dU and io^{5}dU are potent competitive inhibitors (with respect to dT) of thymidine kinase and are effective substrates. This inhibition of thymidine kinase and of other thymidine-processing enzymes is an important aspect of the biochemistry of these analogues, but the potent cellular effects result from subsequent formation of the triphosphates and their incorporation into DNA. Br^{5}dU and io^{5}dU are lethal or toxic to cells, are teratogenic (inhibit embryonic development) and mutagenic, induce oncogenic viruses, and affect cellular differentiation (Goz, 1978). Of particular interest is the fact that br^{5}dU can inhibit cellular differentiation without altering the rate of proliferation or viability of cells (Rutter *et al.*, 1973). Biswas *et al.* (1979) have classified the effects of br^{5}dU on differentiation into two types—the suppression of synthesis of specific proteins in some cells, and the stimulation of production of cell-specific proteins in other cells. Unfortunately, br^{5}dU and io^{5}dU are incorporated selectively, but not specifically, into DNA of infected cells, and these same effects on normal cells complicate systemic use. For example, attempts to treat herpetic encephalitis, a lethal herpes infection, caused severe neurological toxicity.

The effectiveness of io^{5}dU and br^{5}dU *in vivo* is limited also by the rapid cleavage of the ribosyl bond by nucleoside phosphorylase and subsequent catabolic degradation or dehalogenation of the free base (Prusoff and Goz, 1975). Br^{5}dUMP and io^{5}dUMP are also dehalogenated by thymidylate synthase. This dehalogenation involves activation of the inert vinyl halide by the addition of an enzyme-bound nucleophile to C-6, forming a transient dihydropyrimidine. Displacement of the halide by a thiol group, followed by oxidative loss of this substituent and elimination of the C-5 nucleophile, completes the process (Fig. 5-14). Unlike the reaction with dUMP, or inhibition by fl^{5}dUMP, this reaction does not require reduced folate cofactor (Garrett *et al.*, 1979).

As with fl^{5}ura and fl^{5}dU, several strategies have been used to increase the therapeutic effectiveness of br^{5}dU and io^{5}dU (Prusoff and Goz, 1975).

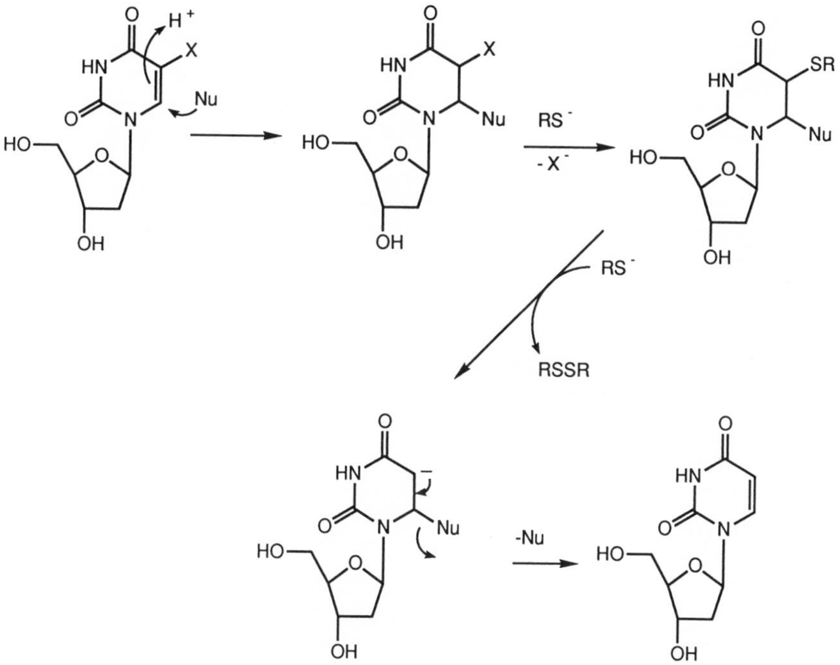

Figure 5-14. Mechanism of dehalogenation of br^{5}dU and io^{5}dU by thymidylate synthase (Garrett *et al.*, 1979).

For example, since io^{5}dUTP competes with dTTP for incorporation into DNA, fl^{5}dU has been coadministered to deplete dTTP stores to favor incorporation of io^{5}dUTP. In addition, inhibition of thymidylate synthase by fl^{5}dUMP blocks deiodination of io^{5}dUMP, deiodination that otherwise would increase dUMP levels. Increased dUMP levels, in turn, would reverse thymidylate synthase inhibition and increase dTMP production. The combination of increased io^{5}dUTP and decreased dTTP levels resulting from fl^{5}dU administration causes a significant increase in incorporation of io^{5}dUTP into DNA and increased lethality (Benson *et al.*, 1985).

5.2.3 Cellular Effects of Br^{5}dU and Io^{5}dU

5.2.3.1 Effects of Br^{5}dU and Io^{5}dU on Cellular Control Mechanisms

The mechanisms by which br^{5}dU and io^{5}dU produce their many effects on cellular functions have been studied extensively and have been related in many cases to alterations in DNA structure and function. Functioning normally, the eucaryotic cell responds to molecular signals by synthesizing the appropriate amount of individual proteins required to carry out the assignment of the particular cell type. Since, in a given organism, the gene content of all cells is the same, this requires precise activation and repression of individual genes. Gene control occurs at the level of transcription through regulation of the levels of synthesis of primary RNA transcripts encoded by a given DNA. In addition, post-transcriptional control is mediated by regulation of the processing of primary transcripts and by regulation of metabolic stability of mRNA.

Many of the effects of incorporation of br^{5}dU and io^{5}dU into DNA are caused by aberrant regulation of cellular function. Goz (1978) and Rutter et al. (1973) have discussed mechanistic possibilities in detail, and many interrelated, complex, and subtle factors are involved. DNA-linked effects include mutagenesis related to increased sensitivity to light and radiation, induction of excision and repair at specific sites, modification of transcription, and alteration of binding of regulatory proteins (Rutter et al., 1973). Increased efficiency of binding of regulatory proteins to the analogue-containing DNA has been implicated by the results of several studies. Analogue incorporation occurs preferentially into intermediate repetitive sequences (repeated 10^3–10^5 per genome) which are believed to be important in gene regulation. Regulation at the level of transcription is mediated in part by binding of repressor or activator proteins to these DNA sequences, and chromosomal proteins have been shown to be bound more efficiently to DNA sequences containing the br^{5}dUMP or io^{5}dUMP residues (Rutter et al., 1973, and references therein).

5.2.3.2 Br^{5}dU as a Molecular Probe of the Mechanism of Cell Differentiation

The ability of br^{5}dU to arrest differentiation of cells has made this analogue a valuable tool for the study of mechanisms of differentiation. For example, Biswas and Hanes (1982) have demonstrated that amplification of the prolactin gene sequence in rat pituitary tumor cell clones is responsible for the reversible induction of prolactin synthesis by br^{5}dU. Biswas et al. (1984) have identified a DNA sequence (an amplicon) adjacent to the prolactin gene which is responsible for this amplification. Cloned 10.3-kb segments from the 5′-end of the prolactin gene containing this sequence were ligated to the thymidine kinase gene of HSV-1 and transferred to thymidine kinase-deficient mouse fibroblasts. Treatment of these cells with br^{5}dU caused amplification of both the rat prolactin gene and the HSV-1 thymidine kinase gene, demonstrating that the information for br^{5}dU-induced gene amplification contained in the rat prolactin gene causes amplification of another gene adjacent to this sequence. In a subsequent study, this 10.3-kb segment was linked by the 5'- and the 3'-end to HSV-1 thymidine kinase and to human growth hormone genes. After transfer to mouse L cells, induction of amplification of both genes by br^{5}dU again was demonstrated. The amplification information is associated with a 4-kb DNA sequence in the 5'-flanking region of the rat prolactin gene (Pasion et al., 1987). While the mechanism of this induction of amplification is still unknown, it had been shown previously to be related directly to incorporation of br^{5}dU into DNA.

The discussion above illustrates the induction of cell-specific processes by br^{5}dU. Included among many examples of the inhibition of cell-specific processes is the dedifferentiation induced by incorporation of br^{5}dU into DNA of chondrocytes. As a consequence of this incorporation, a cell-specific product, type II collagen, is no longer synthesized (Fernandez et al., 1985).

5.2.3.3 Br^{5}dU as a Tool for Cytokinetics

The incorporation of br^{5}dU into DNA has also been exploited in the development of a convenient and accurate procedure for the study of cellular kinetics. Measurement of the rate of cell division and the determination of the timing of the phases of cell growth are important for a number of reasons, for example, in the design of antitumor chemotherapeutic schemes. Earlier procedures that relied on measurement of tritiated thymidine incorporation required time-consuming autoradiographic procedures or liquid scintillation spectroscopy for detection. The former measurement technique is not applicable to such problems as rapid monitoring of tumor growth, while the latter gives no information on the distribution of labeled cells in a given population sampled. As a solution to these and other complications associated with the existing procedures, Gratzner (1982) developed monoclonal antibodies for br^{5}dU and used these antibodies to detect, by immunofluorescent measurements, the incorporation of br^{5}dU into DNA. The utility of this method was increased by the development of a staining procedure that permits simultaneous determination of total DNA and amount of incorporation of br^{5}dU. This detection procedure permits quantification of very low levels of br^{5}dU incorporation and, combined with flow cytometric measurement of total DNA content of cells, provides such information as the fraction of cells in the S-phase (DNA-synthesizing) (Dolbeare et al., 1983). This powerful technique is now used extensively, especially in the field of cell cycle kinetics (Gray and Mayall, 1985, and references therein).

5.2.3.4 Mutagenicity of Br⁵dU

Br⁵dU is well established as an effective mutagen in both procaryotic and eucaryotic systems. To rationalize these mutagenic effects, Freese (1959) proposed that the presence of bromine induces rare enolization of br⁵dUTP, or of br⁵ura residues in DNA, causing this base, in comparison to thymidine, to pair more frequently with guanine (in place of adenine). Thus, two mechanisms are possible. Errors of incorporation are produced by the pairing of br⁵dUTP with a gua residue in replicating DNA, while errors of replication are caused by the pairing of dGTP with br⁵ura present in replicating DNA. Data from mammalian cells indicated that mutagenicity was determined by the concentration of br⁵dU to which the cells were exposed but was independent of the degree of incorporation of br⁵dU, suggesting that errors of incorporation predominate (Kaufman and Davidson, 1978). This "pool-dependent" mutagenesis is more effective when the ratio of br⁵dUTP to dCTP is high, since this high ratio increases the chances that guanine residues will mispair with br⁵dUTP. Recently, Kaufman (1984) has shown that replication-dependent mutagenesis can be demonstrated if br⁵dUMP-containing DNA is allowed to replicate in the presence of a high ratio of dGTP to dATP, in the absence of br⁵dUTP ("DNA-dependent mutagenesis").

5.2.3.5 Br⁵dU in the Detection and Induction of Sister Chromatid Exchange

Chromosome or chromatid breaks and rearrangements are important indices of chromosomal damage caused by such agents as chemicals and irradiation. The development of a microfluorometric method for the detection of exchange of base sequences between sister chromatids [sister chromatid exchange (SCE)], based on the incorporation of br⁵dU into DNA (Latt, 1973), has revealed that this lesion occurs at least 200 times more frequently than chromosome or chromatid breaks (Latt, 1974), and the analysis of SCE is now considered to provide a good indication of genotoxic damage leading to cancer and mutations. This br⁵dU procedure, which now is used widely to detect SCEs, subsequently was itself found to induce SCEs (Kato, 1974). Using br⁵dU, Kaufman (1987) has studied the relationship of nucleotide pool imbalance, mutagenesis, and induction of SCEs.

5.2.3.6 Teratogenic Properties of Br⁵dU and Io⁵dU

The influence of br⁵dU and io⁵dU on embryonic development has been studied extensively. Altered gene sequences and DNA replication due to incorporation of the halogenated analogue into regulatory and structural gene loci have been implicated (Biggers *et al.*, 1987, and references therein).

5.2.3.7 *Cellular Effects: Summary*

As structural analogues of thymidine, br^{5}dU and io^{5}dU can participate in and alter profoundly many of the cellular processes involving DNA. In addition, io^{5}dU and especially br^{5}dU have become extremely valuable tools for a number of molecular biological studies. The studies described above are only a few examples of an enormous volume of research in which these analogues are used, either to alter cellular processes or as molecular probes. While io^{5}dU is an effective topical antiviral agent, studies in this area recently have focused on the development of new halogenated analogues showing greater selectivity of action (see below). Finally, while the effectiveness of cl^{5}dU, br^{5}dU, and io^{5}dU as antitumor agents—the purpose for which the compounds were developed originally—has been limited, there has been recent renewed interest in their application as radiosensitizers.

5.2.4 5-Halodeoxyuridines as Radiosensitizers

While using X-ray sensitization in studies of mutagenicity of br^{5}dU and io^{5}dU, Djordjevic and Szybalski (1960) made the important discovery that cells which had incorporated these analogues into their DNA were rendered much more sensitive to radiation. Szybalski (1974) has reviewed his seminal work on the applications of this radiosensitization in several areas of molecular biology. Selective sensitization of tumor cells was expected to occur as a result of increased DNA synthesis in the generally more rapidly proliferating tumor cells, relative to that in normal cells. Unacceptable levels of damage to normal tissue in early attempts to exploit this tactic in radiation therapy led Szybalski to conclude that this was "a good idea which failed" (Szybalski, 1974). However, in a resurgence of interest in this strategy, factors such as improved procedures for drug delivery and more judicious choice of tumors for study have led to more encouraging results (Kinsella *et al.*, 1984; Mitchell *et al.*, 1986). As a further aid to this work, the anti-br^{5}dU/io^{5}dU monoclonal antibody labeling technique discussed in Section 5.2.3.3 has provided improved data on the labeling index and cell-cycle times of tumors compared to those of surrounding tissue. In addition, several strategies have been employed to enhance the selective tumor cell incorporation of the halogenated pyrimidines. For example, the presence of fl^5dU enhances analogue incorporation and radiosensitization as a result of inhibition of thymidylate synthase (lower dTMP levels; see also Section 5.2.2). *In vitro* and *in vivo* studies have demonstrated sensitization by 5-chlorodeoxycytidine (cl^5dC) coadministered with tetrahydrouridine, a further exploitation of the higher tumor levels of C deaminase and C-dUMP deaminase (Perez and Greer, 1986; Russell *et al.*, 1986). In this regard, the use of cl^5dC has advantages because it is a better substrate for dC kinase than is br^5dC or io^5dC. Inhibition of the ability of tumor DNA repair mechanisms to reverse radiation damage has also been found to potentiate the radiosensitization produced by halogenated pyrimidines in a human carcinoma cell culture (Boothman *et al.*, 1987c). As discussed by Mitchell *et al.* (1986), further improvements in this approach appear likely. Results of phase I and II studies with br^5dU have been very promising, particularly with respect to unresectable sarcomas.

5.2.5 5-[^{77}Br]Bromodeoxyuridine and 5-[^{125}I]Iododeoxyuridine

Deoxyuridine analogues labeled with short-lived radionuclides have been used to deliver sources of damaging radiation to DNA chains. Thus, when [^{77}Br]br^5dU and [^{125}I]io^5dU are incorporated into DNA, the emission of Auger electrons by the radionuclides produces local damage to DNA strands (Kassis *et al.*, 1987, and references therein). Review of the theory and applications of this research is beyond the scope of this chapter.

5.2.6 Development of New Antiviral Drugs Based on Structural Analogues of 5-Halopyrimidines

5.2.6.1 5-Iodo-5′-amino-2′,5′-dideoxyuridine (AIdU), a Case of Serendipity

The profound effects on normal cell function caused by the incorporation of br^5ura and io^5ura into DNA make clear the need for an improved selectivity for incorporation into target cells before these effects can be used effectively in systemic therapeutic schemes. Prusoff *et al.* (1979) prepared 5-iodo-5′-amino-2′,5′-dideoxyuridine. (AIdU) (Fig. 5-15) with the hope that this analogue would block the incorporation of thymidine preferentially into DNA of virus-infected cells, without itself being incorporated. This effort was based on earlier reports that 5′-amino-2′,5′-deoxythymidine is a potent inhibitor of thymidine kinase, which is, as noted previously, an enzyme with enhanced activity in many tumor cells and virus-infected cells.

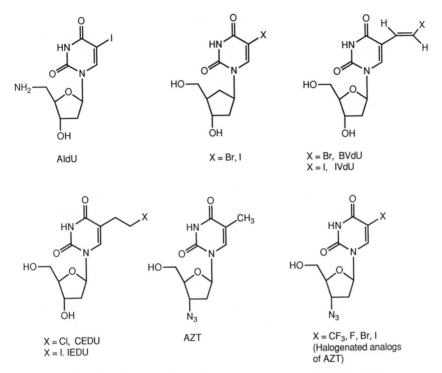

Figure 5-15. Further examples of 5-substituted pyrimidines as antiviral agents.

In HSV-1-infected cells, AIdU was found to be less potent than io^{5}dU, CF$_3$dU, and ara-C, but also it was found to be totally devoid of toxicity to normal cells. Further studies with a variety of cultured normal cell lines confirmed this lack of toxicity, and other effects of io^{5}dU, such as mutagenicity, also were absent. The toxicity of AIdU to virus-infected cells is dependent on phosphorylation by virus-induced thymidine kinase and incorporation of AIdUTP into viral and host DNA. The selectivity arises from the unexpected finding that AIdU is not a substrate for thymidine kinase in uninfected cells. AIdUTP is a substrate for DNA polymerase and is incorporated into DNA through a phosphoramidate bond, producing toxicity through strand breaks and impaired functioning (Prusoff *et al.*, 1979, 1985).

5.2.6.2 Carbocyclic Analogues of 5-Halo-2'-deoxyuridines

Of several carbocyclic analogues of uracil and cytosine nucleosides prepared, the 5-bromo- and 5-iodouridine (Fig. 5-15) analogues were the most active, having activity against HSV-1 comparable to that of io^{5}dU. A dependence on thymidine kinase activity suggests a similar mechanism of activation as for io^{5}dU and br^{5}dU (Shealy *et al.*, 1983).

5.2.6.3 (E)-5-Bromovinyldeoxyuridine (BVdU) and (E)-5-Iodovinyldeoxyuridine (IVdU)

The examination of a series of 5-vinyl-2'-deoxyuridine derivatives for antiherpes activity revealed that (*E*)-5-bromovinyldeoxyuridine (BVdU) and (*E*)-5-iodovinyldeoxyuridine (IVdU) (Fig. 5-15) both were more selective and more potent than all previously described antiviral agents against HSV-1 in the test system used, primary rabbit kidney cells (De Clercq *et al.*, 1979). The effectiveness of these compounds again is dependent on their selective phosphorylation by virus-induced dT kinase. The triphosphates produced by further phosphorylation are competitive inhibitors and excellent substrates for DNA polymerase and thus are incorporated into the DNA of virus-infected cells. An increase in strand breaks results, along with impaired template activity for RNA synthesis. The low activity against HSV-2 was explained by the inability of HSV-2-encoded dT kinase to carry out phosphorylation (De Clercq, 1984). BVdU and IVdU have also shown promise against varicella–zoster virus (VZV), responsible for chicken pox (varicella) and shingles (zoster) (Shigata *et al.*, 1985). Early phase I clinical trials for the topical and systemic treatment of immunodepressed patients suffering from severe HSV-1 and VZV infections have produced rapid cures with no evidence of toxicity (De Clercq and Walker, 1984). The report by Bubley *et al.* (1987) provides a discussion of more detailed biochemical studies and references to other recent work with these analogues, while extensive reviews on these and related pyrimidine analogues are presented by De Clercq (1985a) and by De Clercq and Walker (1984).

5.2.6.4 5-(2-Haloalkyl)-2'-deoxyuridines

The 5-alkyl-substituted pyrimidine series was extended to include analogues having 2-haloalkyl substituents (Fig. 5-15). Among these, CEdU, the most active, was found to be a potent and selective inhibitor of herpes viruses both *in vitro* and *in vivo* (Rosenwirth *et al.*, 1985).

5.2.6.5 5-Halo-3'-azido Analogues of Pyrimidine Deoxynucleosides

3'-Azido-3'-deoxythymidine (AZT) (Fig. 5-15) is a potent inhibitor of the human immunodeficiency virus (HIV) and has been found to produce clinical improvement and prolong life of some AIDS patients. Replacement of the 5-methyl group of AZT with the trifluoromethyl group abolished activity, while analogues in which the 5-methyl group was replaced by a fluoro, bromo, or iodo group retained significant anti-HIV activity (Fig. 5-15). None of the compounds evaluated was as active as AZT (Lin *et al.*, 1988).

5.3 Biochemistry of Halogenated Purines and Their Nucleosides

5.3.1 Halogenated Purine Nucleosides as Antitumor and Antiviral Agents

5.3.1.1 2-Haloadenosines

The first fluorinated purine analogue developed as a potential antitumor agent, 2-fluoroadenosine (fl^{2}A, Fig. 5-16), was found to be highly cytotoxic (Montgomery and Hewson, 1957). Unfortunately, this extreme toxicity was present in both tumor and normal cells, so this analogue had no therapeutic potential. However, the anabolic and catabolic behavior of fl^{2}A proved to be very relevant to the development of another fluoroadenine nucleoside, 9-β-D-arabinofuranosyl-2-fluoro-adenine (ara-fl^{2}A; Fig. 5-16). Fl^{2}A is readily phosphorylated by adenosine kinase to form 2-fluoroadenosine monophosphate (fl^{2}AMP) and is further phosphorylated to the triphosphate (fl^{2}ATP) [reviewed by Montgomery (1982)]. Inhibition of *de novo* purine biosynthesis by the action of fl^{2}AMP

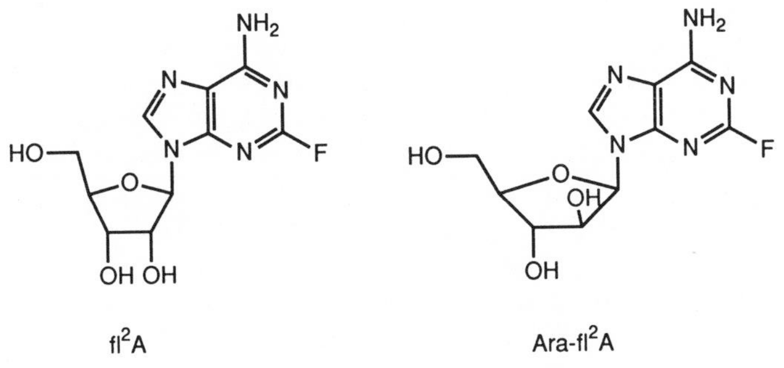

fl^{2}A   Ara-fl^{2}A

Figure 5-16. Fluorinated purine nucleosides.

as a feedback inhibitor, the inhibition of RNA polymerase by fl²ATP, and the incorporation of fl²ATP into RNA all have been suggested as contributing to the cytotoxicity of fl²A (Shigeura *et al.*, 1965). In contrast, fl²A is resistant to the action of adenosine deaminase, the major catabolic enzyme of adenosine. It is on this property that the therapeutic potential of ara-fl²A is based (Section 5.3.1.2). The cytotoxicity of fl²A is dependent on the small size of fluorine, since the activities of the other haloadenosines decrease markedly with increasing size of the halogen, suggesting that adenosine kinase cannot tolerate steric bulk at the 2 position of adenosine (Montgomerey, 1982).

5.3.1.2 9-β-ᴅ-Arabinofuranosyl-2-fluoroadenine (Ara-fl²A)

9-β-ᴅ-Arabinofuranosyladenine (ara-A) is phosphorylated *in vivo* to the triphosphate and, as such, is a potent inhibitor of DNA synthesis and has both antiviral and antitumor activity. The altered furanose ring contributes to this activity by changing reactivity at the 3′-hydroxyl group, and incorporation of ara-A into DNA inhibits DNA chain elongation. Ara-A also inhibits ribonucleotide reductase and DNA polymerase α. In contrast to the related, clinically useful 1-β-ᴅ-arabinofuranosylcytosine (ara-C), ara-A has been of limited clinical effectiveness because of rapid deamination by adenosine deaminase to the inactive arabinosylhypoxanthine. Attempts to circumvent this problem by administration of deaminase inhibitors such as deoxycoformycin have been complicated by undesirable side effects (Spriggs *et al.*, 1986, and references therein).

As an alternative approach, ara-fl²A (Fig. 5-16) was synthesized with the expectation that this would be a metabolically stable analogue of ara-A, based on the resistance of fl²A to the action of adenosine deaminase (Brockman *et al.*, 1977). Ara-fl²A is a substrate for adenosine deaminase, but it is deaminated very slowly and, after phosphorylation to the triphosphate, is selectively toxic to tumor cells, both *in vivo* and *in vitro* (Montgomery, 1982). Despite its structural similarity to adenosine, ara-fl²A is phosphorylated not by adenosine kinase, but by deoxycytidine kinase. Ara-fl²A is a potent inhibitor of ribonucleotide reductase and of DNA polymerase α. The combined effects of these inhibitions have been termed a "self-potentiating" inhibition of DNA synthesis. Thus, the effect of a reduced level of deoxyribonucleotide pools is potentiated by the further inhibition of the incorporation of these nucleotides into DNA (Tseng *et al.*, 1981). A recent study has shown that ara-fl²A can also be incorporated into RNA (Spriggs *et al.*, 1986).

In considering the mechanism for the selective cytotoxicity of ara-fl²A to tumor cells, the importance of potential differences in nucleoside transport by tumor and normal proliferating cells has been established. A combination of more active transport and more active metabolic processing in the tumor cell appears to be important in producing drug sensitivity. The broader implications of this observation with respect to other tumor cells are under investigation (Barrueco *et al.*, 1987; Sirotnak *et al.*, 1983).

Ara-fl^{2}A has activity against L1210 and P388 leukemias and is also effective against solid tumors. However, phase I clinical trials of ara-fl^{2}A, in the form of its water-soluble monophosphate, revealed unexpectedly severe toxic effects, including central nervous system toxicity and dose-limiting granulocytopenia and thrombocytopenia (e.g., Hutton *et al.*, 1984). Although ara-fl^{2}A had been shown to be resistant to phosphorolytic cleavage by nucleoside phosphorylase from mammalian cells (Avramis and Plunkett, 1983), the severe toxicity of 2-fluoroadenine (fl^{2}ade) and its detection in experimental animals and in the urine of patients during phase I trials suggest formation of this metabolite. Huang and Plunkett (1987) recently have shown that extracts of bacterial cells catalyze the formation of fl^{2}ade from ara-fl^{2}A. On the basis of this observation, they proposed that the formation of fl^{2}ade *in vivo* may occur as a result of the metabolism of ara-fl^{2}A by intestinal flora.

5.3.1.3 *Effect of Furanose Structure on Cytotoxicity of Haloadenine Nucleosides*

In contrast to the 2-haloadenosine series, 2-fluoro-, 2-chloro-, and 2-bromodeoxyadenosine are cytotoxic and are curative in a leukemia L1210 system. From this, Montgomery (1982) concluded that, unlike adenosine kinase, deoxycytidine kinase can tolerate bulk at the 2 position of adenine. The opposite trends were found with 8-halo-substituted adenine nucleosides, suggesting that adenosine kinase, but not deoxycytidine kinase, can tolerate bulk at this position (Montgomery, 1982).

5.3.2 Activity of Haloadenosine Analogues at Adenosine Receptors

Extracellular adenosine receptors, coupled to cAMP, mediate a host of physiological effects of adenosine, such as hypotensive and negative chronotropic actions on the heart. Halogenated adenosines, for example 2-chloroadenosine, are among active analogues that have been used extensively in the pharmacological characterization of adenosine receptors [for a recent review, see Jacobson (1988)].

5.4 Nucleoside Analogues Containing Halogen on the Sugar Moiety

5.4.1 2'-Deoxy 2'-Halogenated Nucleosides

5.4.1.1 Effect of 2'-Substituent on Nucleoside and Nucleotide Conformation

Nucleosides can exist in either of two conformations, the C-3' endo (N) or C-2' endo (S) forms (Fig. 5-17), and the different preference of deoxyribose and ribose moieties in nucleic acids for these forms is now believed to be the basis for much of the structural and functional diversity of nucleic acids. At the monomer nucleoside level, interconversion between the two forms is quite rapid. In polynucleic acids, RNA is confined to a 3'-endo structure, while the more flexible DNA has predominantly the 2'-endo configuration (B-DNA) but can also assume the 3'-endo configuration to give A-DNA under certain conditions. Several studies have been conducted to determine the effects of C'-2 substitution on conformation, using a variety of physical methods and *ab initio* calculations (Bergstrom and Swartling, 1988, and references therein). Of particular significance with regard to halogen substitution are the results of Uesugi *et al.* (1979), who found a linear relationship between the electronegativity of 2'-substituents and sugar conformation in adenine nucleosides. Thus, 2'-deoxy-2'-fluoroadenosine (2'-FdA) has an unusually high population of N configuration (67%) compared to that of dA (19%) and the 2'-iodo analogue (2'-IdA) (7%), as well as to that of A (36%).

5.4.1.2 Biochemistry of 2'-Deoxy-2'-fluororibonucleosides

Guschlbauer and co-workers have studied extensively the effects of fluorine in the 2' position of nucleosides with respect to nucleoside conformation, polynucleotide structure, and biological activity (Wohlrab *et al.*, 1985, and references therein). While studies with polynucleotides confirm

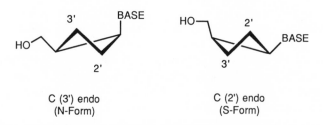

C (3') endo   C (2') endo
(N-Form)   (S-Form)

Figure 5-17. The N- and S-forms of nucleosides.

that poly(2'-fluoronucleotides) behave as polyribonucleotides (see below, Section 5.4.1.3), the monomeric analogues, in many systems, behave biochemically as deoxyribonucleosides, despite the preference for the N configuration. Thus, of several 2'-substituted analogues of dUMP tested, only 2'-deoxy-2'-fluorouridine monophosphate (2'-FdUMP; Fig. 5-18) was a substrate for thymidylate synthase (Wohlrab *et al.*, 1978). In addition, 2'-deoxy-2'-fluoronucleoside triphosphates have been shown to be substrates for RNA polymerases, certain DNA polymerases, and reverse transcriptases (Aoyama *et al.*, 1985, and references therein). Because this activity in DNA-directed systems indicated possible antiviral activity, the inhibitory activity of 2'-deoxy-2'-fluorocytidine (2'FdC; Fig. 5-18) against a number of herpes viruses was investigated. 2'FdC is phosphorylated by viral thymidine kinase with comparable efficiency to that for thymidine. Activity comparable to that of ara-C was found against HSV-1, HSV-2, pseudorabies virus, and equine abortion virus, while cytotoxicity was found to be only $\frac{1}{10}$th as great. Consistent with the behavior toward polymerases noted above, 2'FdC is incorporated into both viral DNA and RNA (Wohlrab *et al.*, 1985).

5.4.1.3 Oligo(2'-deoxy-2'-halogenoribonucleotides) as Templates and Inhibitors of DNA Biosynthesis

The observation that 2'-FdA has an unusually high preference for the 3'-endo conformation is significant with respect to polynucleotide structure and function since a nucleoside residue in well-stacked ribonucleotides also favors 3'-endo puckering. A series of poly(2'-substituted-2'-deoxyadenylic acids), including poly(2'-FdA), poly(2'-CldA), and poly(2'-BrdA), were found to have physical properties consistent with poly(ribonucleic acids)

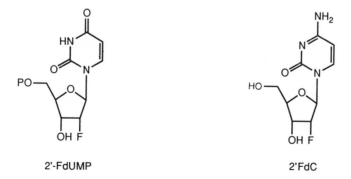

2'-FdUMP   2'FdC

Figure 5-18. Examples of 2'-fluorinated pyrimidine nucleosides.

rather than poly(deoxyribonucleic acids) (Fukui *et al.*, 1982a, and references therein). Similarly, three dinucleoside monophosphates containing 2'-FdA (2'-FdA–2'-FdA, 2'-FdA–A, and A–2'-FdA; Fig. 5-19) were prepared to examine the effects of the strong preference of the fluorinated ribose for the 3'-endo configuration on dimer stability and structure. UV, CD, and NMR spectral data all confirmed a stacked structure for the three dimers, but with greater base overlap than with A–A. Similarly, the three dimers formed complexes with poly(U) that were more stable than the A–A · poly(U) complex. The orders of stability and base stacking were 2'-FdA–2'-FdA > 2'-FdA–A > A–2'-FdA (Uesugi *et al.*, 1981).

From the physicochemical data noted above, which extended similar data obtained from the monomeric nucleosides, it is reasonable to expect that poly(2'-deoxy-2'-fluororibonucleosides) would behave as the polyribonucleosides rather than the polydeoxyribonucleosides. In fact, in some systems, the polymer containing the 2'-fluoronucleoside is more effective than that containing the natural base. Thus, poly(2'-FdA) and other poly(2'-halo-2'-deoxyadenylic acids) function *in vitro* as messenger RNAs in protein-synthesizing systems, causing incorporation of lysine into polypeptides. Moreover, poly(2'-FdA) produces a more rapid initial incorporation and is more stable in the system than poly(riboadenylic acid) [poly(rA)] (Fukui *et al.*, 1982a). In another example, poly(2'-FdA) was shown to resemble closely poly(rA) as a functional template in viral

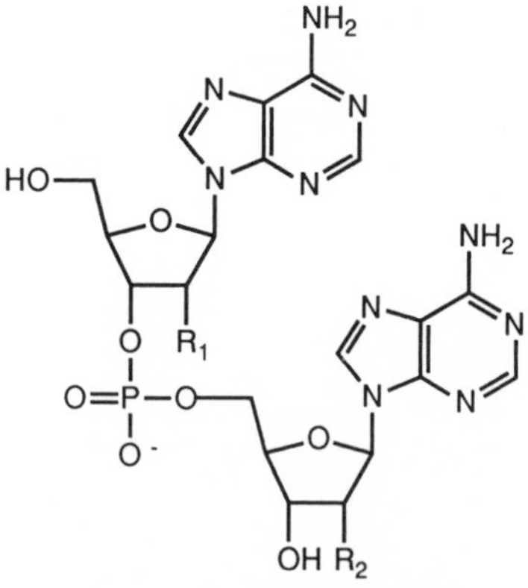

$R_1 = R_2 = F$ (2'FdA-2'FdA)
$R_1 = F; R_2 = OH$ (2'FdA-A)
$R_1 = OH; R_2 = F$ (A-2'FdA)

Figure 5-19. Three fluorinated dinucleoside monophosphates (Uesugi *et al.*, 1981).

RNA-directed polymerase reactions, such as $(rA)_n \cdot (dT)_{12}$-catalyzed reverse transcriptase from Rauscher leukemia virus. In contrast, the DNA polymerase β-activity catalyzed by $(dA)_n \cdot (dT)_{12}$ was inhibited by poly(2'-FdA) (Chandra et al., 1981). Similarly, poly(2'-deoxy-2'-fluoroinosinic acid) [poly(2'-FdI)] functioned as an effective template for the reverse transcriptase from Moloney murine leukemia virus (Fukui et al., 1982b).

Template analogues for polymerases compete with the normal genome for the template-binding site for the enzyme and thus have the potential to inhibit DNA synthesis. Furthermore, the development of effective inhibitors of RNA-directed DNA polymerase (reverse transcriptase) represents an attractive approach for the development of chemotherapeutic agents for the treatment of RNA virus-infected cells, including the HIV infection associated with AIDS. The inhibition of vesicular stomatitis virus in mouse cells and avian myeloblastosis virus (AMV) RNA-dependent DNA polymerase by poly(2'-deoxy-2'-fluorouridylic acid) [poly(2'-FdU)] was an early indication of this potential (Erickson and Grosch, 1974). Fukui and De Clercq (1982) examined a series of 2'- and 2-substituted polynucleotide derivatives as reverse transcriptase inhibitors and found poly(2-fluoroadenylic acid) [poly(2-FA)] to be a potent inhibitor.

5.4.1.4 Interferon Induction by 2'-Deoxy 2'-Halogenated Nucleoside-Containing Double-Helical RNAs

Recent interest in the mode of action of interferon as a biological response modifier and in the potential of exploiting antiviral properties of interferon has prompted syntheses of polynucleotide complexes as artificial interferon inducers. Included in structural requirements for such complexes is the presence of a double-helical RNA configuration. In a series of $(I)_n \cdot (C)_n$ and $(A)_n \cdot (U)_n$ analogues containing 2'-deoxy-2'-halogenoribonucleotide analogues, $(2'\text{-FdI})_n \cdot (C)_n$ and $(2'\text{-CldI})_n \cdot (C)_n$ had unexpectedly high interferon-inducing potency, while other 2'-substituted analogues lacked activity. Greater thermal stability of $(2'\text{-FdI})_n \cdot (C)_n$ and a higher resistance to nucleases suggest that this analogue should have a longer biological half-life than $(I)_n \cdot (C)_n$ (De Clercq et al., 1980).

The several studies discussed above give clear indication of the ability of a 2'-halogen, particularly fluorine, to impose a ribose configuration on polynucleotides containing this substitution. The overall geometry of the polynucleotide, rather than any specific interaction of the 2'-OH group, is critical in determining activity. As will be seen in the next section, substitution of halogen in the 2'-arabino- or "up"-configuration has altogether different consequences for the biochemical behavior of nucleosides.

5.4.1.5 5-Substituted 1-(2'-Deoxy-2'-halogeno-β-D-arabinosyl)cytosines and -uracils—New Potent Antiviral Agents

The effectiveness of ara-C as an anticancer drug and as an inhibitor of DNA virus replication is among evidence that structural modifications at the 2' position of nucleosides can lead to useful altered biological behavior. In 1979, Fox and his co-workers initiated studies of 2'-fluoro-2'-deoxy-arabinofuranosylpyrimidine nucleosides as antiviral agents [reviewed by Fox *et al.* (1988)]. In addition to nucleosides containing the natural pyrimidine bases, analogues also were prepared which were variously substituted on the pyrimidine ring (Fig. 5-20), in particular to take advantage of the potent biological activity of certain 5-substituted pyrimidine nucleosides. The results of this research have led to potential antitumor drugs and agents effective against such viral strains as HSV-2 and cytomegalovirus viruses, strains which previously had been relatively insensitive to chemotherapy (Watanabe *et al.*, 1979). Among these analogues, 1-(2'-deoxy-2'-fluoro-β-D-arabinosyl)-5-iodocytosine (FIAC), the 5-bromo analogue (FBrAC), and 1-(2'-deoxy-2'-fluoroarabinosyl)-5-

R = CH₃; FMAU
R = C₂H₅; FEAU

R = I; FIAC
R = Br; FBrAC

ClMAC

X = Br ; 2'-Deoxy-2'-fluoro-
arabinosyl-5-bromocytosine

X = I; 2'-Deoxy-2'-fluoro
arabinosyl-5-iodocytosine

2'F-ara-fl⁵U

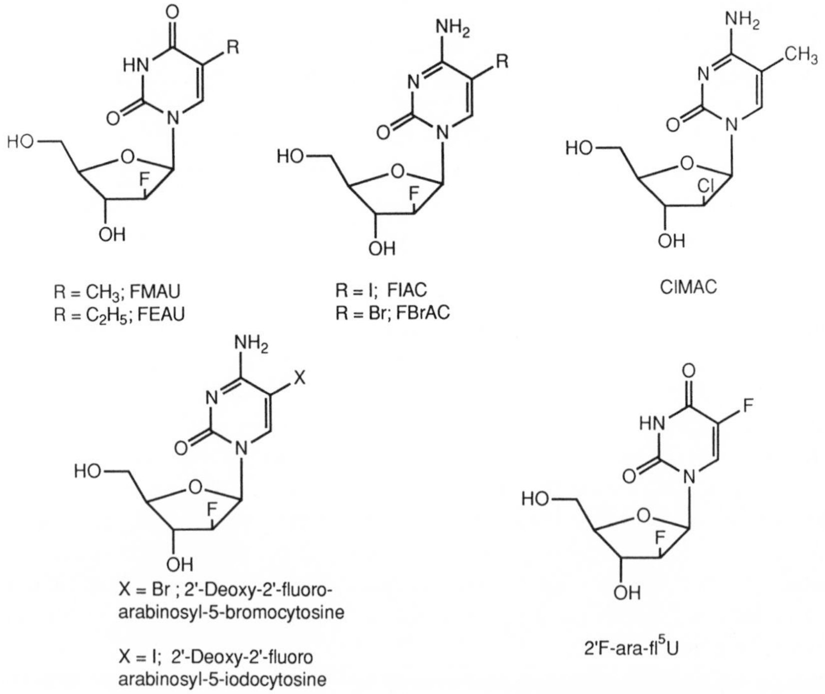

Figure 5-20. Examples of 2'-deoxy-2'-haloarabinosyl nucleosides.

methyluracil (FMAU) were highly potent inhibitors of HSV-1 and HSV-2 in Vero cell monolayers. FIAC was the most active, with an ED_{50} of 0.01 μM for both HSV-1 and HSV-2. Toxicity toward uninfected Vero cells was observed only at significantly higher doses. 1-(2'-Deoxy-2'-fluoro-β-D-arabinosyl)-5-ethyluracil (FEAU) was less potent but also less toxic and had a very favorable therapeutic index. FEAU may be clinically effective against hepatitis B virus (Fox et al., 1988). In general, a 2'-fluoro substituent was the most effective in a given series, with the exception that 1-(2'-chloro-2'-deoxyarabinosyl)-5-methylcytosine (ClMAC) was more potent ($ED_{50} = 0.22\ \mu M$) than the corresponding 2'-fluoro analogue ($ED_{50} = 0.8\ \mu M$) against HSV-2. In dramatic contrast to such antiviral agents as $br^5 dU$ and $io^5 dU$, the arabino nucleosides ClMAC, FBrAC, and 1-(2'-chloro-2'-deoxyarabinosyl)-5-iodocytosine had considerably higher activity against HSV-2 than against HSV-1, possibly reflecting different substrate specificities for the virus-induced kinases. In further contrast, these analogues have been shown not to be mutagenic (Watanabe et al., 1983).

The essentiality of having the 2'-halo substituent "up" is seen from the 10^3 times lower activity shown by 1-(2'-deoxy-2'-fluororibosyl)-5-iodocytidine relative to that shown by FIAC (Watanabe et al., 1983). Based on ab initio calculations of preferred geometries, Sapse and Snyder (1985) proposed that the requirement for fluorine "up" in FMAU is caused by a fluorine-induced restriction of rotation of the base about the sugar–base bond when fluorine is in the arabino position. The origin of such a restriction is unlikely to be steric in nature. Dipole–dipole repulsions between the $C^{\delta+} - F^{\delta-}$ bond and a $C^{\delta+} = O^{\delta-}$ bond in the base may play a role. Sapse and Snyder suggested that such a locked conformation may expose the analogue more effectively to the action of DNA polymerase. The selectivity of these analogues, as in previous examples, is dependent on virus-encoded enzymes which have different biochemical properties than those in normal cells. For example, HSV-1- and HSV-2-encoded thymidine kinase preferentially phosphorylates FMAU, and FMAU triphosphate is a selective inhibitor of virus-encoded DNA polymerase (Cheng et al., 1981; Ruth and Cheng, 1981).

Subsequent in vivo studies with FIAC and FMAU have confirmed their antiviral activity, as well as their selective toxicity toward tumor cells. FIAC has been shown to have clinical efficacy in immunosuppressed patients suffering from HSV infections (Young et al., 1983). In phase I evaluation of FMAU in treatment of advanced cancer, central nervous system toxicity proved to be dose limiting. This neurotoxicity, which was unexpected based on preclinical toxicity studies, appears to be related to DNA effects in the brain (Fanucchi et al., 1985).

In addition to their effectiveness against HSV-1 and HSV-2, FIAC, FIAU, FMAC, and FMAU are also active against VZV and human cytomegalovirus (HCMV) (Mar *et al.*, 1984; Colacino and Lopez, 1983, and references therein). HCMV infection is life threatening and has become a much more serious problem with the increased use of immunosuppression following transplantation procedures. It has also been associated with Kaposi's sarcoma, prevalent in AIDS patients. This virus does not specify an HCMV-thymidine kinase for its replication and thus lacks the target enzyme exploited in herpes and VZV chemotherapy. The action of purified HCMV-induced DNA polymerase and cellular DNA polymerase α on the triphosphates of FMAU, FMAC, and FIAC confirmed earlier indications that the analogues are better substrates for viral DNA polymerase than cellular DNA polymerase and thus are preferentially incorporated into viral DNA. The viral DNA polymerase also is more effectively inhibited by the analogues (Mar *et al.*, 1985).

1-β-D-Arabinofuranosyl-5-fluorouracil (ara-fl^{5}U; Fig. 5-8) (Section 5.1.10) has recently been shown to have selective action against HCMV replication. This appears to be another case of a "self-potentiating" activity. Although HCMV does not encode its own thymidine kinase, cellular kinase activity in infected cells is increased, leading to increased phosphorylation of ara-fl^{5}U. Formation of the monophosphate (ara-fl^{5}UMP), an effective inhibitor of thymidylate synthase, causes a decrease in dTTP pools. The effect of this is potentiated further by the inhibition of HCMV DNA polymerase by the triphosphate, ara-fl^{5}UTP. Ara-fl^{5}UTP, however, does not replace dTTP in viral DNA synthesis. The corresponding arabinosyl-5-bromo-, -5-chloro-, and -5-iodouracil analogues did not inhibit thymidylate synthase and thus were less effective anti-HCMV agents, even though they were better inhibitors of HCMV DNA polymerase (Suzuki *et al.*, 1985).

1-(2'-Deoxy-2'-fluoro-β-D-arabinosyl)-5-fluorouracil   (2'-F-ara-fl^{5}U; Fig. 5-20) combines biochemical properties caused by a 2'-fluorine in the arabino configuration with the potent effects of the 5'-fluorine of fl^{5}ura. 2'-F-ara-fl^{5}U is a kinase substrate and, as the monophosphate, inhibits thymidylate synthase by the same mechanism as does fl^{5}dUMP. The 2'-fluoro group confers resistance to phosphorolytic cleavage, causing prolonged thymidylate synthase inhibition in comparison to that produced by fl^{5}dU. While 2'-F-ara-fl^{5}U is moderately cytotoxic, the expected absence of RNA effects such as seen with fl^{5}dU could be advantageous in chemotherapy (Coderre *et al.*, 1983). The selective inhibition of human HCMV replication by 2'-F-ara-fl^{5}U has been attributed to its functioning as a substrate for viral, but not host, DNA synthesis (Suzuki *et al.*, 1987).

The aforementioned compounds are examples of a new generation of highly potent and selective antiviral agents. They can also be considered new lead compounds in the design of even more effective drugs. For example, recognition that a 2'-fluorine in the arabino configuration can improve antiviral activity has led to the synthesis of additional analogues in this series (see, e.g., Su *et al.*, 1986; Griengl *et al.*, 1987; Biggadike *et al.*, 1987).

5.4.1.6 1-(2',3'-Dideoxy-2'-fluoro-β-D-arabinofuranosyl)adenine (2'-F-ara-ddA)

The electronegativity of fluorine has been exploited to solve a problem of drug delivery (Marquez *et al.*, 1987). 2',3'-Dideoxynucleosides, including dideoxyadenosine (ddA), have shown promise in the inhibition of the cytopathic effects of the HIV virus that causes AIDS. Since viral chemotherapeutic approaches to the management of AIDS will require a long period of continuous treatment, oral administration of the drug is desirable. However, ddA is known to be particularly acid sensitive ($t_{1/2}$ at pH 1 = 35 s), and exposure to stomach acid likely would hydrolyze the glycosidic bond. In contrast, the presence of fluorine in the 2'-arabino configuration in 2'-F-ara-ddA (Fig. 5-21) completely blocks hydrolysis at pH 1, as a result of destabilization of the oxonium ion intermediate involved in the hydrolysis. 2'-F-ara-ddA was as effective as AZT in protecting cultured ATH8 cells against the effects of HIV infection (Marquez *et al.*, 1987).

5.4.1.7 2'-Deoxy 2'-Halogenated Nucleosides as Inhibitors and Mechanistic Probes of Ribonucleotide Reductase

Ribonucleotide reductases (RTR) catalyze the reduction of ribo-nucleotides to deoxyribonucleotides, and thus they play a critical role

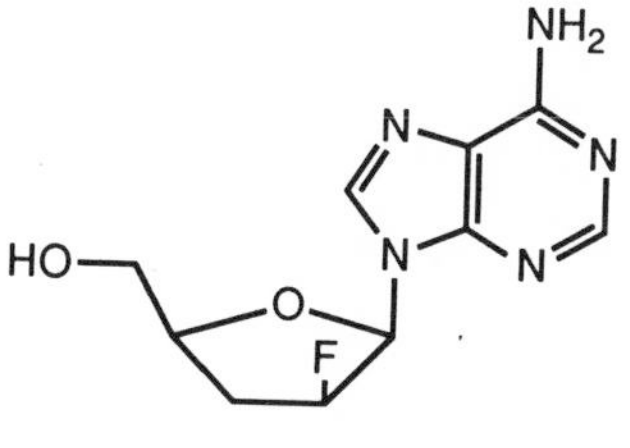

2'-F-ara-ddA

Figure 5-21. 2'F-ara-ddA. Through its inductive effect, fluorine stabilizes the otherwise very hydrolytically labile glycosidic bond (Marquez *et al.*, 1987).

in *de novo* DNA synthesis in all types of organisms. One class of these enzymes, which includes ribonucleoside diphosphate reductases (RDPR) from *E. coli* and from mammalian cells, uses deoxyribonucleoside diphosphates as substrates and contains a tyrosyl radical associated with a binuclear iron(III) center. A second class, predominantly from procaryotes, are ribonucleoside triphosphate reductases (RTPR) and have adenosyl-cobalamin as a cofactor. A third class, requiring Mn^{3+}, has recently been identified. Despite dissimilarities in structure and cofactor requirements, the mechanisms of reduction for these three classes are thought to be similar [reviewed by Ashley and Stubbe (1985)].

Thelander *et al.* (1976) made the important observation that 2'-chloro-2'-deoxycytidine diphosphate (2'-CldCDP) and 2'-chloro-2'-deoxyuridine diphosphate (2'-CldUDP) (Fig. 5-22) irreversibly inhibit the RTR from *E. coli* with release of chloride. The demonstration by Subbe and Kozarich (1980) that 2'-deoxy-2'-fluoroadenosine diphosphate (2'-FdADP) and 2'-deoxy-2'-fluorocytidine diphosphate (2'-FdCDP) behaved quite similarly to the chloro analogues indicated that a radical process was involved in the inhibition, consistent with the presence of the tyrosyl radical at the active site. Based on these and other observations, Stubbe and Kozarich have formulated mechanisms both for inhibition by 2'-deoxy-2'-halonucleosides as well as for the reduction of the natural substrates (Fig. 5-23). According to this proposal, homolytic elimination of halogen gives 2-methylene-3(2*H*)-furanone. Addition of an enzyme-bound nucleophile to this reactive Michael acceptor results in inactivation. In both RDPR- and RTPR-catalyzed reductions of the natural substrates, the 2'-hydroxyl of ribose is replaced by solvent hydrogen, and reduction is coupled to oxidation of a dithiol. During reduction, the hydroxyl group must be protonated by a

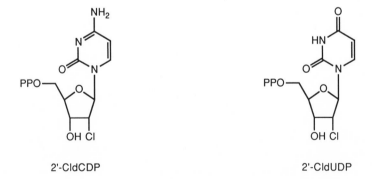

2'-CldCDP   2'-CldUDP

Figure 5-22. Examples of 2'-halogenated nucleosides that function as ribonucleotide reductase inhibitors.

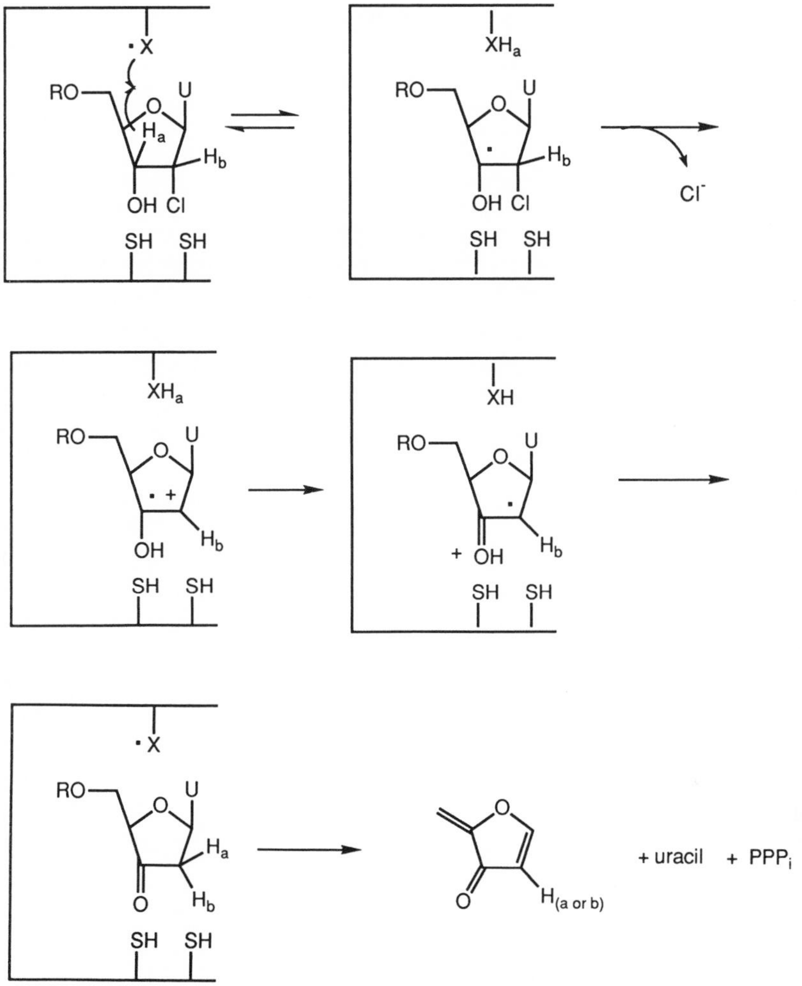

Figure 5-23. Ribonucleotide reductase inhibition by 2'-halogenated nucleosides.

thiol group, producing the thiolate anion, which is required for rapid cyclization to the disulfide with concomitant reduction of the cation-radical. In contrast with the halogenated analogues, elimination of halide occurs without protonation, and this results in a slower oxidative cyclization of the dithiol and, thus, to slower reduction of the cation-radical (Ashley and Stubbe, 1985; Stubbe and Kozarich, 1980; Harris *et al.*, 1984).

5.4.2 Biochemistry of 3′-Deoxy-3′-halonucleosides

5.4.2.1 2′,3′-Dideoxy-3′-fluororibonucleoside Triphosphates as Substrates for DNA Polymerase

A significant difference in the biochemistry of 2′-deoxy-2′-fluoro- and 3′-deoxy-3′-fluoronucleosides is seen in the ability of the latter, as the tri-phosphates, to act as efficient chain terminators in DNA biosynthesis. Waehnert and Langen (1979) reported that the triphosphate (3′-FddTTP) of 2′,3′-dideoxy-3′-fluorothymidine (3′-FddT; Fig. 5-24) is a potent competitive inhibitor, with respect to dTTP, of DNA polymerase from *Micrococcus luteus* and *Streptomyces hygroscopicus.* 3′-FddTTP is incor-porated into the DNA by both microbial DNA polymerases and, after incorporation, produces a 3′-FddT chain terminus. In contrast, Schroeder

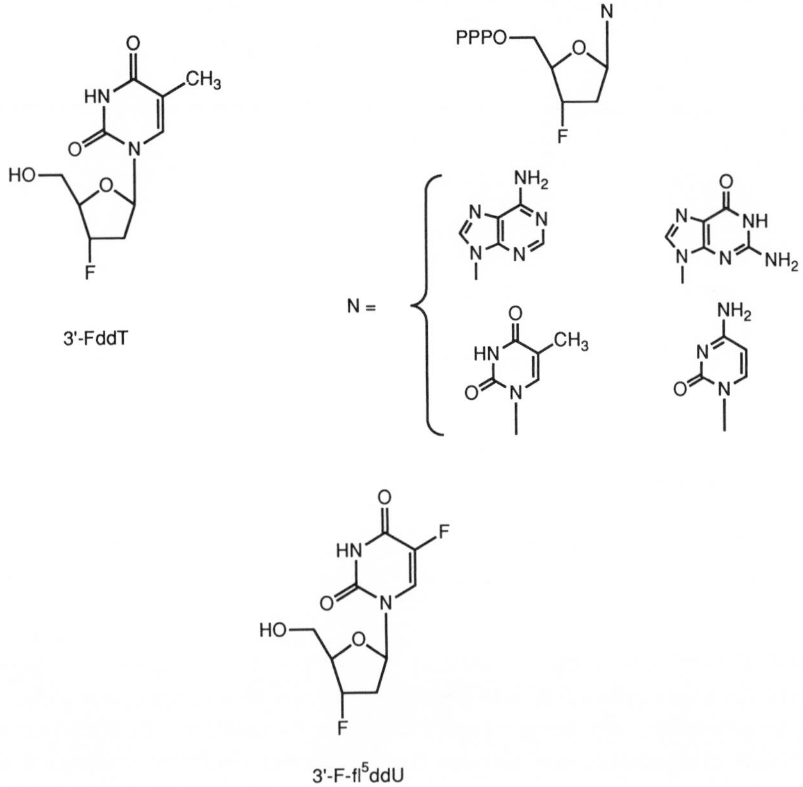

Figure 5-24. Examples of 3′-deoxy-3′-fluororibonucleosides and triphosphates.

and Jantschak (1980) reported that 3'-FdTTP is not incorporated into DNA by phage T4 wild-type and mutant DNA polymerase, but is a weak competitive inhibitor.

The effect of the presence of the 2',3'-dideoxy-3'-fluororibose moiety on DNA synthesis has been extended to the remaining natural substrates for DNA polymerase. Chidgeavadze et al. (1985, 1986) prepared 2',3'-dideoxy-3'-fluororibonucleoside triphosphates (3'-FddNTP = 3'-FddTTP, 3'-FddCTP, 3'-FddATP, and 3'-FddGTP; Fig. 5-24) and evaluated their effectiveness as substrates for a variety of DNA polymerases and as chain terminators. The 3'-ddNTPs were not substrates for DNA polymerase α but were chain-terminating substrates for E. coli DNA polymerase I, avian myeloblastosis virus reverse transcriptase, and calf thymus terminal deoxyribonucleotidyltransferase. The clearer termination patterns derived from these nucleosides suggest that they may have advantages over the nonfluorinated ddNTPs in a sequencing procedure developed by Sanger et al. (1977). Their selectivities for certain polymerase enzymes further suggest possible targets for selective toxicity (Beabealashvilli et al., 1986).

5.4.2.2 5-Fluoro-2',3'-dideoxy-3'-fluorouridine and Its 5'-Phosphate

The DNA-selective and potent activity frequently seen with fl^{5}dU in cell culture is not shown when used in vivo because of rapid phosphorolytic cleavage to fl^{5}ura. 5-Fluoro-2',3'-dideoxy-3'-fluorouridine (3'-F-fl^{5}ddU; Fig. 5-24) was synthesized with the expectation that substitution of fluorine for hydroxyl at C-3' would inhibit the action of dT phosphorylase. Indeed, 3'-F-fl^{5}dU was not a substrate for dT phosphorylase but was a weak competitive inhibitor. Unfortunately, 3'-F-fl^{5}ddUMP only weakly inhibited

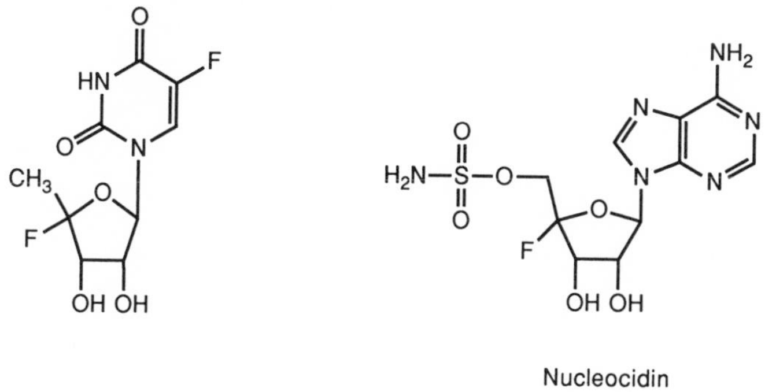

Nucleocidin

Figure 5-25. Two examples of 4'-fluoro nucleosides: 5'-deoxy-4',5-difluorouridine and nucleocidin.

thymidylate synthase and was four orders of magnitude less potent than fl^{5}dU in inhibiting the growth of L1210 tumor cells (Ajmera *et al.*, 1984).

5.4.3 4'-Halogenated Nucleosides

5'-Deoxy-4',5-difluorouridine (Fig. 5-25) was prepared to serve as a prodrug for fl^{5}ura, with the expectation that the 4'-fluorine would render this analogue a better substrate for uridine phosphorylase than is the well-studied 5'-deoxy-fl^{5}U (cf. Section 5.1.8). These expectations were met, and the 4'-fluoro analogue was an order of magnitude more effective in inhibition of L1210 cells in culture (Ajmera *et al.*, 1988).

The antibiotic nucleocidin (Fig. 5-25) is another example of a nucleoside containing a C-4' halogen. It is also one of a very few naturally occurring fluorine-containing compounds.

5.4.4 5'-Halogenated Nucleosides

5'-Deoxy-5'-methylthioadenosine (MTA) is derived from *S*-adenosyl-methionine during the biosynthesis of spermidine and spermine. The first

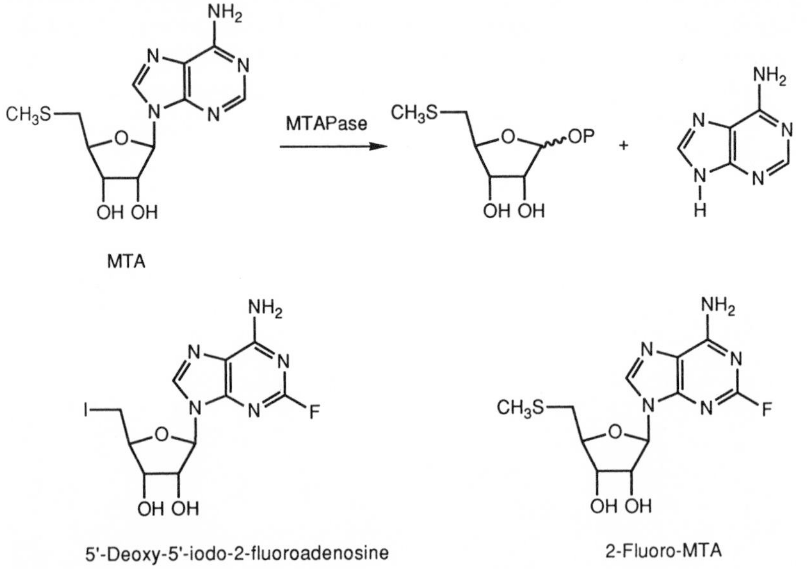

Figure 5-26. MTAPase-catalyzed reaction and halogenated adenosine derivatives as MTAPase inhibitors.

step in a process which converts MTA to adenine nucleotides and methionine is the phosphorolytic cleavage of MTA by MTAPase (Fig. 5-26). Savarese *et al.* (1987, and references therein) have developed MTA-like compounds which, through the action of MTAPase, serve as prodrugs for the release of antimetabolites *in vivo*. Doubly halogenated drugs, such as 5'-deoxy-5'-iodo-2-fluoroadenosine, were found to be less effective than the singly substituted analogue, 5'-deoxy-5'-methylthio-2-fluoroadenine (2-fluoro-MTA). It appears that the potent cytotoxicity of the fl^{2}ade that is formed masks any effects of the 5-methylthioribose-1-phosphate analogues, likewise produced by the action of MTAPase (Savarese *et al.*, 1987).

5.4.5 6'-Halogenated Nucleosides

An example of carbocyclic analogues of nucleosides substituted at C'-6 with halogen is found in the fluorinated analogue of ($\pm$) aristeromycin (Fig. 5-27).

Adding significance to the replacement of CH$_2$ in carbocyclic nucleosides with the CHF group is the evidence that the latter replacement may resemble more closely the ether oxygen of the furanose (Blackburn and Kent, 1986; also see Section 5.5). Biggadike *et al.* (1987) have prepared C'-6-fluorinated carbocyclic analogues of the potent antiviral agent io^{5}dU with fluorine in the α and β configuration (Fig. 5-27). While the α-fluoro analogue was extremely active against HSV-1 (approximately one-half the activity of acyclovir), the β-fluoro analogue was two orders of magnitude less active.

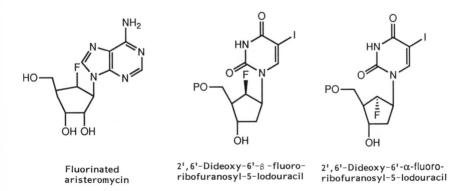

Fluorinated aristeromycin

2',6'-Dideoxy-6'-β-fluoro-ribofuranosyl-5-Iodouracil

2',6'-Dideoxy-6'-α-fluoro-ribofuranosyl-5-Iodouracil

Figure 5-27. Examples of 6'-deoxy-6'-fluorinated nucleosides.

5.5 Nucleosides Containing Halogenated Phosphate Esters

5.5.1 Halophosphate Analogues of Nucleotides

Phosphate hydroxyl groups have been replaced with fluorine as a means to investigate the importance of phosphate ionization and biochemical behavior, since the fluorophosphate will have only one anionic charge at physiological pH, in contrast to the doubly charged natural relative. For example, the inability of adenosine monophosphofluoridate (Fig. 5-28) to activate glycogen phosphorylase b was interpreted as indicating an absolute requirement for a dianionic phosphate at the nucleotide activator site of the phosphorylase (Withers and Madsen, 1980). Other examples are cited in the review by Bergstrom and Swartling (1988).

5.5.2 Fluoroalkylphosphonate Analogues of Nucleotides

Phosphate esters are ubiquitous and important biological structures, and, accordingly, research with analogues of phosphate esters has been extensive. Many phosphate ester analogues are produced by an "isosteric" replacement of oxygen with nitrogen, sulfur, or methylene functions. Because the methylene replacement present in phosphonates introduces hydrolytic stability, these have received considerable attention in the design of hydrolytically stable nucleotide analogues. For example, oligo(deoxyribonucleoside methylphosphonates) are being studied as nonionic sequence-specific inhibitors of mRNA expression as an approach to viral chemotherapy (Smith et al., 1986). While the replacement of oxygen with a methylene group introduces only minimal changes in stereochemistry, electronic factors are perturbed significantly because of the greater electronegativity of oxygen relative to carbon. Blackburn and co-

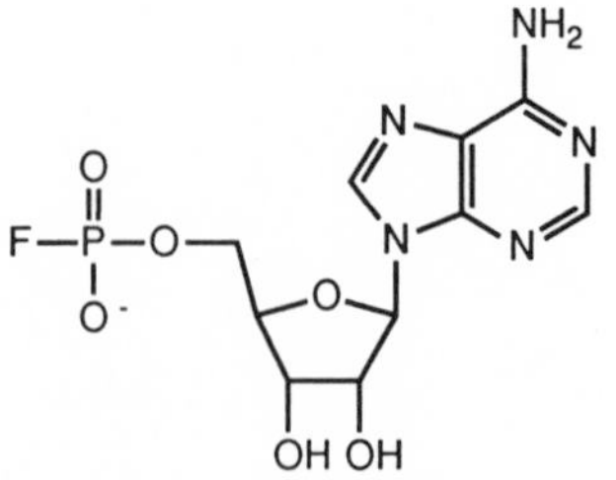

Adenosine monophosphofluoridate

Figure 5-28. An example of a nucleoside fluorophosphate analogue.

Table 5.1. Acidity of Pyrophosphate and
Phosphonates, $H_2O_3P-X-PO_3H_2$[a]

X	pK_{a2}	pK_{a3}
O	5.77	8.22
CH_2	7.45	10.96
CHF	6.15	9.35
CF_2	5.80	8.00

[a] Blackburn et al., 1986.

workers, stressing the importance of giving equal weight to polar considerations in these molecules, have studied the more electronegative halomethylene groups as oxygen replacements in phosphonates. In this strategy, the α-fluoromethylene replacement has advantages over use of other halogens both for steric reasons and because of the relative inertness of alkyl fluorides relative to the other alkyl halides (Blackburn et al., 1986, and references therein). The effectiveness of the difluoromethylene group in

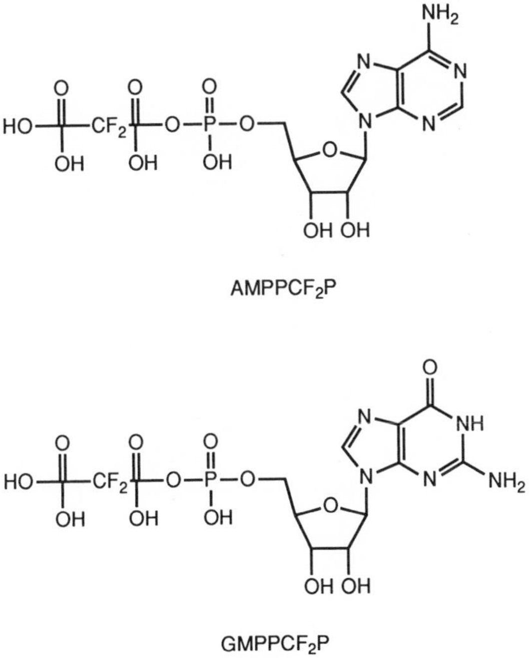

AMPPCF$_2$P

GMPPCF$_2$P

Figure 5-29. Difluoromethylene phosphonate analogues of ATP and GTP.

mimicking oxygen is indicated by the ionization constants found for pyrophosphate and the methylene, fluoromethylene, and difluoromethylene phosphonate analogues (Table 5-1).

Based on the above considerations, Blackburn *et al.* (1986, and references therein) prepared a series of β,γ-bridged analogues of ATP and studied these in a number of biological systems. Of these analogues, the β,γ-difluoromethylene analogue (AMPPCF$_2$P; Fig. 5-29) was found to be, relative to ATP, a better inhibitor of hexokinase and a better substrate for RNA polymerase. AMPPCF$_2$P was a good substrate for adenylate cyclase, as was the difluoromethylene GTP analogue (GMPPCF$_2$P; Fig. 5-29).

Incubation of extracts of interferon-treated cells with double-stranded RNA and ATP produces a low-molecular-weight inhibitor of protein synthesis. The identification of this inhibitor as 5'-*O*-triphosphoryladenyl(2' → 5')adenyl(2' → 5')adenosine (2-5A; Fig. 5-30) has aided in the understanding of the biochemistry of interferon. 2-5A is also the first natural oligonucleotide shown to contain a 2',5'-linkage. There has been much recent interest in the study of analogues of 2-5A, both as potential interferon-inducing agents and as antagonists to the action of 2-5A (Torrence, 1985). (2'-5')Oligoadenylate synthase, the enzyme which catalyzes the synthesis of 2-5A from ATP, uses both AMPPCF$_2$P and AMPPCl$_2$P as substrates, forming 2-5A oligomers with three to eight

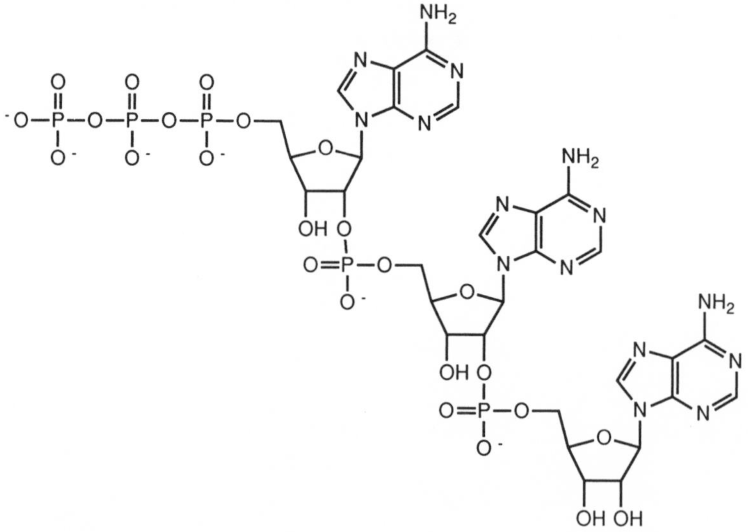

2-5A

Figure 5-30. The structure of 2-5A.

adenylated units, albeit in lower yields than seen with ATP. The hydrolytic stability of these 2-5A analogues and their lack of toxicity suggest that they may prove to be valuable stable antagonists of the 2-5A system (Blackburn et al., 1986).

Bergstrom et al. (1987) have prepared difluoromethylphosphonate analogues of thymidine 3'- and 5'-monophosphate, as well as the 3',5'-dimer (Fig. 5-31). The dimer was separated into the two diastereomic forms, an important consideration in the construction of oligonucleotides

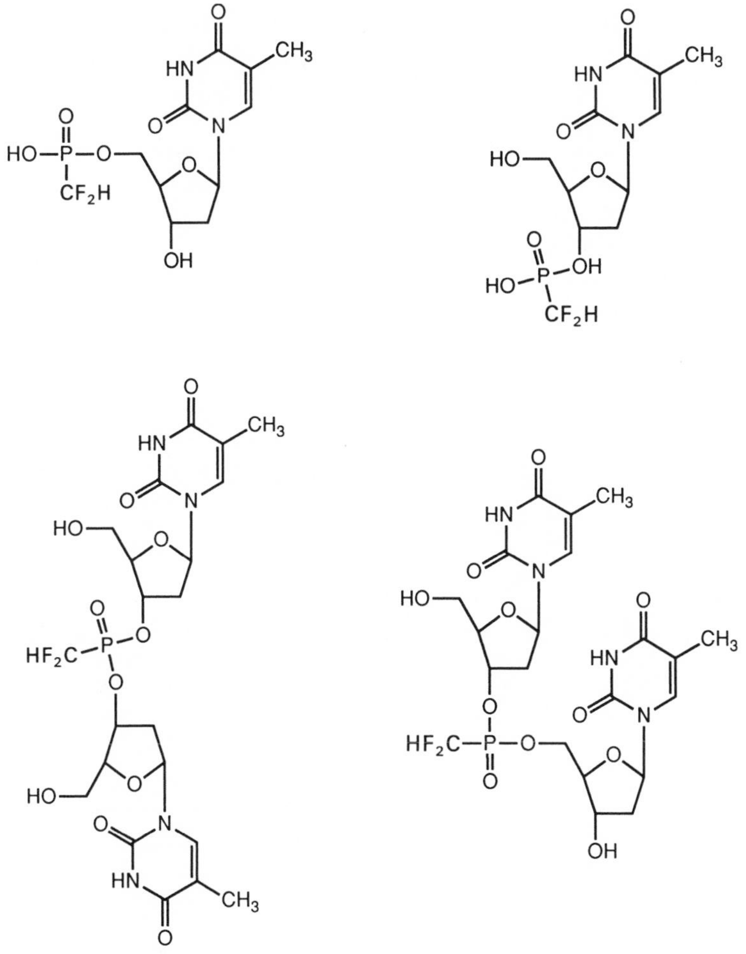

Figure 5-31. Difluoromethylene analogues of thymidine 5'- and 3'-monophosphates and of the 3',3'- and 3',5'-dimers.

with a bridging chiral phosphorus. These analogues should provide interesting biochemical data, particularly with regard to complementary sequence binding.

5.6 Conclusion

Progress in the development of antitumor and antiviral agents continues at a rapid pace. A major part of this work has been based on the leads provided by simple halogenated pyrimidines, purines, and their nucleosides. In this chapter, an attempt has been made to describe some of the directions in which this research is going, placed in the framework of halogenated nucleosides and nucleotides, and to provide specific examples in each area. Because of the enormous choice, selection of these examples has been difficult, and many important contributions have not been included. Considerable attention was given to the development of new antiviral compounds, both because of important new breakthroughs in this field and because of the urgent clinical need for these agents. New generations of compounds, together with improved combination chemotherapeutic schemes, should provide even more successes in treatment.

References

Ajmera, S., Bapat, A. R., Danenberg, K., and Danenberg, P. V., 1984. Synthesis and biological activity of 5-fluoro-2',3'-dideoxy-3'-fluorouridine and its 5'-phosphate, *J. Med. Chem.* 27:11–14.

Ajmera, S., Bapat, A. R., Stephanian, E., and Danenberg, P. V., 1988. Synthesis and interaction with uridine phosphorylase of 5'-deoxy-4',5-difluorouridine, a new prodrug of 5-fluorouracil, *J. Med. Chem.* 31:1094–1098.

Aoyama, H., Sarih-Cottin, L., Tarrago-Litvak, L., Kakiuchi, N., Litvak, S., and Guschlbauer, W., 1985. 2'-Fluoro-2'-deoxypolynucleosides as templates and inhibitors for RNA- and DNA-dependent DNA polymerases, *Biochim. Biophys. Acta* 824:225–232.

Armstrong, R. D., Lewis, M., Stern, S. G., and Cadman, E. C., 1986a. Acute effect of 5-fluorouracil on cytoplasmic and nuclear dihydrofolate reductase messenger RNA metabolism, *J. Biol. Chem.* 261:7366–7371.

Armstrong, R. D., Takimoto, C. H., and Cadman, E. D., 1986b. Fluoropyrimidine-mediated changes in small nuclear RNA, *J. Biol. Chem.* 261:21–24.

Ashley, G. W., and Stubbe, J., 1985. Current ideas on the chemical mechanism of ribonucleotide reductases, *Pharmacol. Ther.* 30:301–329.

Avramis, V. I., and Plunkett, W., 1983. 2-Fluoro-ATP: A toxic metabolite of 9-β-D-arabinosyl-2-fluoroadenine, *Biochem. Biophys. Res. Commun.* 113:35–43.

Barrueco, J. R., Jacobsen, D. M., Chang, C.-H., Brockman, R. W., and Sirotnak, F. M., 1987. Proposed mechanism of therapeutic selectivity for 9-β-D-arabinofuranosyl-2-fluoroadenine against murine leukemia based upon lower capacities for transport and phosphorylation in poliferative intestinal epithelium compared to tumor cells, *Cancer Res.* 47:700–706.

Beabealashvilli, R. S., Scamrov, A. V., Kutateladze, T. V., Mazo, A. M., Krayevsky, A. A., and Kukhanova, M. K., 1986. Nucleoside 5'-triphosphates modified at sugar residues as substrates for calf thymus terminal deoxynucleotidyl transferase and for AMV reverse transcriptase, *Biochim. Biophys. Acta* 868:136–144.

Beltz, R. E., and Visser, D. W., 1955. Growth inhibition of *Escherichia coli* by new thymidine analogues, *J. Am. Chem. Soc.* 77:736–738.

Benson, A. B., III, Trump, D. L., Cummings, K. B., and Fischer, P. H., 1985. Modulation of 5-iodo-2'-deoxyuridine metabolism and cytotoxicity in human bladder cancer cells by fluoropyrimidines, *Biochem. Pharmacol.* 34:3925–3931.

Benz, C., and Cadman, E., 1981. Modulation of 5-fluorouracil metabolism and cytotoxicity by antimetabolite pretreatment in human colorectal adenocarcinoma HCT-8, *Cancer Res.* 41:994–999.

Bergstrom, D., and Swartling, D. J., 1988. Fluorine substituted analogues of nucleic acid components, in *Fluorine-Containing Molecules* (J. F. Liebman, A. Greenberg, and W. R. Dolbier, Jr., eds.), VCH Publishers, New York, pp. 259–308.

Bergstrom, D., Romo, E., and Shum, P., 1987. Fluorine substituted analogues of nucleosides and nucleotides, *Nucleosides Nucleotides* 6:53–63.

Biggadike, K., Borthwick, A. D., Evans, D., Exall, A. M., Kirk, B. E., Roberts, S. M., Stephenson, L., Youds, P., Slawin, M. Z., and Williams, D. J., 1987. Synthesis of fluorinated carbocyclic nucleosides: Preparation of carbocyclic 1-(2'-deoxy-6'-fluoro-ribofuranosyl)-5-iodouracils, *J. Chem. Soc., Chem. Commun.* 1987:255–256.

Biggers, W. J., Barnea, E. R., and Sanyal, M. K., 1987. Anomalous neural differentiation induced by 5-bromo-2'-deoxyuridine during organogenesis in the rat, *Teratology* 35:63–75.

Biswas, D. K., and Hanes, S. D., 1982. Increased level of prolactin gene sequences in bromodeoxyuridine treated GH cells, *Nucleic Acids Res.* 10:3995–4008.

Biswas, D. K., Abdullah, K. T., and Brennessel, B. A., 1979. On the mechanism of 5-bromodeoxyuridine induction of prolactin synthesis in rat pituitary tumor cells, *J. Cell Biol.* 81:1–9.

Biswas, D. K., Hartigen, J. A., and Pichler, M. H., 1984. Identification of DNA sequence responsible for 5-bromodeoxyuridine-induced gene amplification, *Science* 225:941–943.

Blackburn, G. M., and Kent, D. E., 1986. Synthesis of α- and γ-fluoroalkylphosphonates, *J. Chem. Soc., Perk. in Trans. 1* 1986:913–917.

Blackburn, G. M., Perree, T. D., Rashid, A., Bisbal, C., and Lebleu, B., 1986. Isosteric and isopolar analogues of nucleotides, *Chem. Scr.* 26:21–24.

Boothman, D. A., Briggle, T. V., and Greer, S., 1985. Metabolic channeling of fluoro-2'-deoxycytidine utilizing inhibitors of its deamination in cell culture, *Mol. Pharmacol.* 27:584–594.

Boothman, D. A., Briggle, T. V., and Greer, S., 1987a. Tumor-selective metabolism of 5-fluoro-2'-deoxycytidine coadministered with tetrahydrouridine compared to 5-fluoro-uracil in mice bearing Lewis lung carcinoma, *Cancer Res.* 47:2354–2362.

Boothman, D. A., Briggle, T. V., and Greer, S., 1987b. Protective, tumor-selective dual pathway activation of 5-fluoro-2'-deoxycytidine provided by tetrahydrouridine in mice bearing mammary adenocarcinoma-755, *Cancer Res.* 47:2344–2353.

Boothman, D. A., Greer, S., and Pardee, A. B., 1987c. Potentiation of pyrimidine radiosensitizers in human carcinoma cells by β-lapachone (3,4-dihydro-2,2-dimethyl-2H-naphtho-[1,2-b]pyran-5,6-dione), a novel DNA repair inhibitor, *Cancer Res.* 47:5361–5366.

Brockman, R. W., Schabel, F. M., Jr., and Montgomery, J. A., 1977. Biological activity of

9-β-D-arabinofuranosyl-2-fluoroadenine, a metabolically stable analogue of 9-β-D-arabinofuranosyladenine, *Biochem. Pharmacol.* 26:2193–2196.

Bubley, G. J., Balzarini, J., Crumpacker, C. S., De Clercq, E., and Schnipper, L. E., 1987. The effect of (*E*)-5-(2-bromovinyl)-2'-deoxyuridine on DNA repair and mutagenesis of herpes simplex virus type 1, *Virology* 161:242–244.

Caradonna, S. J., and Cheng, Y.-C., 1980. The role of deoxyuridine triphosphate nucleotidohydrolase, uracil-DNA glycosylase, and DNA polymerase α in the metabolism of FUdR in human tumor cells, *Mol. Pharmacol.* 18:513–520.

Carrico, C. K., and Glazer, R. I., 1979. Augmentation by thymidine of the incorporation and distribution of 5-fluorouracil in ribosomal RNA, *Biochem. Biophys. Res. Commun.* 87:664–670.

Chandra, P., Demirhan, I., and De Clercq, E., 1981. A study of antitemplate inhibition of mammalian, bacterial, and viral DNA polymerases by 2- and 2'-substituted derivatives of polyadenylic acid, *Cancer Lett.* 12:181–193.

Chaudhuri, N. K., Montag, B. J., and Heidelberger, C., 1958. Studies on fluorinated pyrimidines III: The metabolism of 5-fluorouracil-2-C^{14} and 5-fluoroorotic-2-C^{14} acid *in vivo*, *Cancer Res.* 18:318–328.

Cheng, Y.-C., and Nakayama, K., 1983. Effects of 5-fluoro-2'-deoxyuridine on DNA metabolism in HeLa cells, *Mol. Pharmacol.* 23:171–174.

Cheng, Y.-C., Dutschman, G., Fox, J. J., Watanabe, K. A., and Machida, H., 1981. Differential activity of potential antiviral nucleoside analogues on herpes simplex virus-induced and human cellular thymidine kinases, *Antimicrob. Agents Chemother.* 20:420–423.

Chidgeavadze, Z. G., Scamrov, A. V., Beabealashvilli, R. Sh., Kvasyuk, E. I., Zaitseva, G. V., Mikhailopulo, I. A., Kowollik, G., and Langen, P., 1985. 3'-Fluoro-2',3'-dideoxyribonucleoside 5'-triphosphate: Terminators of DNA synthesis, *FEBS Lett.* 183:275–278.

Chidgeavadze, Z. G., Beabealashvilli, R. Sh., Krayevsky, A. A., and Kukhanova, M. K., 1986. Nucleoside 5'-triphosphates with modified sugars as substrates for DNA polymerases, *Biochim. Biophys. Acta* 868:145–152.

Clark, J. L., Berger, S. H., Mittelman, A., and Berger, F. G., 1987. Thymidylate synthase gene amplification in a colon tumor resistant to fluoropyrimidine chemotherapy, *Cancer Treat. Rep.* 71:261–265.

Coderre, J. A., Santi, D. V., Matsuda, A., Watanabe, K. A., and Fox, J. J., 1983. Mechanism of action of 2',5-difluoro-1-arabinosyluracil, *J. Med. Chem.* 26:1149–1152.

Cohen, S. S., Flaks, J. G., Barner, H. D., Loeb, M. R., and Lichtenstein, J., 1958. The mode of action of 5-fluorouracil and its derivatives, *Proc. Natl. Acad. Sci. USA* 44:1004–1012.

Colacino, J. M., and Lopez, C., 1983. Efficacy and selectivity of some nucleoside analogues as anti-human cytomegalovirus agents, *Antimicrob. Agents Chemother.* 24:505–508.

Cory, J. G., Breland, J. C., and Carter, G. L., 1979. Effect of 5-fluorouracil on RNA metabolism in Novikoff hepatoma cells, *Cancer Res.* 39:4905–4913.

Danenberg, P. V., 1977. Thymidylate synthetase—a target enzyme in cancer chemotherapy, *Biochim. Biophys. Acta* 473:73–92.

Danenberg, P. V., Heidelberger, C., Mulkins, M. A., and Peterson, A. R., 1981. The incorporation of 5-fluoro-2'-deoxyuridine into DNA of mammalian tumor cells, *Biochem. Biophys. Res. Commun.* 102:654–659.

Darnell, J., Lodish, H., and Baltimore, D., 1986. *Molecular Cell Biology*, Scientific American Books, distributed by W. H. Freeman and Company, New York, pp. 316–321.

De Clercq, E., 1984. Biochemical aspects of selective antiherpes activity of nucleoside analogues, *Biochem. Pharmacol.* 33:2159–2169.

De Clercq, E., 1985a. Recent trends in antiviral chemotherapy, in *Proceedings of the 1st International TNO Conference on Antiviral Research, Antiviral Research, Suppl. 1* (A. Billiau,

E. De Clercq, and H. Schellekens, eds.), Elsevier Science Publishers, Rotterdam, pp. 11–19.

De Clercq, E., 1985b. Synthetic pyrimidine nucleoside analogues, in *Approaches to Antiviral Agents* (M. E. Harnden, ed.), VCH Publishers, New York, pp. 57–99.

De Clercq, E., and Walker, R. T., 1984. Synthesis and antiviral properties of 5-vinyl pyrimidine nucleoside analogues, *Pharmacol. Ther.* 26:1–44.

De Clercq, E., Descamps, J., De Somer, P., Barr, P. J., Jones, A. S., and Walker, R. T., 1979. (*E*)-5-(2-Bromovinyl)-2′-deoxyuridine: A potent and selective antiherpes agent, *Proc. Natl. Acad. Sci. USA* 76:2947–2951.

De Clercq, E., Stollar, B. D., Hobbs, J., Fukui, T., Kakiuchi, N., and Ikehara, M., 1980. Interferon induction by two 2′-modified double-helical RNAs, poly(2′-fluoro-2′-deoxyinosinic acid)poly(cytidylic acid) and poly(2′-chloro-2′-deoxyinosinic acid)poly(cytidylic acid), *Eur. J. Biochem.* 107:279–288.

Dexter, D. L., Woberg, W. H., Ansfield, F. J., Helson, L., and Heidelberger, C., 1972. The clinical pharmacology of 5-trifluoromethyl-2′-deoxyuridine, *Cancer Res.* 32:247–253.

Djordjevic, B., and Szybalski, W., 1960. Genetics of human cell lines. III. Incorporation of 5-bromo- and 5-iododeoxyuridine into the deoxyribonucleic acid of human cells and its effect on radiation sensitivity, *J. Exp. Med.* 112:509–531.

Dolbeare, F., Gratzner, H., Pallavicini, M. G., and Gray, J. W., 1983. Flow cytometric measurement of total DNA content and incorporated bromodeoxyuridine, *Proc. Natl. Acad. Sci. USA* 80:5573–5577.

Dollinger, M. R., Burchenal, J. H., Kries, W., and Fox, J. J., 1967. Analogues of 1-β-D-arabinofuranosylcytosine. Studies on mechanisms of action in Burkitt's cell culture and mouse leukemia, and *in vitro* deamination studies, *Biochem. Pharmacol.* 16:689–706.

Dolnick, B. J., and Pink, J. J., 1983. 5-Fluorouracil modulation of dihydrofolate reductase RNA levels in methotrexate-resistant KB cells, *J. Biol. Chem.* 258:13299–13306.

Dolnick, B. J., and Pink, J. J., 1985. Effects of 5-fluorouracil on dihydrofolate reductase and dihydrofolate reductase mRNA from methotrexate-resistant KB cells, *J. Biol. Chem.* 260:3006–3014.

Douglas, K. T., 1987. The thymidylate synthesis cycle and anticancer drugs, *Med. Res. Rev.* 7:441–475.

Erickson, R. J., and Grosch, J. C., 1974. The inhibition of avian myeloblastosis virus deoxyribonucleic acid polymerase by synthetic polynucleotides, *Biochemistry* 13:1987–1993.

Fanucchi, M. P., Leyland-Jones, B., Young, C. W., Burchenal, J. H., Watanabe, K. A., and Fox, J. J., 1985. Phase I trial of 1-(2′-deoxy-2′-fluoro-1-β-D-arabinofuranosyl)-5-methyluracil (FMAU), *Cancer Treat. Rep.* 69:55–59.

Fernandez, M. P., Young, M. F., and Sobel, M. E., 1985. Methylation of type II and type I collagen genes in differentiated and dedifferentiated chondrocytes, *J. Biol. Chem.* 260:2347–2378.

Filler, R., and Naqvi, S. M., 1982. Fluorine in biomedicinal chemistry. An overview of recent advances and selected topics, in *Biomedicinal Aspects of Fluorine Chemistry* (R. Filler and Y. Kobayashi, eds.), Kodansha Ltd., Tokyo; Elsevier Biomedical Press, New York, pp. 1–16.

Fox, J. J., Falco, E. A., Wempen, I., Pomeroy, D., Dowling, M. D., and Burchenal, J. H., 1972. Oral and parenteral activity of 2,2′-anhydro-1-β-D-arabinofuranosyl-5-fluorocytosine against both intraperitoneally and intracerebrally inoculated mouse leukemia, *Cancer Res.* 32:2269–2272.

Fox, J. J., Watanabe, K. A., Chou, T. C., Schinazi, R. F., Soike, K. F., Fourel, I., Gantz, G., and Trepo, C., 1988. Antiviral activities of 2′-fluorinated arabinosyl-pyrimidine nucleosides, in *Fluorinated Carbohydrates, Chemical and Biochemical Aspects* (N. F. Taylor, ed.), ACS Symposium Series, No. 374, American Chemical Society, Washington, D.C., pp. 176–190.

Freese, E., 1959. The specific mutagenic effect of base analogues on phage T_4, *J. Mol. Biol.* 1:87–105.

Fukui, T., and De Clercq, E., 1982. Inhibition of murine leukemia virus reverse transcriptase by 2-halogenated polyadenylic acids, *Biochem. J.* 203:755–760.

Fukui, T., De Clercq, E., Kakiuchi, N., and Ikehara, M., 1982a. Template activity of poly-(2′-fluoro-2′-deoxyinosinic acid) for murine leukemia virus reverse transcriptase, *Cancer Lett.* 16:129–135.

Fukui, T., Kakiuchi, N., and Ikehara, M., 1982b. Protein synthesis using poly(2′-halogeno-2′-deoxyadenylic acids) as messenger, *Biochim. Biophys. Acta* 697:174–177.

Garrett, C., Wataya, Y., and Santi, D. V., 1979. Thymidylate synthetase. Catalysis of dehalogenation of 5-bromo- and 5-iodo-2′-deoxyuridylate, *Biochemistry* 18:2798–2804.

Glazer, G. I., and Hartman, K. D., 1983. *In vivo* translation of messenger RNA following exposure of human colon carcinoma cells in culture to 5-fluorouracil and 5-fluorouridine, *Mol. Pharmacol.* 23:540–546.

Glazer, R. I., and Lloyd, L. S., 1982. Association of cell lethality with incorporation of 5-fluorouracil and 5-fluorouridine into nuclear RNA in human colon carcinoma cells in culture, *Mol. Pharmacol.* 21:468–473.

Glazer, R. I., and Peale, A. L., 1979. The effect of 5-fluorouracil on the synthesis of nuclear RNA in L1210 cells in vitro, *Mol. Pharmacol.* 16:270–277.

Goulian, M., Bliele, B., and Tseng, Y. B., 1980. Methotrexate-induced misincorporation of uracil into DNA, *Proc. Natl. Acad. Sci. USA* 77:1956–1960.

Goz, B., 1978. The effects of incorporation of 5-halogenated deoxyuridines into the DNA of eukaryotic cells, *Pharmacol. Rev.* 29:249–271.

Gratzner, H. G., 1982. Monoclonal antibody to 5-bromo- and 5-iodouridine: A new reagent for detection of DNA replication, *Science* 218:474–475.

Gray, J. W., and Mayall, B. H. (eds.), 1985. Monoclonal antibodies against bromo-deoxyuridine, *Cytometry* 6:499–673.

Griengl, H., Wanek, E., Schwarz, W., Streicher, W., Rosenwirth, B., and De Clercq, E., 1987. 2′-Fluorinated arabinonucleosides of 5-(2-haloalkyl)uracil: Synthesis and antiviral activity, *J. Med. Chem.* 30:1199–1204.

Harnden, M. R. (ed.), 1985. *Approaches to Antiviral Agents*, VCH Publishers, New York.

Harris, G., Ator, M., and Stubbe, J., 1984. Mechanism of inactivation of *Escherichia coli* and *Lactobacillus leichmannii* ribonucleotide reductases by 2′-chloro-2′-deoxynucleotides: Evidence for generation of 2-methylene-3-(2H)-furanone, *Biochemistry* 23:5214–5225.

Heidelberger, C., 1975a. Fluorinated pyrimidines and their nucleosides, in *Antineoplastic and Immunosuppressive Agents, Part II, Handbood of Experimental Pharmacology*, Vol. XXXVIII/2 (A. C. Sartorelli and D. G. Johns, eds.), Springer-Verlag, New York, pp. 193–231.

Heidelberger, C., 1975b. On the molecular mechanisms of the antiviral activity of trifluoro-thymidine, *Ann. N.Y. Acad. Sci.* 255:317:325.

Heidelberger, C., and King, D. H., 1979. Trifluorothymidine, *Pharmacol. Ther.* 6:427–442.

Heidelberger, C., Chaudhuri, N. K., Danneberg, P., Mooren, D., Griesbach, L., Duschinsky, R., Schnitzer, R. J., Pleven, E., and Scheiner, J., 1957. Fluorinated pyrimidines, a new class of tumor-inhibitory compounds, *Nature* 179:663–666.

Heidelberger, C., Parsons, D., and Remy, D. C., 1962. Synthesis of trifluoromethyluracil and 5-trifluoromethyluracil-2′-deoxyuridine, *J. Am. Chem. Soc.* 84:3597–3598.

Heidelberger, C., Danenberg, P., and Moran, R. G., 1983. Fluorinated pyrimidines and their nucleosides, *Adv. Enzymol.* 54:58–119.

Houghton, J. A., Weiss, K. D., Williams, L. G., Torrance, P. M., and Houghton, P. J., 1986. Relationship between 5-fluoro-2'-deoxyuridylate, 2'-deoxyuridylate, and thymidylate synthase activity subsequent to 5-fluorouracil administration, in xenografts of human colon adenocarcinomas, *Biochem. Pharmacol.* 35:1351–1358.

Huang, P., and Plunkett, W., 1987. Phosphorolytic cleavage of 2-fluoroadenine from 9-β-D-arabinofuranosyl-2-fluoroadenine by *Escherichia coli*. A pathway for 2-fluoro-ATP production. *Biochem. Pharmacol.* 36:2945–2950.

Hutton, J. J., Von Hoff, D. D., Kuhn, J., Philips, J., Hersh, M., and Clark, G., 1984. Phase I clinical investigation of 9-β-D-arabinofuranosyl-2-fluoroadenine-5'-phosphate (NSC 312887), a new purine antimetabolite, *Cancer Res.* 44:4183–4186.

Ingraham, H. A., Tseng, B. Y., and Goulian, M., 1982. Nucleotide levels and incorporation of 5-fluorouracil and uracil into DNA cells treated with 5-fluorodeoxyuridine, *Mol. Pharmacol.* 21:211–216.

Jacobson, K. A., 1988. Chemical approaches to the definition of adenosine receptors, in *Adenosine Receptors* (D. M. F. Cooper and C. Londos, eds.), Alan R. Liss, New York, pp. 1–26.

Kassis, A. I., Sastry, K. S., and Adelstein, S. J., 1987. Kinetics of uptake, retention, and radiotoxicity of [125]IUdR in mammalian cells: Implications of localized energy deposition by Auger processes, *Radiat. Res.* 109:78–89.

Kato, H., 1974. Spontaneous sister chromatid exchanges detected by a BUdR-labeling method, *Nature* 251:70–72.

Kaufman, E. R., 1984. Replication of DNA containing 5-bromouracil can be mutagenic in Syrian hamster cells, *Mol. Cell. Biol.* 4:2449–2454.

Kaufman, E. R., 1987. Uncoupling of the induction of mutations and sister-chromatid exchanges by the replication of 5-bromouracil-substituted DNA, *Mutat. Res.* 176:133–144.

Kaufman, E. R., and Davidson, R. L., 1978. Bromodeoxyuridine mutagenesis in mammalian cells: Mutagenesis is independent of the amount of bromouracil in DNA, *Proc. Natl. Acad. Sci. USA* 75:4982–4986.

Kaysen, J., Spriggs, D., and Kufe, D., 1986. Incorporation of 5-fluorodeoxycytidine and metabolites into nucleic acids of human MCF-7 breast carcinoma cells, *Cancer Res.* 46:4534–4538.

Kessel, D., Hall, T. C., and Wodinsky, I., 1966. Nucleotide formation as a determinant of 5-fluorouracil response in mouse leukemia, *Science* 154:911–913.

Keyomarsi, K., and Moran, R. G., 1986. Folinic acid augmentation of the effects of fluoropyrimidines on murine and human leukemic cells, *Cancer Res.* 46:5229–5235.

Kinsella, T. J., Mitchell, J. B., Russo, A., Morstyn, G., and Glatstein, E., 1984. The use of halogenated thymidine analogues as clinical radiosensitizers: Rationale, current status, and future prospects: Non-hypoxic cell sensitizers, *Int. J. Radiat. Oncol. Biol. Phys.* 10:1399–1406.

Kremer, A. B., Mikita, T., and Beardsley, G. P., 1987. Chemical consequences of incorporation of 5-fluorouracil into DNA as studied by NMR, *Biochemistry* 26:391–397.

Kufe, D. W., Major, P. P., Egan, E. M., and Loh, E., 1981. 5-Fluoro-2'-deoxyuridine incorporation in L1210 DNA, *J. Biol. Chem.* 256:8885–8888.

Langenbach, R. J., Dannenberg, P. V., and Heidelberger, C., 1972. Thymidylate synthetase: Mechanism of inhibition by 5-fluoro-2'-deoxyuridylate, *Biochem. Biophys. Res. Commun.* 48:1565–1571.

Latt, S. A., 1973. Microfluorometric detection of deoxyribonucleic acid replication in human metaphase chromosomes, *Proc. Natl. Acad. Sci. USA* 70:3395–3399.

Latt, S. A., 1974. Sister chromatid exchanges, indices of human chromosome damage and repair: Detection by fluorescence and induction by mitomycin C, *Proc. Natl. Acad. Sci. USA* 71:3162–3166.

Lin, T.-S., Guo, J.-I., Schinazi, R. F., Chu, C. K., Xiang, J.-N., and Prusoff, W. H., 1988. Synthesis and antiviral activity of various 3′-azido analogues of pyrimidine deoxyribonucleosides against human immunodeficiency virus (HIV-1, HTLV-III/LAV), *J. Med. Chem.* 31:336–340.

Lonn, U., and Lonn, S., 1986. DNA lesions in human neoplastic cells and cytotoxicity of 5-fluoropyrimidines, *Cancer Res.* 46:3866–3870.

Mar, E.-C., Patel, P. C., Cheng, Y.-C., Fox, J. J., Watanabe, K. A., and Huang, E.-S., 1984. Effects of certain nucleoside analogues on human cytomegalovirus replication *in vitro*, *J. Gen. Virol* 65:47–53.

Mar, E.-C., Chiou, J.-F., Cheng, Y.-C., and Huang, E.-S., 1985. Human cytomegalovirus-induced DNA polymerase and its interaction with the triphosphates of 1-(2′-deoxy-2′-fluoro-β-D-arabinofuranosyl)-5-methyluracil, -5-iodocytosine, and -5-methylcytosine, *J. Virol.* 56:846–851.

Marquez, V. E., Tseng, C. K.-H., Kelley, J. A., Mitsuya, H., Broder, S., Roth, J. S., and Driscoll, J. S., 1987. 2′,3′-Dideoxy-2′-fluoro-ara-A. An acid stable purine nucleoside active against human immunodeficiency virus (HIV), *Biochem. Pharmacol.* 36:2719–2722.

Mekras, J. A., Boothman, D. A., Perez, L. M., and Greer, S., 1984. Use of 5-fluorodeoxycytidine and tetrahydrouridine to exploit high levels of deoxycytidylate deaminase in tumors to achieve DNA- and target-directed therapies, *Cancer Res.* 44:2551–2560.

Mekras, J. A., Boothman, D. A., and Greer, S. B., 1985. Use of 5-trifluoromethyldeoxycytidine and tetrahydrouridine to circumvent catabolism and exploit high levels of cytidine deaminase in tumors to achieve DNA- and target-directed therapies, *Cancer Res.* 45:5270–5280.

Mini, E., Moroson, B. A., and Bertino, J. R., 1987. Cytotoxicity of floxuridine and 5-fluoro-uracil in human T-lymphoblast leukemia cells: Enhancement by leucovorin, *Cancer Treat. Rep.* 71:381–389.

Mitchell, J. B., Russo, A., Kinsella, T. J., and Glatstein, E., 1986. The use of nonhypoxic cell sensitizers in radiobiology and radiotherapy, *Int. J. Radiat. Oncol. Biol. Phys.* 12:1513–1518.

Montgomery, J. A., 1982. Has the well gone dry? The first Cain memorial award lecture, *Cancer Res.* 42:3911–3917.

Montgomery, J. A., and Hewson, K., 1957. Synthesis of potential anticancer agents X. 2-Fluoroadenosine, *J. Am. Chem. Soc.* 79:4559.

Myers, C. E., 1981. The pharmacology of the fluoropyrimidines, *Pharmacol. Rev.* 33:1–15.

Nakayama, C., Wataya, Y., and Santi, D. V., 1981. Interaction of 1-(5-phospho-β-D-arabinofuranosyl)-5-substituted-uracils with thymidylate synthetase: Mechanism-based inhibition by 1-(5-phospho-β-D-arabinosyl)-5-fluorouracil, *J. Med. Chem.* 24:1161–1165.

Newman, E. M., and Santi, D. V., 1982. Metabolism and mechanism of action of 5-fluoro-deoxycytidine, *Proc. Natl. Acad. Sci. USA* 79:6419–6423.

Osterman, D. G., DePillis, G. D., Wu, J. C., Matsuda, A., and Santi, D. V., 1988. 5-Fluorocytosine in DNA is a mechanism-based inhibitor of *Hha*I methylase, *Biochemistry* 27:5204–5210.

Paison, S. G., Hartigan, J. A., Kumar, V., and Biswas, D. K., 1987. DNA sequence responsible for the amplification of adjacent genes, *DNA* 6:419–428.

Parker, W. B., Kennedy, K. A., and Klubes, P., 1987. Dissociation of 5-fluorouracil-induced

DNA fragmentation from either its incorporation into DNA or its cytotoxicity in murine T-lymphoma (S-49) cells, *Cancer Res.* 47:979–982.

Perez, L. M., and Greer, S., 1986. Sensitization to X ray by 5-chloro-2'-deoxycytidine co-administered with tetrahydrouridine in several mammalian cell lines and studies of 2'-chloro derivatives, *Int. J. Radiat. Oncol. Biol. Phys.* 12:1523–1527.

Prusoff, W. H., 1959. Synthesis and biological activity of iododeoxyuridine, an analogue of thymidine, *Biochim. Biophys. Acta* 32:295–296.

Prusoff, W. H., and Goz, B., 1975. Halogenated pyrimidine deoxyribonucleosides, in *Antineoplastic and Immunosuppressive Agents, Part II, Handbook of Experimental Pharmacology*, Vol. XXXVIII/2 (A. C. Sartorelli and D. G. Johns, eds.), Springer-Verlag, New York, pp. 272–347.

Prusoff, W. H., Chen, M. S., Fischer, P. H., Lin, T.-S., Shiau, G. T., Schinaze, R. F., and Walker, J., 1979. Antiviral iodinated pyrimidine deoxyribonucleosides: 5-Iodo-2'-deoxyuridine; 5-iodo-2'-deoxycytidine; 5-iodo-5'-amino-2',5'-dideoxyuridine, *Pharmacol. Ther.* 7:1–34.

Prusoff, W. H., Mancini, W. R., Lin, T.-S., Lee, J.-J., Siegel, S. A., and Otto, M. J., 1984. Physical and biological consequences of incorporation of antiviral agents into virus DNA, *Antiviral Res.* 4:303–315.

Prusoff, W. H., Zucker, M., Mancini, W. R., Otto, M. J., Lin, T.-S., and Lee, J.-J., 1985. Basic biochemical and pharmacological aspects of antiviral agents, in *Proceedings of the 1st International TNO Conference on Antiviral Research* (A. Billau, E. De Clercq, and H. Schellekens, eds.), *Antiviral Research, Suppl. 1*, Elsevier Science Publishers, Rotterdam, pp. 1–10.

Rosenwirth, B., Griengl, H., Wanek, E., and De Clercq, E., 1985. 5-(2-Chloroethynyl)-2'-deoxyuridine: A potent and selective inhibitor of herpes viruses, in *Proceedings of the 1st International TNO Conference on Antiviral Research* (A. Billau, E. De Clercq, and H. Schellekens, eds.), *Antiviral Research, Suppl. 1*, Elsevier Science Publishers, Rotterdam, pp. 21–28.

Rossler, K., Meyers, G.-J., and Stocklin, G., 1977. Labeling and animal distribution studies of 5-astatouracil and 5-astatodeoxyuridine (211 At). *J. Labeled Compd. Radiopharm.* 13:271.

Russell, K. J., Rice, G. C., and Brown, J. M., 1986. *In vitro* and *in vivo* radiation sensitization by the halogenated pyrimidine 5-chloro-2'-deoxycytidine, *Cancer Res.* 46:2883–2887.

Ruth, J. L., and Cheng, Y.-C., 1981. Nucleoside analogues with clinical potential in antivirus chemotherapy: The effect of several thymidine and 2'-deoxycytidine analogue 5'-triphosphates on purified human (α, β) and herpes simplex virus (types 1, 2) DNA polymerases, *Mol. Pharmacol.* 20:415–422.

Rutter, W. J., Pictet, R. L., and Morris, P. W., 1973. Toward molecular mechanisms of developmental processes, *Annu. Rev. Biochem.* 42:601–645.

Sanger, F., Nicklen, S., and Coulson, A. R., 1977. DNA sequencing with chain-terminating inhibitors, *Proc. Natl. Acad. Sci. USA* 74:5463–5467.

Santi, D. V., and McHenry, C. S., 1972. 5-Fluoro-2'-deoxyuridylate: Covalent complex with thymidylate synthetase, *Proc. Natl. Acad. Sci. USA* 69:1855–1857.

Santi, D. V., Pogolotti, A. L., James, T. L., Wataya, Y., Ivanetich, K. M., and Lam, S. S. M., 1976. Thymidylate synthase: Interaction with 5-fluoro and 5-trifluoromethyl-2'-deoxyuridylic acid, in *Biochemistry Involving Carbon–Fluorine Bonds* (R. Filler, ed.), ACS Symposium Series, No. 28, American Chemical Society, Washington, D.D., pp. 57–76.

Santi, D. V., Pogolotti, A. L., Jr., Newman, E. M., and Wataya, Y., 1982. Aspects of the biochemistry and biochemical pharmacology of 5-fluorinated pyrimidines, in *Biomedicinal Aspects of Fluorine Chemistry* (R. Filler and J. Kobayashi, eds.), Kodansha Ltd., Tokyo; Elsevier Biomedical, Amsterdam, pp. 123–142.

Sapse, A. M., and Snyder, A. G., 1985. *Ab Initio* studies of the antiviral drug 1-(2-fluoro-2-deoxy-β-D-arabinofuranosyl)thymine, *Cancer Invest.* 3:115–121.

Savarese, T. M., Cannistra, A. J., Parks, R. E., Jr., Secrist, J. A., III, Shortnacy, A. T., and Montgomery, J. A., 1987. 5′-Deoxy-5′-methylthioadenosine phosphorylase—IV. Biological activity of 2-fluoroadenine-substituted 5′-deoxy-5′-methylthioadenosine analogues, *Biochem. Pharmacol.* 36:1881–1893.

Schildkraut, I., Cooper, G. M., and Greer, S., 1975. Selective inhibition of the replication of herpes simplex virus by 5-halogenated analogues of deoxycytidine, *Mol. Pharmacol.* 11:153–158.

Schroeder, C., and Jantschak, J., 1980. Inhibitor studies of phage T4 wild-type and mutant DNA polymerase. IV. The substrate analogue 3′-fluorothymidine 5′-triphosphate, *Z. Allg. Mikrobiol.* 20:657–662.

Schuetz, J. D., Collins, J. M., Wallace, H. J., and Diasio, R. B., 1986. Alteration of the secondary structure of newly synthesized DNA from murine bone marrow cells by 5-fluorouracil, *Cancer Res.* 46:119–123.

Shealy, Y. F., O'Dell, C. A., Shannon, W. M., and Arnett, G., 1983. Carbocyclic analogues of 5-substituted uracil nucleosides: Synthesis and antiviral properties, *J. Med. Chem.* 26:156–161.

Shigata, S., Yokota, T., and De Clercq, E., 1985. Therapy of varicella-zoster virus infection—mechanism of action of (*E*)-5-(2-bromovinyl)-2′-deoxyuridine, in *Proceedings of the 1st International TNO Conference on Antiviral Research* (A. Billau, E. De Clercq, and H. Schellekens, eds.), *Antiviral Research, Suppl. 1*, Elsevier Science Publishers, Rotterdam, pp. 35–44.

Shigeura, H. T., Boxer, G. E., Sampson, S. D., and Meloni, M. L., 1965. Metabolism of 2-fluoroadenine by Ehrlich ascites cells, *Arch. Biochem. Biophys.* 111:713–719.

Sirotnak, F. M., Chello, P. L., Dorick, D. M., and Montgomery, J. A., 1983. Specificity of systems mediating transport of adenosine, 9-β-D-arabinofuranosyl-2-fluoroadenine, and other purine nucleoside analogues in L1210 cells, *Cancer Res.* 43:104–109.

Smith, C. C., Aurelian, L., Reddy, M. P., Miller, P. S., and Ts'o, P. O. P., 1986. Antiviral effect of an oligo(nucleoside methylphosphonate) complementary to the splice junction of herpes simplex virus type 1 immediate early pre-mRNAs 4 and 5, *Proc. Natl. Acad. Sci. USA* 83:2787–2791.

Sowers, L. C., Eritja, R., Kaplan, B. E., Goodman, M. F., and Fazakerley, G. V., 1987. Structural and dynamic properties of a fluorouracil–adenine base pair in DNA studied by proton NMR, *J. Biol. Chem.* 262:15436–15442.

Spears, C. P., Shani, J., Shahinian, A. H., Wolf, W., Heidelberger, C., and Danenberg, P. V., 1985. Assay and time course of 5-fluorouracil incorporation into RNA of L1210/0 ascites cells *in vivo*, *Mol. Pharmacol.* 27:302–307.

Spector, S. A., Tyndall, M., and Kelley, E., 1983. Inhibition of human cytomegalovirus by trifluorothymidine, *Antimicrob. Agents Chemother.* 23:113–118.

Spriggs, D., Robbins, R., Mitchell, T., and Kufe, D., 1986. Incorporation of 9-β-D-arabino-furanosyl-2-fluoroadenine into HL-60 cellular RNA and DNA, *Biochem. Pharmacol.* 35:247–252.

Stubbe, J., and Kozarich, J. W., 1980. Fluoride, pyrophosphate, and base release from 2′-deoxy-2′-fluoronucleoside 5′-diphosphates by ribonucleoside-diphosphate reductase, *J. Biol. Chem.* 255:5511–5513.

Su, T.-L., Watanabe, K. A., Schinazi, R. F., and Fox, J. J., 1986. Nucleosides. 136. Synthesis and antiviral effects of several 1-(2-deoxy-2-fluoro-β-D-arabinofuranosyl)-5-alkyluracils. Some structure–activity relationships, *J. Med. Chem.* 29:151–154.

Suzuki, S., Saneyoshi, M., Nakayama, C., Yukihirl, N., and Yoshida, S., 1985. Mechanism of selective inhibition of human cytomegalovirus replication by 1-β-D-arabinofuranosyl-5-fluorouracil, *Antimicrob. Agents Chemother.* 28:326–330.

Suzuki, S., Misra, H. K., Wiebe, L. I., Knaus, E. E., and Tyrrell, L. J., 1987. A proposed mechanism for the selective inhibition of human cytomegalovirus replication by 1-(2'-deoxy-2'-fluoro-β-D-arabinofuranosyl)-5-fluorouracil, *Mol. Pharmacol.* 31:301–306.

Szybalski, W., 1974. X-ray sensitization by halopyrimidines, *Cancer Chemother. Rep.* 58:539–557.

Tanaka, M., Kimura, K., and Yoshida, S., 1983. Enhancement of the incorporation of 5-fluorodeoxyuridylate into DNA of HL-60 cells by metabolic modulations, *Cancer Res.* 43:5145–5150.

Thelander, L., Larsson, B., Hobbs, J., and Eckstein, F., 1976. Active site of ribonucleoside diphosphate reductase from *Escherichia coli.* Inactivation of the enzyme by 2'-substituted ribonucleoside diphosphates. *J. Biol. Chem.* 251:1398–1405.

Torrence, P. F., 1985. How interferon works, in *Biological Response Modifiers* (P. F. Torrence, ed.), Academic Press, Orlando, Florida, pp. 77–105.

Tseng, W.-C., Derse, D., Cheng, Y.-C., Brockman, R. W., and Bennett, L. L., 1981. *In vitro* biological activity of 9-β-D-arabinofuranosyl-2-fluoroadenine and the biochemical actions of its triphosphate on DNA polymerases and ribonucleotide reductase from HeLa cells, *Mol. Pharmacol.* 21:474–477.

Uesugi, S., Miki, H., Ikehara, M., Iwahashi, H., and Kyogoku, Y., 1979. A linear relationship between electronegativity of 2'-substituents and conformation of adenine nucleosides, *Tetrahedron Lett.* 42:4073–4076.

Uesugi, S., Takatsuka, Y., Ikehara, M., Cheng, D. M., Kan, L. S., and Ts'o, P. O. P., 1981. Synthesis and characterization of the dinucleoside monophosphates containing 2'-fluoro-2'-deoxyadenosine, *Biochemistry* 20:3056–3062.

Visser, D. W., Frisch, D. M., and Huang, B., 1960. Synthesis of 5-chlorodeoxyuridine and a comparative study of 5-halodeoxy uridines in *E. coli, Biochem. Pharmacol.* 5:157–164.

Waehnert, U., and Langen, P., 1979. Incorporation of 3'-deoxy-3'-fluorothymidylate into DNA *in vitro*, 19th Hungarian Annual Meeting of Biochemistry, pp. 27–28.

Watanabe, K., Reichman, U., Hirota, K., and Fox, J. J., 1979. Nucleosides. 110. Synthesis and antiherpes virus activity of some 2'-fluoro-2'-deoxyarabinofuranosylpyrimidine nucleosides, *J. Med. Chem.* 22:21–24.

Watanabe, K. A., Reichman, U., Chu, C. K., Hollenberg, D. H., and Fox, J. J., 1980. Nucleosides. 116. 1-(β-Xylofuranosyl)-5-fluorocytosines with a leaving group on the 3'-position. Potential double-barreled masked precursors of anticancer nucleosides, *J. Med. Chem.* 23:1088–1094.

Watanabe, K., Su, T.-L., Klein, R. S., Chu, C. K., Matsude, A., Chun, M. W., Lopez, C., and Fox, J. J., 1983. Nucleosides. 123. Synthesis of antiviral nucleosides: 5-Substituted 1-(2-deoxy-2-halogeno-β-D-arabinosyl)cytosines and -uracils. Some structure–activity relationships, *J. Med. Chem.* 26:152–156.

Weckbecker, G., and Keppler, D. O., 1984. Substrate properties of 5-fluorouridine diphospho sugars detected in hepatoma cells, *Biochem. Pharmacol.* 33:2291–2298.

Wilkinson, D. S., and Crumley, J., 1977. Metabolism of 5-fluorouracil in sensitized and resistant Novikoff hepatoma cells, *J. Biol. Chem.* 252:1051–1056.

Wilkinson, D. S., Tlsty, T. D., and Hanas, R. J., 1975. The inhibition of ribosomal RNA synthesis and maturation in Novikoff hepatoma cells by 5-fluorouridine, *Cancer Res.* 35:3014–3020.

Will, C. L., and Dolnick, B. J., 1986. 5-Fluorouracil augmentation of dihydrofolate reductase gene transcripts containing intervening sequences in methotrexate resistant KB cells, *Mol. Pharmacol.* 29:643–648.

Wingard, J. R., Stuart, R. K., Saral, R., and Burns, W. H., 1981. Activity of trifluorothymidine against cytomegalovirus, *Antimicrob. Agents Chemother.* 20:286–290.

Withers, S. G., and Madsen, N. B., 1980. Nucleotide activation of phosphorylase *b* occurs only when the nucleotide phosphate is in a dianionic form, *Biochem. Biophys. Res. Commun.* 97:513–519.

Wohlrab, F., Haertle, T., Trichtinger, T., and Guschlbauer, W., 1978. 2'-Deoxy-2'-fluorouridine-5'-phosphate: An alternate substrate for thymidylate synthetase from *Escherichia coli* K12, *Nucleic Acids Res.* 5:4753–4759.

Wohlrab, F., Jamieson, A. T., Hay, J., Mengel, R., and Guschlbauer, W., 1985. The effect of 2'-fluoro-2'-deoxycytidine on herpes virus growth, *Biochim. Biophys. Acta* 824:233–242.

Young, C. W., Schneider, R., Leyland-Jones, B., Armstrong, D., Tan, C., Lopez, C., Watanabe, K. A., Fox, J. J., and Philips, F. S., 1983. Phase I evaluation of 2'-fluoro-5-iodo-β-D-arabinofuranosylcytosine in immunosuppressed patients with herpes virus infection, *Cancer Treat. Rep.* 43:5006–5009.

Yung, N. C., Burchenal, J. H., Fecher, R., Duschinsky, R., and Fox, J. J., 1961. Nucleosides. XI. Synthesis of 1-β-D-arabinofuranosyl-5-fluorouracil and related nucleosides. *J. Am. Chem. Soc.* 83:4060–4065.

Biochemistry of Halogenated Carbohydrates

6.1 Introduction

Halogenated carbohydrates received early attention as tools for the study of glycolysis, glyconeogenesis, and sugar transport [for reviews, see Barnett (1972), Taylor (1972), and Taylor *et al.* (1976)]. This research has been extended to many areas wherein halogenated carbohydrates have been used effectively as mechanistic probes and as leads for chemotherapeutic agents. For example, the successes realized in the use of halogenated sugars as components of nucleosides, nucleotides, and nucleic acids in the development of antitumor and antiviral agents have been discussed at length in the preceding chapter. Cell-surface glycoproteins are involved in immune response and other cellular recognition phenomena, and strategies to alter glycoprotein biosynthesis based on use of carbohydrate analogues have received much recent attention as a basis for chemotherapeutic drug design. Fluorine-18-labeled sugars are now clinically important positron emission tomography (PET) scanning agents. Halogenated sugars have also been used to study the mechanism of "sweetness." These are among the topics that will be reviewed in this chapter. A recent concise overview of several of these topics related to fluorinated carbohydrates has been given by Kent (1988).

Introduction of a carbon–halogen bond into a fully hydroxylated carbohydrate structure $(C_nH_{2n}O_n)$ requires the formal replacement of a hydroxyl group with halogen in order to produce a stable analogue. This raises the important question of whether the product should be considered functionally as a halogenated analogue of a deoxysugar (halogen as a hydrogen replacement) or as an analogue of the parent sugar (halogen as a hydroxyl replacement). With respect to fluorinated carbohydrates, the physicochemical similarities between the $C-F$ and $C-OH$ bond in several respects are closer than those between the $C-F$ and the $C-H$ bond (Barnett, 1972). Several examples confirm that the $C-F$ replacement in many instances mimics a C−OH, an observation that has proven quite useful in mechanistic studies. Reviews covering the syntheses of fluorinated carbohydrates include those prepared by Card (1985), Penglis (1981), and Kent (1972). Several topics were covered in a recent ACS symposium (Taylor, 1988).

6.2 Transport and Metabolism of Halogenated Carbohydrates

In this section, topics will be considered by proceeding from the more complex polysaccharides to simpler monosaccharides and from hexoses by natural progression to trioses. Thus, the processing of glycogen to glucose (or uptake of glucose) and subsequent metabolism through available pathways to triose precursors of the Krebs cycle is an example of the biochemical processing of carbohydrates. (The incorporation of monosaccharides into macromolecules, for example, during the biosynthesis of glycoproteins and glyconeogenesis, will be discussed separately in Section 6.3.) For reference purposes, a schematic summary of carbohydrate interconversions is given in Fig. 6-1.

6.2.1 Fluorodeoxy Sugars as Probes for the Specificity of the Glycogen Phosphorylase—Glucose Complex

Glycogen is a polysaccharide made up of glucose residues linked by α-1,4-glycosidic linkages branched by α-1,6 linkages (Fig. 6-2). As glycogen is the principal storage depot of glucose in mammalian systems, a precise balance of biosynthesis and hydrolysis is required to maintain proper blood glucose levels. As shown in Fig. 6-2, glycogen phosphorylase cleaves successive 1,4-glycoside bonds from the nonreducing end of the polysaccharide, releasing glucose-1-phosphate which, among other fates, enters the glycolytic pathway after conversion to glucose-6-phosphate by phosphoglucomutase. Phosphorylase is an allosteric enzyme, the inactive form of which is stabilized by glucose. Thus, as part of the control of blood glucose levels, glucose acts to retard the further hydrolysis of glycogen.

The replacement of hydroxyl groups with fluorine has been used by several investigators to map critical hydrogen bonding interactions between carbohydrates and proteins in diverse biological systems. The comparable size of fluorine and the hydroxyl group is a major advantage in this strategy, since this precludes the binding of water in the empty space that would be present in the corresponding deoxy sugar. A recent example of the use of fluorodeoxy sugars as probes for hydrogen bonding, as well as

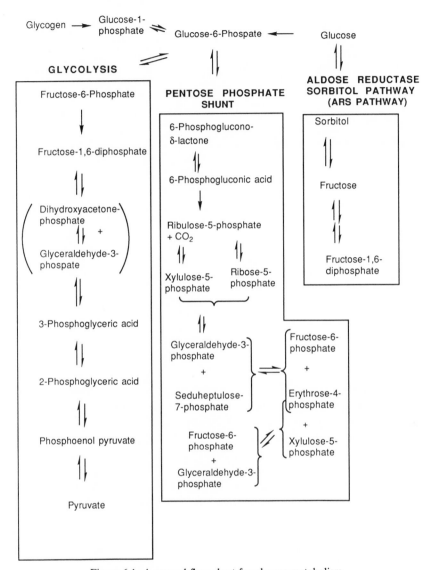

Figure 6-1. A general flow chart for glucose metabolism.

an effective analysis of the principles involved, is provided by a study of hydrogen bonding in the glycogen phosphorylase–glucose complex (Withers *et al.*, 1988; Street *et al.*, 1986). In this investigation, glucose analogues were studied wherein the hydroxyl groups were successively replaced by hydrogen, replaced by fluorine, or inverted. From affinity data,

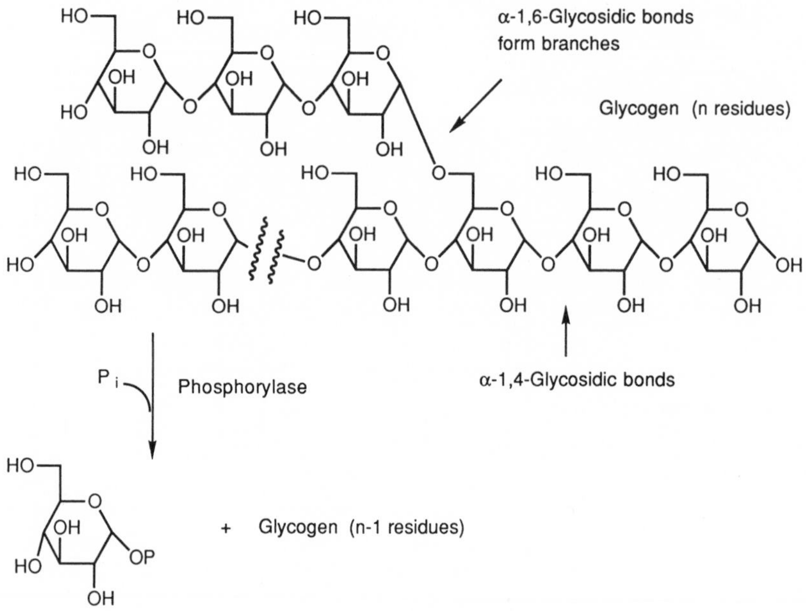

Figure 6-2. Glycogen structure and phosphorylase cleavage.

the importance of each hydroxyl group of glucose as a hydrogen bond donor or acceptor was evaluated. In interpretation of the data, the ability of fluorine to act as a weak hydrogen bond acceptor was assumed, particularly if the $C-F$ bond was positioned suitably in an enzyme–ligand complex. From these data, the predicted order of hydrogen bond strengths and the role of the sugar hydroxyl as a donor (D) or acceptor (A) at each position was found to be $3(D) = 6(D + A) > 4(D + A) > 2(A) \approx 1(A)$. These predictions were in good accord with reported X-ray crystallographic data.

6.2.2 1-Deoxy-1-fluorohexoses as Substrates for Glycosidases

1-Fluorohexoses and other nonglycosidic substrates have been found to be excellent substrates for glycosidases, a group of enzymes that catalyze the hydrolytic cleavage of simple glycosides and oligosaccharides. Because of the stability of the $C-F$ bond to S_N2 displacement, the substrate activity of these analogues has supported the idea that such enzymes cleave glycosidic bonds through formation of an oxocarbonium ion by E1 elimination of the C-1 substituent (Fig. 6-3). Furthermore, the demonstra-

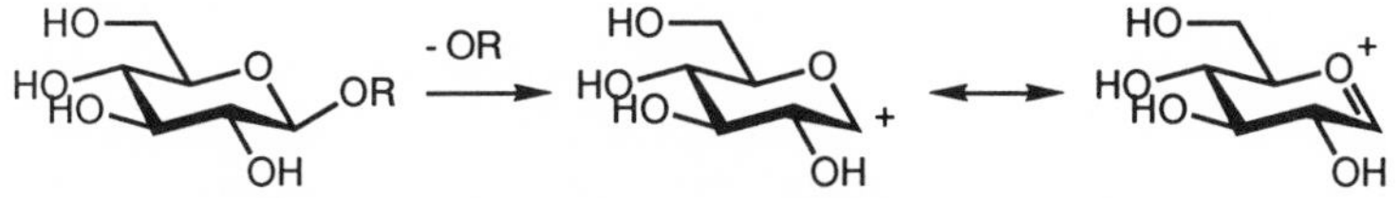

Figure 6-3. E1 elimination in glycosidic bond cleavage.

tion by Hehre and co-workers that both α and β anomers of 1-fluoro-hexoses can serve as substrates for the same enzyme has been pivotal in a recent fundamental reevaluation of the mechanism of these reactions (Chiba *et al.*, 1988, and references therein).

β-Amylase catalyzes the hydrolysis of 1,4-α-D-glycosidic bonds of starch, glycogen, and maltosaccharides, cleaving successive maltosyl residues from the chain ends to produce β-maltose residues (Fig. 6-4). Inversion of configuration at C-1 and a requirement for water as a reactant were long thought to be limiting characteristics of β-amylase-catalyzed reactions. However, catalysis by β-amylase of the synthesis of maltosyl-α-1,4-maltose from β-maltose (the reverse reaction) (Fig. 6-5) suggested that the enzyme might be capable of broader activity (Hehre *et al.*, 1969). β-Amylase subsequently was found to catalyze the hydrolysis of α-maltosyl

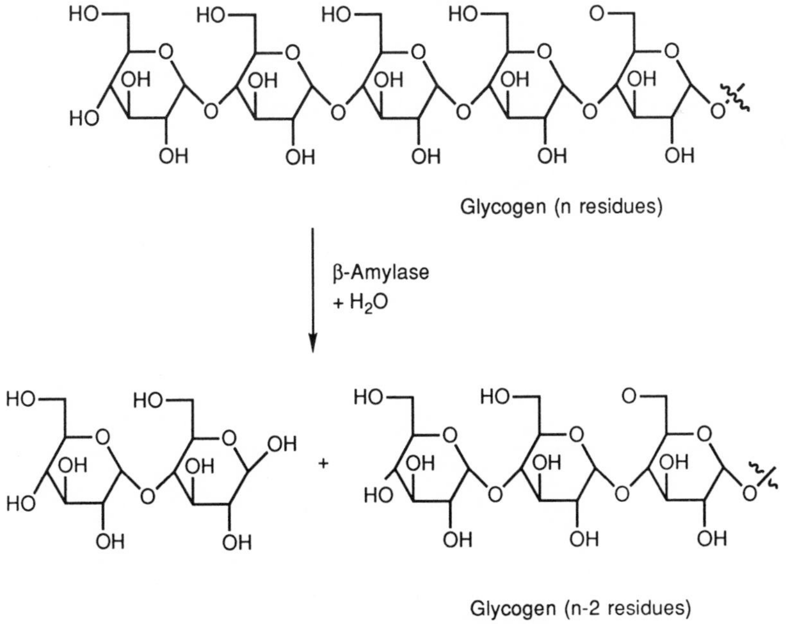

Glycogen (n residues)

β-Amylase
+ H_2O

Glycogen (n-2 residues)

Figure 6-4. Cleavage of glycogen by β-amylase.

fluoride to produce β-maltose (Fig. 6-5) at a rate 30–60 times faster than that for the hydrolysis of the α-1,4-D-glycosidic bonds of malto-triose (Genghof *et al.*, 1978). β-Amylase also catalyzes the hydrolysis of β-maltosyl fluoride, with apparent retention of configuration, to β-maltose, a result difficult to reconcile with previous mechanisms. Kinetic evidence, including the requirement for two moles of substrate and rate enhancement with such compounds as *p*-nitrophenyl-α-D-glucoside, supports a two-step

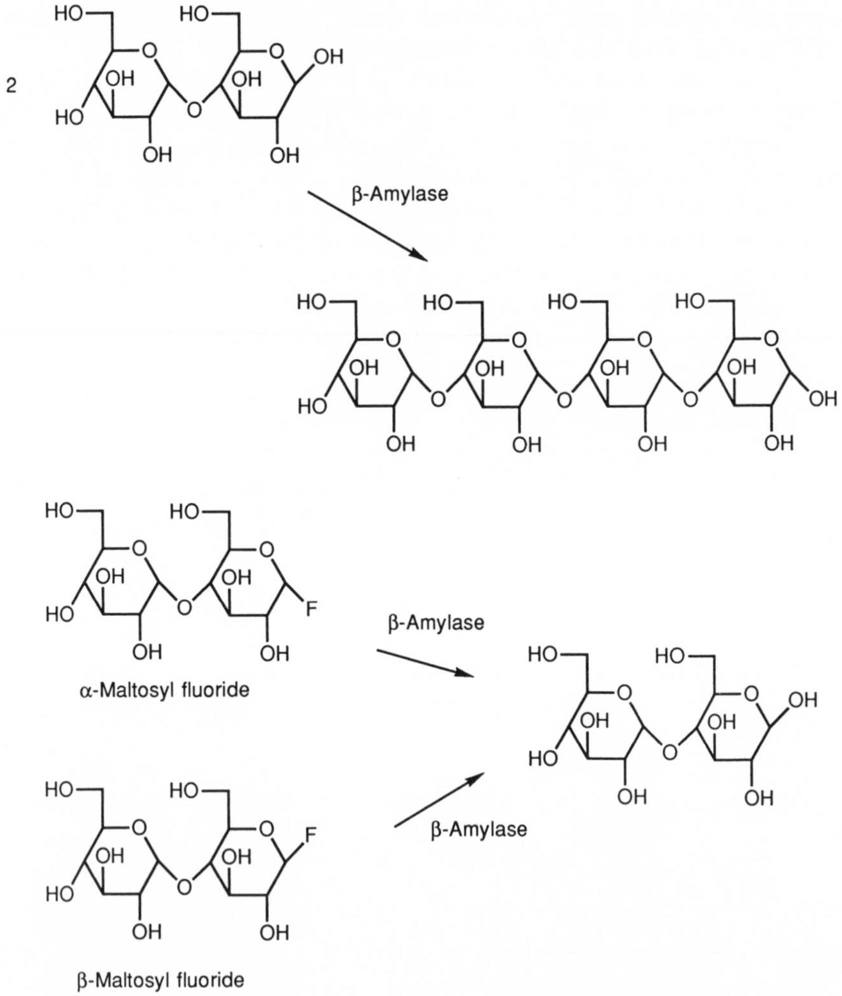

Figure 6-5. β-Amylase-catalyzed synthesis of dimaltose from β-maltose (Hehre *et al.*, 1969) and the formation of β-maltose from either α- or β-maltosyl fluoride (Hehre *et al.*, 1979).

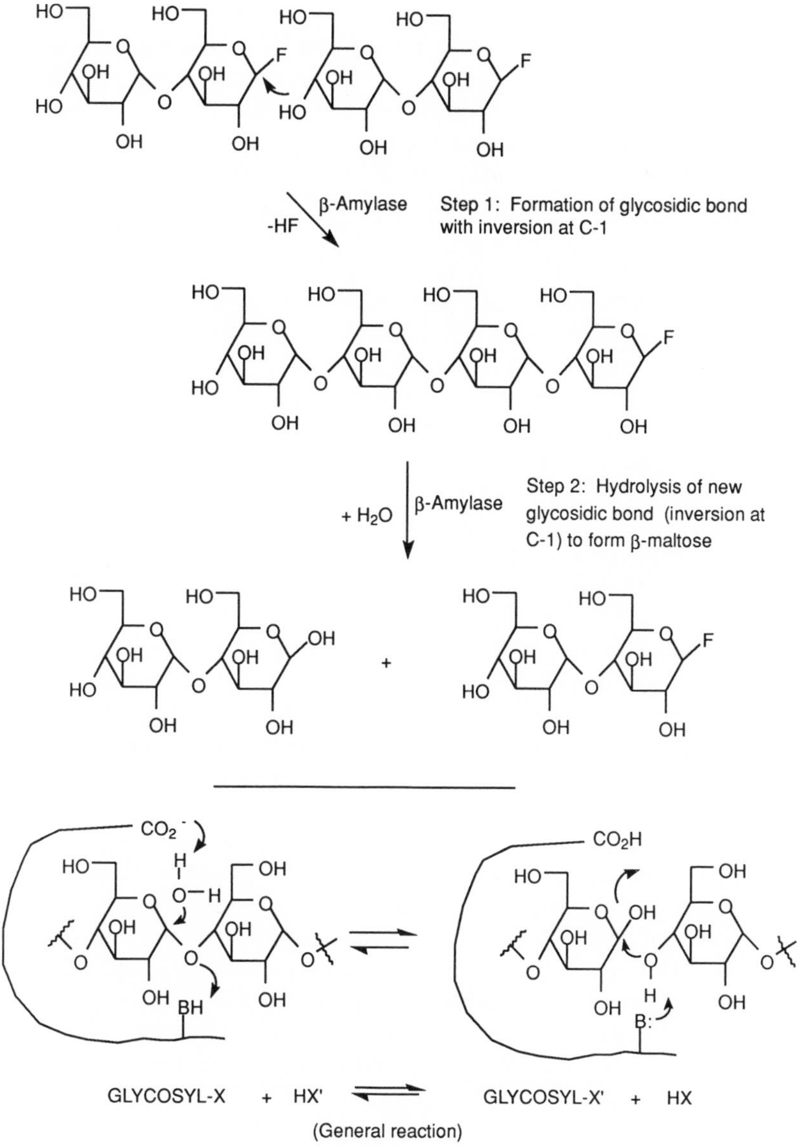

β-Amylase   Step 1: Formation of glycosidic bond
-HF   with inversion at C-1

Step 2: Hydrolysis of new
+ H₂O β-Amylase glycosidic bond (inversion at
C-1) to form β-maltose

GLYCOSYL-X + HX' ⇌ GLYCOSYL-X' + HX

(General reaction)

Figure 6-6. Mechanism of β-amylase-catalyzed formation of β-maltose from β-maltosyl fluoride and the general mechanism for the β-amylase-catalyzed glycosylation reaction (Hehre et al., 1979; see text for details).

process (Fig. 6-6). Thus, the enzyme serves as an initial glycosyltransferase, catalyzing the formation of a tetrasaccharide with loss of fluoride and inversion at C-1 (or formation of a trisaccharide in the presence of such alternate acceptors as p-nitrophenyl-α-D-glucoside). In a second step, the enzyme serves as a hydrolase, catalyzing the normal cleavage of the newly formed 1,4-α-D-glycosidic bond, again with inversion at C-1, to produce β-maltose. From these results, Hehre *et al.* (1979) proposed a general mechanism for the action of glycosidases which emphasizes the catalytic flexibility of the active-site functional groups.

The behavior of several glycosidases extended the generality of these results. The observation that glucoamylase and glucodextranase both hydrolyze α- and β-D-glucosyl fluoride to β-D-glucose parallels the previous observation (Fig. 6-7). In addition, by using methyl α-D-glucopyranoside[^{14}C] as a transferase acceptor, direct evidence was obtained for disaccharide formation when the β anomer was used as substrate (Kitahata *et al.*, 1981). Similar results (Fig. 6-7) were obtained with trehalase, the enzyme which catalyzes the hydrolysis of α,α-trehalase (α-D-glucopyranosyl α-D-glucopyranoside) to α- and β-D-glucose (Hehre *et al.*, 1982). This enzyme also catalyzes the synthesis of α-D-xylopyranoside from β-D-glucosyl fluoride and α-D-xylose (Fig. 6-7) (Kasumi *et al.*, 1986). Providing results which mirrored the previous observations, the β-xylosidase from *Bacillus pumilus* catalyzed rapid hydrolysis of β-D-xylopyranosyl fluoride to produce α-D-xylose. In contrast, the reaction with α-D-xylopyranosyl fluoride was slow, appeared to require two moles of substrate, produced 4-O-β-D-xylopyranosyl-D-xylopyranose as a transient intermediate, and gave α-D-xylose as the final product (Fig. 6-7) (Kasumi *et al.*, 1987).

Based on the divergent paths taken with anomeric pyranosyl fluorides as substrates and on results from other substrate analogues, Hehre and coworkers have concluded that, within the framework of the accepted mechanism, the process of catalysis by glycoside hydrolases and glycosyltransferases (i.e., glycosylases) involves two separately controlled features. A substrate-dependent "plastic" process involves enzyme-catalyzed protonation, the stereochemistry of which may vary from one substrate to the next. A "conserved" aspect concerned with creation of product is independent of substrate (see Fig. 6-6) (Chiba *et al.*, 1988).

6.2.3 2-Deoxy-2-fluoroglucosides as Mechanism-Based Glucosidase Inhibitors

Compounds capable of inhibiting glycosidases have received considerable recent attention as agents for the control of carbohydrate

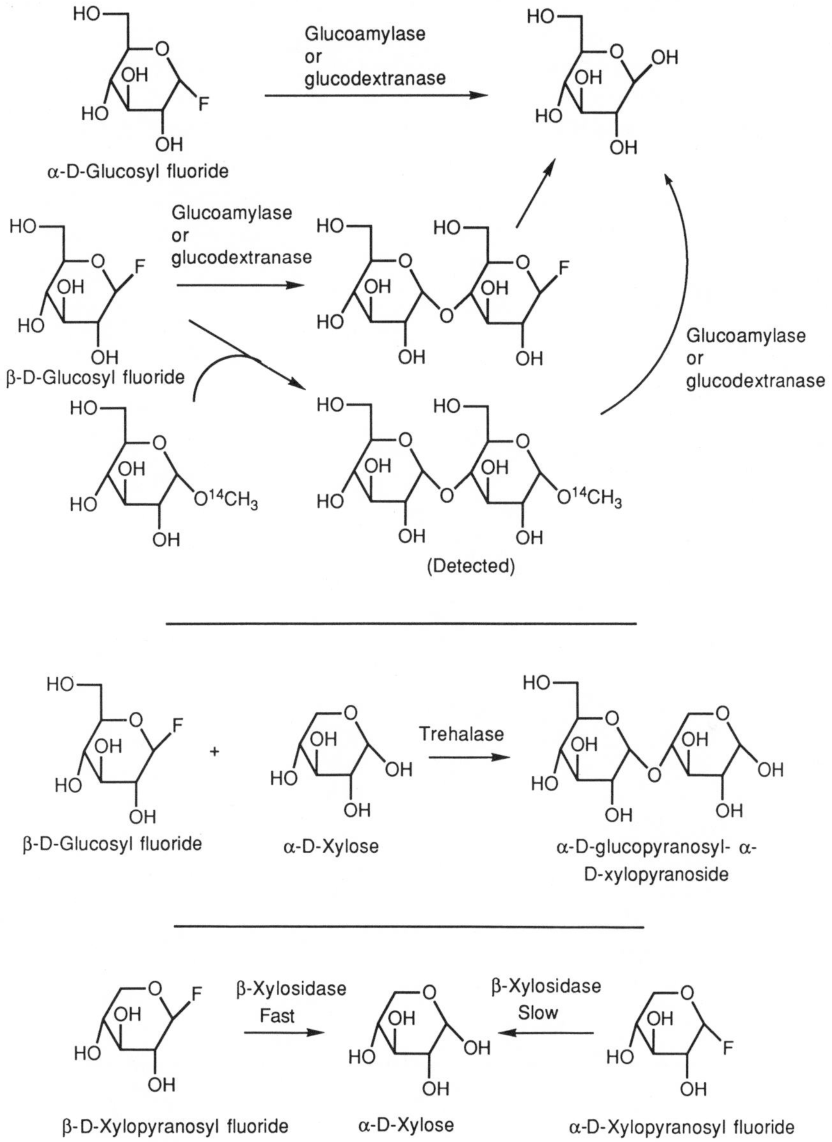

Figure 6-7. Additional glycosidase-catalyzed reactions demonstrating the catalytic versatility of these enzymes (see text for details).

metabolic disorders and as antiviral agents (Tong and Ganem, 1988, and references therein). The oxocarbonium ion-like transition state usually invoked in the glucosidase-catalyzed hydrolyses was the target for the design of mechanism-based inactivators recently reported by Withers *et al.* (1987) (Fig. 6-8). Fluorine present on C-2 slows the rate of C−O bond cleavage by destabilizing this transition state. Incorporation of the highly

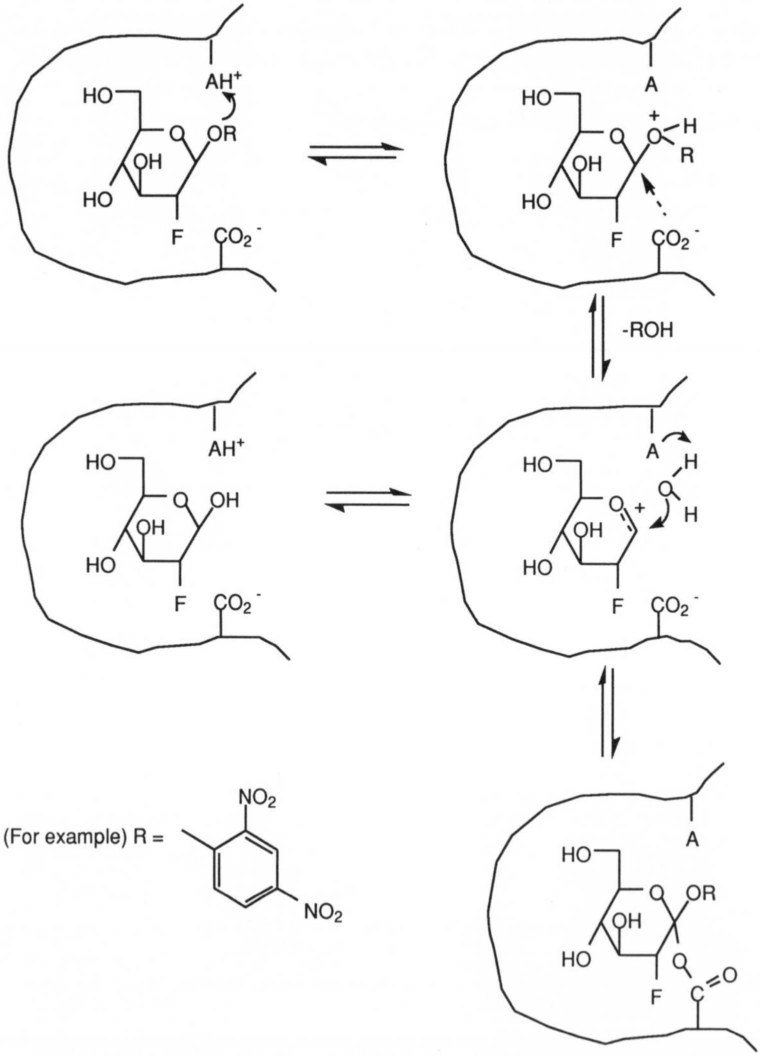

Figure 6-8. A mechanism-based glucosidase inhibitor (Withers *et al.*, 1987).

stabilized 2,4-dinitrophenolate leaving group was expected to increase the rate of ionization sufficiently to allow trapping of this oxocarbonium ion by a suitably situated nucleophile present on the enzyme. Indeed, incubation of 2,4-dinitrophenyl 2-deoxy-2-fluoro-β-D-glucopyranoside (Fig. 6-8) with *Alcaligenes faecalis* β-glucosidase resulted in a rapid time-dependent loss of enzyme activity which followed pseudo-first-order kinetics. The nature of the covalent attachment was expected to permit a slow return of enzyme activity through hydrolysis. An observed substrate dependence of restoration of activity was surprising. This unexpected stability of the 2-fluoroglucosyl–enzyme intermediate in the absence of added glycosides should aid in identification of the active-site nucleophile in ^{19}F-NMR experiments planned by the authors.

6.2.4 Halogenated Hexoses as Probes of the Mechanism and Specificity of Glucose Transport Systems

In mammalian systems, two basic mechanisms are responsible for the transmembrane movement of glucose. Most mammalian cells take up glucose by a facilitated diffusion process catalyzed by membrane proteins, while kidney and intestine use an active Na^+-cotransport mechanism (Wheeler and Whelan, 1988; Wheeler and Hinkle, 1985; Semenza *et al.*, 1984, and references therein). Halogenated hexose analogues have been used effectively in studies of the structural and binding requirements for each of these processes.

In analyzing binding data of fluorinated analogues, the ability of fluorine to serve as a hydrogen bond acceptor usually is assumed (also see Section 6.1). For example, Barnett (1972) systematically replaced hydroxyl groups of glucose by fluorine or hydrogen to explore hydrogen bonding requirements for the active sugar transport of hamster intestine. From the effectiveness of transport of various analogues and the efficiency of inhibition of transport, hydrogen bonds were proposed from the carrier (donor) to C-1, C-3, C-4, and C-6 of the sugar. As an example, while 3-deoxyglucose (3-DG) was poorly transported, 3-deoxy-3-fluoroglucose (3-FDG) was bound as effectively to the carrier as was D-glucose, suggesting that the C-3 fluorine can serve as an effective hydrogen bond acceptor. On the other hand, modification at C-2, including replacement of hydroxyl with fluorine (2-FDG), abolished activity, revealing that interactions at this carbon were more critical and of a different nature than those at the other carbons. From these results, a hypothetical model for the binding of D-glucose to the carrier and a model for the transport process were constructed (Fig. 6-9). For a thorough discussion of the biochemistry of the

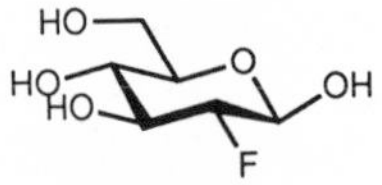

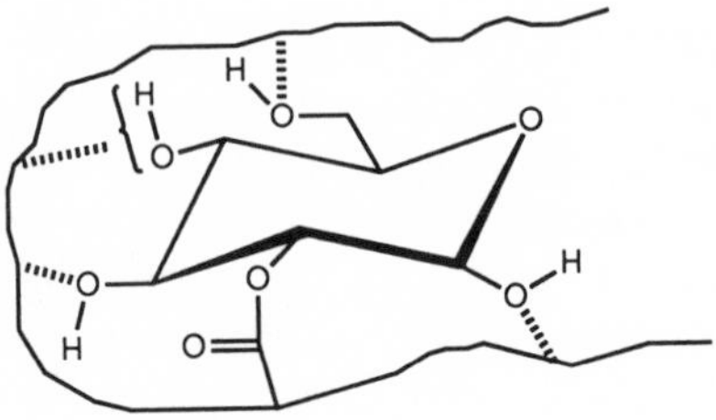

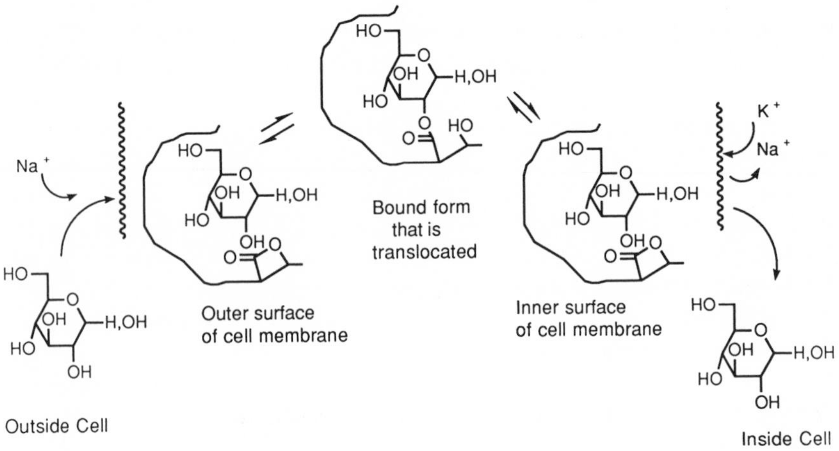

Figure 6-9. A hypothetical model for binding of glucose to the intestine active Na$^+$, glucose cotransport carrier and a proposed mechanism for transport (Barnett, 1972).

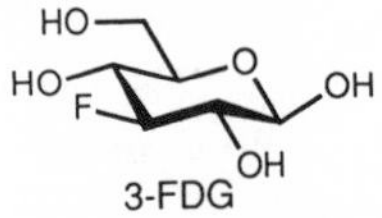

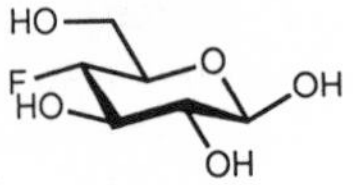

intestinal Na$^+$, D-glucose cotransporter and current models, the review by Semenza *et al.* (1984) is recommended.

The facilitated-diffusion transporter carries glucose alone and is driven by the glucose concentration gradient. The glucose transporter protein of human erythrocytes has been studied extensively and is considered the prototype of a facilitated-diffusion transporter of a single substrate (Wheeler and Whelan, 1988). Using a strategy similar to that employed for investigation of the active-transport system, Barnett *et al.* (1973) measured the effectiveness of specifically substituted glucose analogues of D-glucose as inhibitors of the transport of L-sorbose in order to ascertain the importance of each glucose hydroxyl group. Derivatives having inverted hydroxyl groups, as well as halogenated and deoxy analogues, could bind to the carrier, confirming that no single hydroxyl group is an absolute requirement for binding. However, consideration of the relative inhibition constants of D-glucose for the epimeric, deoxy, and halogenated analogues shows that interactions of the sugar with the carrier protein occur at C-1, C-3, probably C-4, and possibly C-6 (C1 conformation of β-D-glucopyranose). Based on these and other data, Barnett et al. (1975) proposed a model for the carrier protein which also explains the "asymmetric" nature of inhibition, that is, the different affinity for substrate analogues on the inside and the outside of the membrane (Fig. 6-10) [see also the discussion by Stein (1986)].

Riley and Taylor (1973) studied the transport of 3-FDG in human erythrocytes. They obtained results in agreement with those of Barnett and co-workers and determined that, subsequent to transport, there is little metabolism of this analogue. From the structure–affinity relationships observed, they further proposed a mechanism for the mode of action of cytochalasin B, a potent inhibitor of glucose transport ($K_i = 1.2 \times 10^{-7} M$), based on comparable abilities to hydrogen bond to the glucose carrier protein (Taylor et al., 1976). Rees and Holman (1981) carried out a similar investigation of the insulin-sensitive sugar transport system of rat adipocytes and found major similarities in selectivities to those shown for the human erythrocyte carrier. A significant difference was found in that interactions at C-4 appeared to be more critical. Halton et al. (1980) demonstrated another advantage of fluorinated sugar analogues in their study of the uptake of 3-FDG into synaptosomes from rat cerebral cortex. 3-FDG had been shown to be a poor substrate for hexokinase (see Section 6.2.5 below). This fact effectively isolates the transport process from subsequent metabolism, thus avoiding a complicating factor which had been

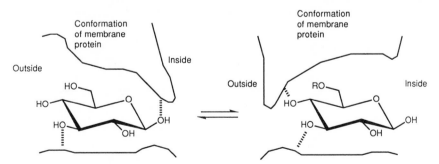

Figure 6-10. A hypothetical model for the membrane protein-catalyzed facilitated glucose transport of the human erythrocyte (Barnett et al., 1975).

present in earlier studies. Kinetic data obtained with this analogue revealed an initial rapid rate of influx which established equilibration of the facilitated transport system; a slower subsequent influx indicated slow removal of 3-FDG from the equilibrium by phosphorylation.

The mechanisms involved in the transport of glucose across the blood–brain barrier (BBB) and within the brain have received increased attention because of the use of [^{18}F]-2-deoxy-2-fluoro-D-glucose ([^{18}F]-2-FDG) as a PET scanning agent (see below, Section 6.2.5.2). A recent paper detailing a study of the distribution of the glucose transporter in the mammalian brain provides a good discussion of these issues (Dick and Harik, 1986).

Halogenated sugars have been used to study transport mechanisms in nonmammalian systems. For example, Card and co-workers have used fluoro- and azidodeoxysucrose analogues to study the topographical surface presented by sucrose to the carrier protein responsible for the transport of sucrose in plants [reviewed by Hitz (1988)]. Studies of this transport had also been complicated by metabolic lability, in this case due to hydrolysis by invertase. Sucrose synthase-catalyzed coupling of 1-deoxy-1-fluorofructose (1; Fig. 6-11) to UDP-glucose had been used to prepare 1′-deoxy-1′-fluorosucrose (2; Fig. 6-11) as an invertase-resistant sucrose analogue for studies in the presence of intracellular invertase (Card and Hitz, 1984). The discovery that this analogue is bound to the carrier protein two times more strongly than sucrose suggested that the 1′-OH in sucrose is intramolecularly hydrogen bonded to the 2′-OH, thus presenting a hydrophobic surface for binding to the protein, and that (1) fluorine at C-1′ presents an even more hydrophobic surface, or (2) fluorine at C-1′ can also accept the hydrogen bond from the 2′-OH. A similar chemoenzymatic approach produced the analogue 3 having the azido group at C-1′. Comparable binding of 3 and 2 showed that the first situation is present. Substitution of the 6′-OH with fluorine similarly resulted in an analogue (4) that binds two times more effectively than sucrose, showing that C-6′ also lies in the hydrophobic cleft of the carrier protein. A similar substitution at C-4′ produced an analogue (5) which was essentially equal to sucrose in binding (Card et al., 1986). Subsequent studies revealed that phenyl α-D-glucopyranoside and phenyl α-D-thioglucopyranoside also could bind effectively to the sucrose carrier, giving more support to the proposal that the fructose moiety interacts with a hydrophobic region of the carrier protein. Binding characteristics of phenyl α-D-thioglycopyranosides in which systematic substitution of the hydroxyls of the glucose was made (e.g., compounds 6a–f; Fig. 6-11) revealed that hydroxyl groups at C-3, C-4, and C-6 are involved in substrate recognition of the carrier protein (Hitz et al., 1986).

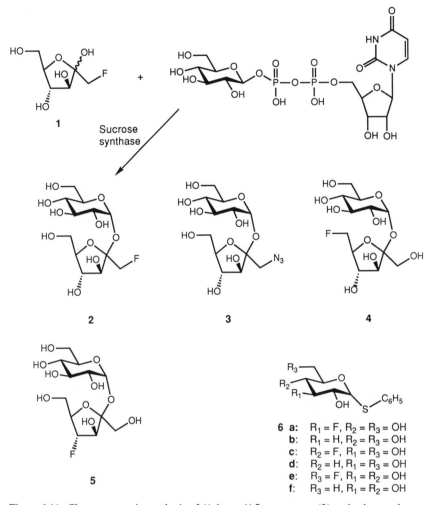

Figure 6-11. Chemoenzymatic synthesis of 1'-deoxy-1'-fluorosucrose (2) and other analogues (3–6) for the study of sucrose transport in plants [reviewed by Hitz, (1988)].

Fluorinated sugars have also been studied in bacterial sugar transport systems. For example, Miles and Pirit (1973) found that 3-FDG is converted to 3-FDG-6-phosphate and taken up by the phosphoenolpyruvate-dependent transferase system of *E. coli*. Taylor and Louie (1977) found that the isomeric 4-FDG also is a substrate for this system and that this analogue was phosphorylated at twice the rate of 3-FDG. Neither sugar was significantly metabolized, but both inhibited the lactose utilization by the cells, probably as a result of repression of β-galactosidase activity.

6.2.5 2-, 3-, 4-, and 6-Halodeoxyhexoses as Substrates and Inhibitors of Metabolic Enzymes

6.2.5.1 Overview of Activity of Halogenated Hexoses with Metabolic Enzymes

A significant portion of the early work with halogenated sugar analogues was concentrated on investigation of effects of halogen substitu-

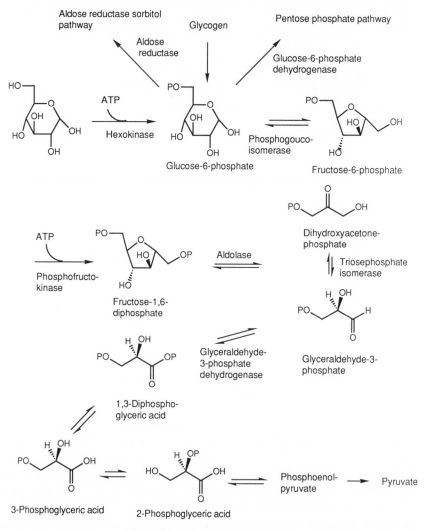

Figure 6-12. The glycolytic pathway.

tion on processing of sugars by glycolytic and other metabolic enzymes. The results of these and more recent studies will be summarized in this section.

Halogen at C-1 of a hexose is at the locus of reactions involving cleavage and formation of glycosidic bonds, as discussed above in Section 6.2.2. During metabolism of glucose by glycolytic enzymes, enzyme-catalyzed transpositioning of functional groups, oxidation–reduction processes, and bond cleavage (Fig. 6-12) are influenced by the presence of halogen substitution on internal carbons. Selected examples of the effects of halogen on these processes will be given below.

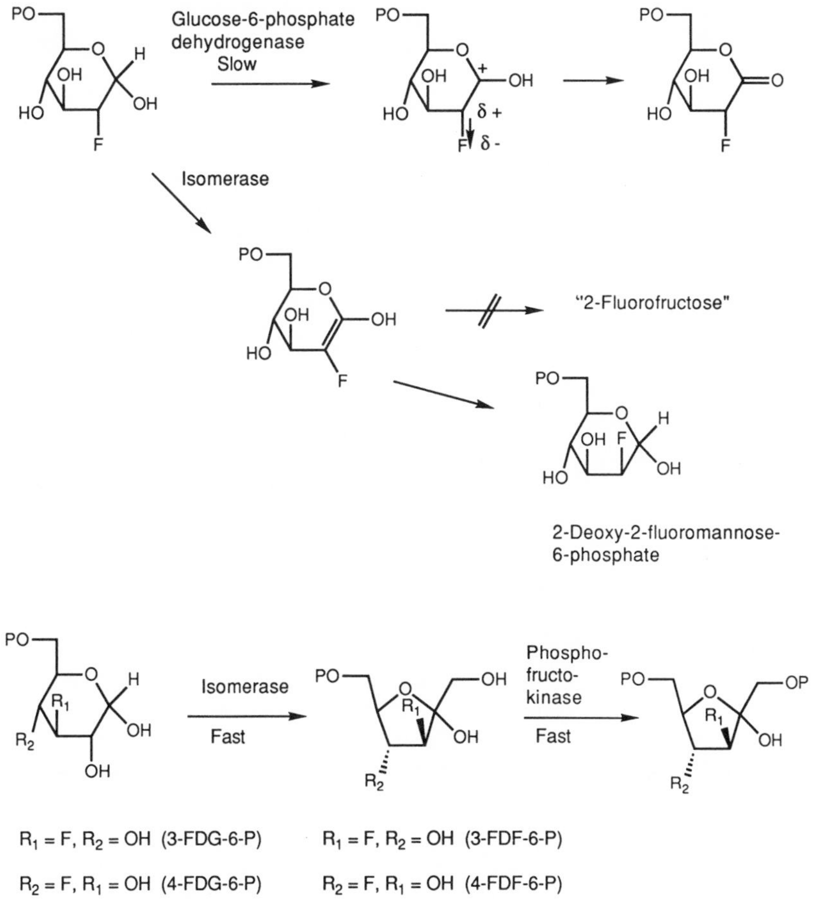

Figure 6-13. Examples of the processing of fluorinated glucose analogues by glycolytic enzymes (see text for details and references).

Early studies by Bessel *et al.* (1972) had shown that 2FDG-6-phosphate is a good substrate for hexokinase but is a very poor substrate (high K_m, low V_{max}) for glucose-6-phosphate dehydrogenase, the enzyme that catalyzes the oxidation of glucose-6-phosphate to 6-phospho-δ-gluconolactone, the initial step in the pentose monophosphate pathway (see below). This latter property was interpreted as consistent with a mechanism for oxidation involving hydride removal from C-1, a process which would be inhibited by the increased electronegativity of C-2 caused by fluorine substitution (Fig. 6-13) (Bessel and Thomas, 1973a). While the 6-phosphate derivatives of 3-FDG and 4-FDG were shown to be good substrates for yeast glucose phosphate isomerase, 2-FDG-6-phosphate, because of the C-2 substituent, cannot form the product of the isomerase reaction. Enolization of 2-FDG is possible (2-deoxy-2-fluoro-D-mannose has been detected as a product of 2-FDG metabolism *in vivo*; see Section 6.2.5.2), and 2-FDG-6-phosphate did exhibit weak competitive inhibition of the isomerase. The fluorinated products of the isomerase reaction, 3-deoxy-3-fluorofructose-6-phosphate (3-FDF-6-phosphate) and 4-deoxy-4-fluorofructose-6-phosphate (4-FDF-6-phosphate), were good substrates for phosphofructokinase (Fig. 6-13) (Bessel and Thomas, 1973b).

Taylor and co-workers have investigated the metabolism of 3-FDG extensively in locusts (Romaschin and Taylor, 1981; Romaschin *et al.*, 1977). While 3-FDG is nontoxic to the rat, even at high doses (up to 5 g/kg), this analogue has a 50% lethal dose (LD_{50}) of 4.8 mg/g in the adult *Locusta migratoria*. Gas chromatographic analyses of metabolites indicated that 3-FDG is converted to 3-fluorosorbitol by aldose reductase with subsequent oxidation to 3-FDF by sorbitol dehydrogenase (Romaschin *et al.*, 1977). [The significance of the aldose reductase sorbitol (ARS) pathway in mammalian systems recently has been noted and will be discussed further in the next section.] The biochemical mode of toxicity of 3-FDG subsequently was ascribed to further metabolism of 3-FDF to the triosephosphate stage, at which point triosephosphate isomerase could be the target of irreversible inhibition (Fig. 6-14) (see Silverman *et al.*, 1975, and Section 6.2.8.3 below). Concomitant release of fluoride would be expected to increase toxicity because of an unusually high sensitivity of locust respiration to fluoride toxicity. The participation of 3-FDG in glycogen biosynthesis of this species will be discussed in Section 6.3.1 below. Taylor and co-workers have also studied the metabolism of 3-FDG and 4-FDG in bacteria. During incubation of [6-^{3}H]-4-FDG with *Pseudomonas putida*, both fluoride and tritiated water are released and 4,5-dihydroxypentanoic acid is produced (Fig. 6-15). Using ^{14}C-labeled 4-FDG, evidence was obtained for the incorporation of radiolabel into cell-surface peptidoglycan [reviewed by Taylor *et al.* (1988)].

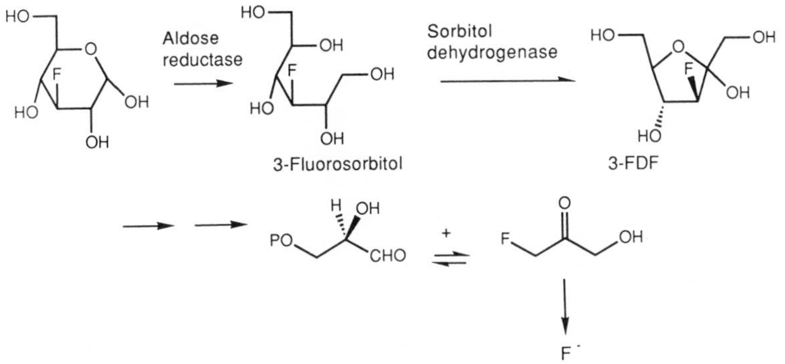

Figure 6-14. Metabolism of 3-FDG by the aldose reductase sorbitol pathway and possible mechanism for the toxicity of 3-FDG in locusts [reviewed by Taylor *et al.* (1988)].

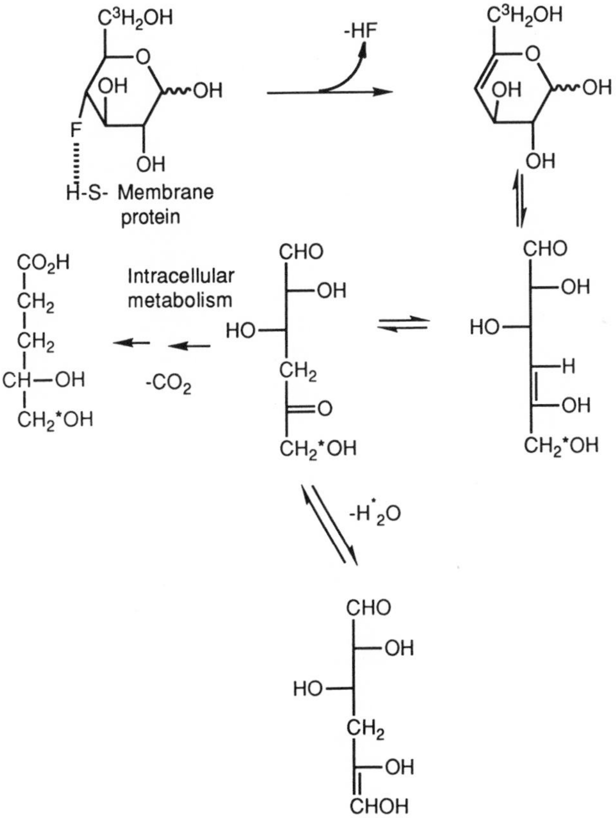

Figure 6-15. Metabolism of 4-FDG in *Pseudomonas putida* [reviewed by Taylor *et al.* (1988)].

6.2.5.2 Application of Deoxyfluoro Sugars for the Noninvasive Quantitation In Vivo of Glucose Metabolism

Under normal conditions, glucose is the sole energy source for the brain. The recognition that measurement of glucose utilization could provide a valuable technique for monitoring brain function, and dysfunction, has led to extensive research on methods for noninvasive procedures to quantitate brain glucose utilization. The development of a [^{14}C]-2-deoxyglucose ([^{14}C]DG) autoradiographic method for measurement of regional cerebral glucose metabolism in living animals was based on the knowledge that 2-deoxyglucose (2-DG) is a substrate for hexokinase, but, once formed, 2-DG-6-phosphate shows no inhibitory or substrate activity for phosphoglucose isomerase or glucose-6-phosphate dehydrogenase (see Fig. 6-12). Thus, after carrier-mediated transport of intravenous pulse-injected [^{14}C]DG into the animal brain, this isolation of the hexokinase reaction effectively traps [^{14}C]DG-6-phosphate in brain tissue over the time course of the measurements. After decapitation, the frozen brain is sectioned, and local tissue concentrations of [^{14}C]DG are determined by autoradiography. Mathematical models using measurable variables are then used to relate these regional concentrations of radioactive tracer to local cerebral glucose consumption (Sokoloff *et al.*, 1977).

While the [^{14}C]DG method represented a major breakthrough for the study of cerebral energy metabolism in animals and is a powerful experimental tool in neuroscience, application to human subjects obviously is impossible. In order to extend this procedure to the human brain, a noninvasive method to measure glucose metabolism was developed based on the used of [^{18}F]-2-FDG as a scanning agent for positron emission tomography (PET) (Phelps *et al.*, 1979). This effort was facilitated by the fact that 2-FDG behaves biochemically quite similarly to 2-DG—that is, 2-FDG is transported efficiently across the blood–brain barrier and is rapidly phosphorylated to 2-FDG-phosphate by hexokinase. 2-FDG-6-phosphate is cleared slowly from tissue because of low membrane permeability, the low activity of glucose-6-phosphatase in the brain, and resistance to further metabolism (however, see below). The [^{18}F]-2-FDG method has proven to be extremely valuable as a research tool and in medical diagnoses in the brain [reviewed by Phelps and Mazziotta (1985)] and in other organs (Schelbert and Phelps, 1984) and is the most widely used of all PET methodologies [for a concise review, see Fowler and Wolf (1986)].

The metabolic stability of 2-FDG-6-phosphate is an important factor in these PET applications. There has been recent evidence, primarily from *in vivo* ^{19}F-NMR investigations, that the metabolism of 2-FDG, at least under certain conditions, can proceed beyond the 2-FDG-6-phosphate stage. Nakada *et al.* (1986) studied the metabolic fate of 2-FDG in the living brain of rats by analysis of ^{19}F-NMR signals 3 h after intravenous injection. In addition to the ^{19}F resonance attributed to 2-FDG and/or 2-FDG-6-phosphate (not resolved), there appeared a signal assigned to 2-deoxy-2-fluoro-6-phospho-δ-gluconolactone   and/or   2-deoxy-2-fluoro-6-phosphogluconate (early fluorinated intermediates of the pentose monophosphate pathway; see Fig. 6-1) and two additional unidentified peaks. The importance of the ARS pathway (Fig. 6-16) (a glucose metabolic pathway that bypasses the hexokinase and phosphofructokinase control points) in the brain (Jeffery and Jörnvall, 1983) suggested that this route could produce metabolites responsible for the two remaining NMR peaks. In a later study, the corresponding fluorinated intermediates of the ARS pathway were synthesized and the ^{19}F-NMR parameters determined. While the resonances of 5-deoxy-5-fluorosorbose (5-FDS) and 5-FDS-1-phosphate could not be separated from those of 2-FDG and 2-FDG-6-phosphate, subsequent *in vivo* NMR experiments clearly indicated the formation of 2-deoxy-2-fluorosorbitol, 2-deoxy-2-fluoro-L-glyceraldehyde, and 2-deoxy-2-fluoroglycerol after injection of 2-FDG, confirming that significant metabolism of 2-FDG occurs by the ARS pathway under these conditions. Thus, while 2-fluorosorbitol formed from 2-FDG by the action of aldose reductase cannot be oxidized at the 2 position to form a fructose derivative, it can be oxidized, as a 5-fluoro carbohydrate, to 5-FDS (Fig. 6-17) (Nakada and Kwee, 1986). The authors have noted, however, that the concentrations used in the NMR experiments are significantly

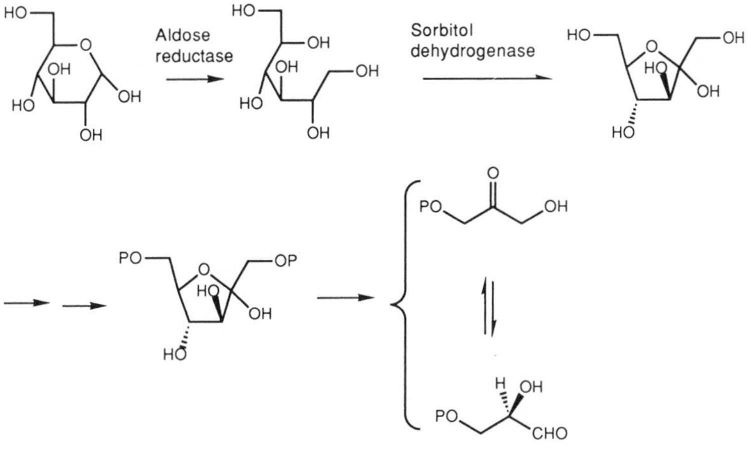

Figure 6-16. The aldose reductase sorbitol (ARS) pathway.

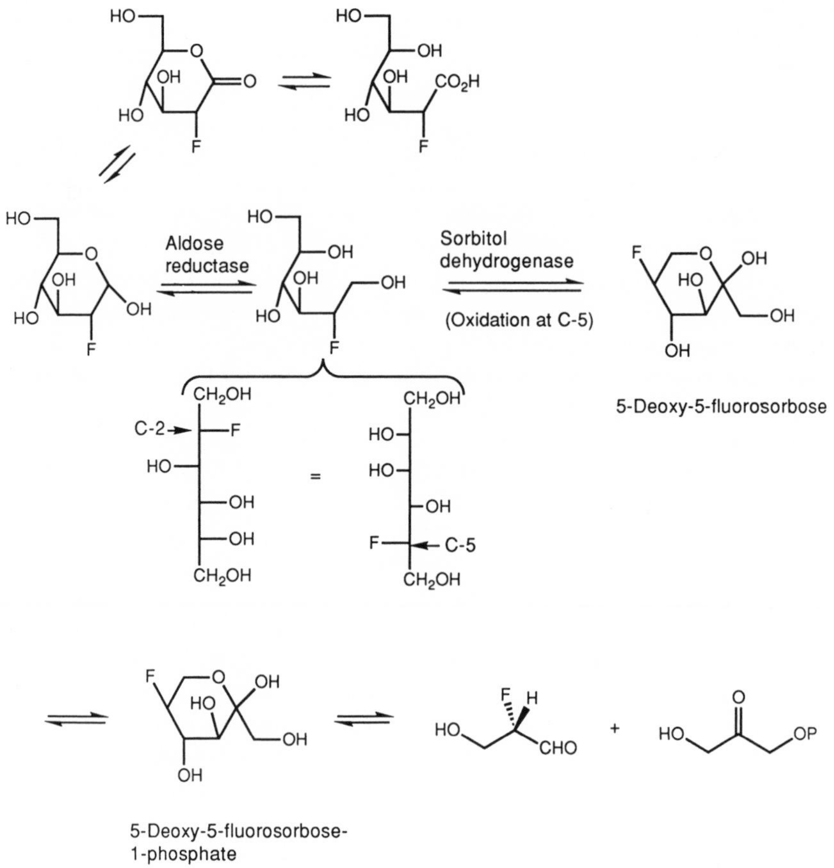

Figure 6-17. Metabolism of 2-FDG in the rat brain (Nakada and Kwee, 1986).

higher than tracer amounts used in the PET procedures and that extrapolation of these results must be done with caution.

In a related study, Kanazawa *et al.* (1986) used ¹⁹F-NMR measurements of isolated tissues to follow the metabolic fate of 2-FDG in the mouse for up to 96 h after injection. Significant amounts of 2-deoxy-2-fluoro-D-mannose (2-FDM) and 2-FDM-6-phosphate were detected, particularly in the heart, and fluorinated compounds were excreted as both glucose and mannose derivatives. A likely biochemical mechanism for this apparent epimerization of C-2 is the formation of a 1,2-enolate by the action of phosphoglucose isomerase on 2-FDG-6-phosphate (see above, Section 6.2.5.1).

The biochemistry of 3-FDG has been studied extensively (see section

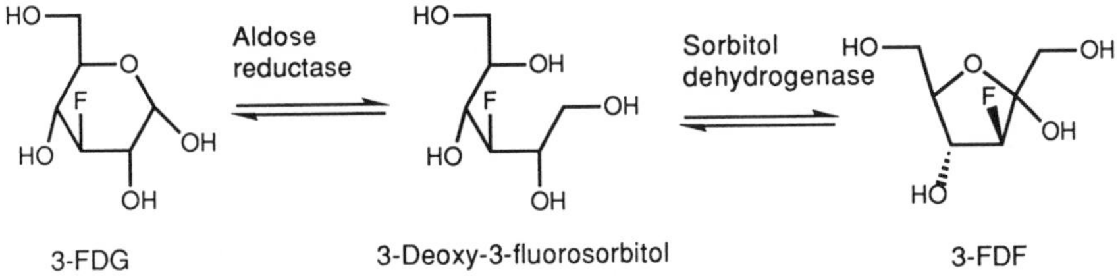

3-FDG   3-Deoxy-3-fluorosorbitol   3-FDF

Figure 6-18. Metabolism of 3-FDG by the ARS pathway (Kwee *et al.*, 1987).

6.2.5.1), and the potential of [^{18}F]-3-FDG as a PET scanning agent has been explored. The ready utilization of 3-FDG by glucose transport systems was confirmed in the rat heart, validating the use of [^{18}F]-3-FDG for the study of glucose transport in the heart (Halama *et al.*, 1984). Consistent with previous results, 3-FDG was found to be a poorer substrate for hexokinase than 2-FDG.

The efficient uptake of 3-FDG into the brain and its efficient reduction by aldose reductase (Romaschin *et al.*, 1977; see section 6.2.5.1 above), combined with its low rate of phosphorylation, make this analogue ideal for the study of the brain ARS pathway, since aldose reductase uses glucose directly, bypassing hexokinase. 3-FDG is also a poor substrate for the phosphoglucose isomerase reaction. Kwee *et al.* (1987) have extended their *in vivo* ^{19}F-NMR investigations to include the study of the metabolism of 3-FDG in the living rat brain. Following intravenous infusion of 3-FDG, peaks assigned to the α and β anomers of 3-FDG, 3-deoxy-3-fluoro-D-sorbitol, and 3-FDF were resolved clearly in the NMR spectrum from the brain, confirming that the analogue is metabolized in this organ primarily by the ARS pathway (Fig. 6-18). The low toxicity of 3-FDG and good resolution of the NMR signals suggest that this analogue will be useful for *in vivo* noninvasive quantitation of the ARS pathway by ^{19}F-NMR spectroscopy and should help in defining the physiological role of this pathway in the brain.

6.2.6 Metabolic Studies of Halogenated Pentoses

The antiviral and antitumor activities of nucleosides derived from halogenated pentoses have been discussed in Chapter 5. In contrast to the extensive research done with these nucleoside analogues, there has been much less exploration of halogenated pentose analogues as inhibitors and substrates of enzymes involved with carbohydrate metabolism. More specifically, a review of the literature has revealed no systematic study of

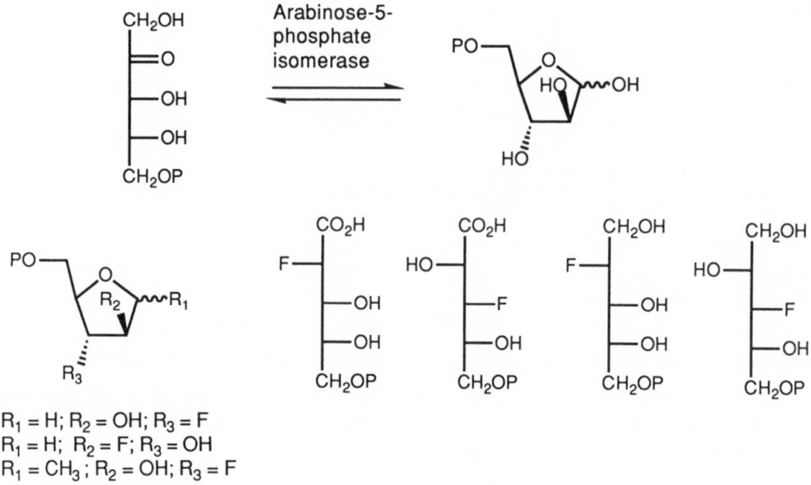

R₁ = H; R₂ = OH; R₃ = F
R₁ = H; R₂ = F; R₃ = OH
R₁ = CH₃ ; R₂ = OH; R₃ = F

Figure 6-19. Fluorinated analogues that were studied as potential inhibitors of arabinose-5-phosphate isomerase (Bigham *et al.*, 1984).

the processing of halogenated pentose analogues by the pentose phosphate shunt.

One example of the development of halogenated pentose analogues for enzyme inhibition is found in the work of Bigham *et al.* (1984). A series of pentose phosphates, aldonic acid phosphates, ribonolactones, and arabinose phosphates, including fluorinated derivatives (Fig. 6-19), were studied as analogues of substrates or intermediates for arabinose 5-phosphate isomerase. This enzyme, which catalyzes the isomerization of ribulose-5-phosphate to arabinose-5-phosphate, is a key enzyme in the biosynthesis of lipopolysaccharide, an essential component of the cell membrane of gram-negative bacteria. In the five-carbon analogues studied, substitution with fluorine at C-2 or at C-3 decreased inhibitory activity in a given series, with the decrease being larger for substitution at C-3.

6.2.7 Halogenated Tetroses

The peculiar preference of brucellae for erythritol was the stimulus for evaluation of halogenated analogues of erythritol and threitol as inhibitors of brucellae as an approach to the treatment of human brucellosis (undulant fever). In this rare example of a study with tetrose analogues, 2-deoxy-2-fluoro-D,L-erythritol (Fig. 6-20) showed erythritol-reversible *in vitro* and *in vivo* activity against brucellae (Smith *et al.*, 1965).

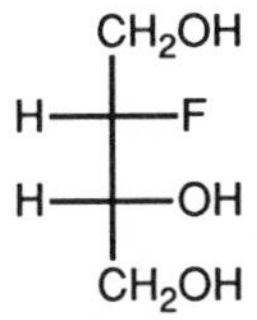

$$CH_2OH$$
$$H \underline{\quad\quad} F$$
$$H \underline{\quad\quad} OH$$
$$CH_2OH$$

Figure 6-20. A fluorinated erythritol derivative active against brucellae (Smith *et al.*, 1965).

2-Deoxy-2-fluoro-D,L-erythritol
(D-isomer shown)

6.2.8 Halogenated Trioses

Halogenated analogues of pyruvate and phosphoenolpyruvate, metabolites immediately downstream from the three-carbon glycolytic intermediates, dihydroxyacetone phosphate, glyceraldehyde phosphate, and phosphoglycerate, have been considered in Chapter 1. Examples of the use of halogenated analogues in the study of triose-processing enzymes will be considered in this section.

6.2.8.1 Glycerol Kinase

An unambiguous demonstration of the ability of fluorine to mimic the hydroxyl group comes from studies done with fluorinated glycerol analogues. Glycerol kinase from *Candida mycoderma* catalyzes phosphorylation of glycerol specifically at C-3. Eisenthal *et al.* (1972) prepared C-1-, C-2-, and C-3-fluorinated analogues of *sn*-glycerol (*s*tereospecifically *n*umbered—in the Fischer projection formulation of molecules having this sort of symmetry, the carbon at the top in the projection that has the 2-OH group to the left is designated C-1 (Fig. 6-21). While the C-1- and C-2-fluorinated analogues were weak substrates for glycerol kinase, the C-3-substituted analogue was a potent competitive inhibitor, establishing C-3 as the site of phosphorylation. While the product from the phosphorylation of 1-deoxy-1-fluoroglycerol is unambiguously 1-deoxy-1-fluoroglycerol-3-phosphate, the product of phosphorylation of the prochiral 2-deoxy-2-fluoroglycerol could be the 1-phosphate or the 3-phosphate, depending upon whether fluorine is recognized by the enzyme as hydrogen or hydroxyl. This ambiguity was resolved by the synthesis of chiral C-1-deuterated 2-deoxy-2-fluoroglycerol (Briley *et al.*, 1975). The action of glycerol kinase on this analogue produced 1-deutero-2-fluoro-2-deoxyglycerol-3-phosphate with complete selectivity, a result which requires fluorine to occupy the site on the enzyme occupied by the 2-hydroxyl group of the natural substrate (Fig. 6-21).

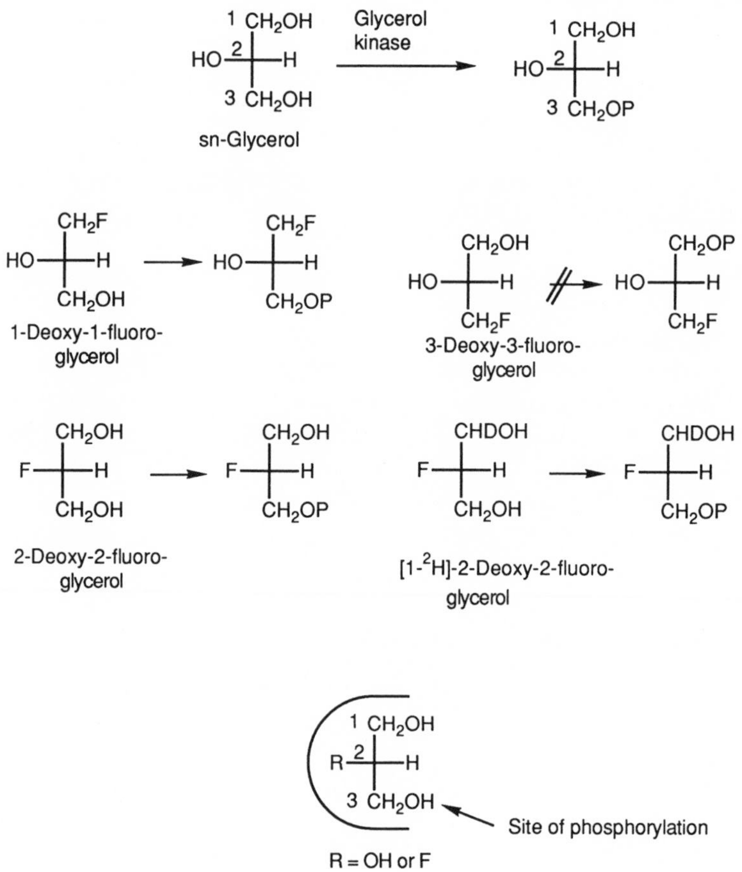

Figure 6-21. Fluoroglycerols as substrates for glycerol kinase (see text for details).

6.2.8.2 Aldolase

Hartman (1970) prepared chloro-, bromo-, and iodoacetol phosphate (Fig. 6-22) as potential active-site-directed inhibitors of fructose-1,6-diphosphate aldolases, enzymes responsible for the cleavage of fructose 1,6-diphosphate to dihydroxyacetone phosphate and glyceraldehyde 3-phosphate. Iodoacetol phosphate rapidly inactivated rabbit muscle aldolase, with a concomitant extensive loss of sulfhydryl groups. Inactivation to the extent of 95% resulted in incorporation of two moles of reagent per mole of protein.

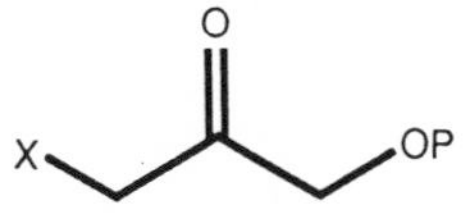

Haloacetol phosphate

Figure 6-22. Haloacetol phosphates as inhibitors of
fructose-1,6-diphosphate aldolases.   X = F,Cl, Br, I

6.2.8.3 Glycerol-3-phosphate Dehydrogenase (GPDH)

Fluorinated substrate and product analogues have been used to investigate several aspects of the mechanism of glycerol-3-phosphate dehydrogenase (GPDH). This enzyme catalyzes the reduction of dihydroxyacetone phosphate to glycerol-3-phosphate as part of the glycerol phosphate shuttle of electrons across the mitochondrial membrane. The toxicity of 1-fluoroglycerol to rats approaches that of fluoroacetate. Fondy *et al.* (1970) have explored strategies to exploit this toxicity in chemotherapeutic schemes, based on the very low to nonexistent levels of cytoplasmic NAD-linked GPDH in a wide variety of human cancer cells. Thus, metabolic oxidation by GPDH of 1-haloglycerol-3-phosphate, or of an appropriate precursor, to 1-halohydroxyacetone-3-phosphate could provide a mechanism for detoxification not available to tumor cells, producing selective toxicity in the latter. Of D,L-1-halo analogues prepared (Fig. 6-23), only the fluoro analogue was active. In a later study, 1-deoxyfluoro-L-glycerol-3-phosphate was found to be a substrate for GPDH with a K_m of 3.8 mM, compared to 0.15 mM for the natural substrate. However, *in vivo* formation of fluoroacetate was apparent by the observed toxicity in mice, comparable to that of fluoroacetate, and a similar induction of hypothermia was observed. The significant reduction of these effects by pyrazole, an inhibitor of alcohol dehydrogenase but not of GDPH,

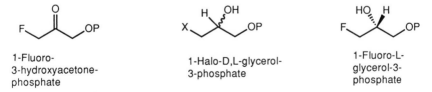

1-Fluoro-
3-hydroxyacetone-
phosphate

1-Halo-D,L-glycerol-
3-phosphate

1-Fluoro-L-
glycerol-3-
phosphate

Figure 6-23. 1-Fluorohydroxyacetone-3-phosphate and 1-haloglycerol-3-phosphate studied as inhibitors of glycerol-3-phosphate dehydrogenase (GPDH).

suggested that metabolism mediated by this enzyme is more significant *in vivo* than oxidation by GPDH (Fondy *et al.*, 1974).

As a substrate for NAD-linked GPDH-catalyzed reduction, 1-deoxy-1-fluoro-3-hydroxyacetone phosphate (Fig. 6-23) has a K_m 333-fold greater than that of dihydroxyacetone phosphate (Silverman *et al.*, 1975). A 15-fold greater K_m for 1-deoxy-1-fluoro-L-glycerol-3-phosphate compared to that for glycerophosphate suggested that, with both substrates, the hydroxyl group of C-1 is required for hydrogen bond donation to the enzyme active site. The slow turnover observed in the oxidation step similarly suggested that hydride transfer from C-2 is involved in the rate-limiting step, since this transfer would be disfavored by the electron-withdrawing effects of fluorine at C-1. Similarly to fluoroglycerol phosphate, 1-fluoro-3-hydroxyacetone phosphate induced hypothermia in mice. However, in contrast to the case of fluoroglycerol, the hypothermic activity of the fluoroketone was not blocked by pyrazole (Silverman *et al.*, 1975).

Transport restrictions of phosphorylated analogues *in vivo* present potential problems for their effective use in chemotherapeutic schemes. Accordingly, studies were extended to include nonphosphorylated esters of halogenated hydroxyacetone. In the 3-halo-1-benzoyloxypropan-2-one series (Fig. 6-24), toxicity to mice decreased in the order bromo > fluoro > chloro, with toxicity of the chloro- and bromoketones being related to their alkylating ability. In contrast, activity *in vivo* against L1210 mouse leukemia and Ehrlich ascites carcinoma was greatest in the chloroketone series. The benzoate and nitrobenzoate esters of 3-chloro-1-hydroxy-propanone completely cured mice bearing Ehrlich ascites carcinoma at doses well below the LD_{50}. The fluoroketone esters were less active and showed greater toxicity (Pero *et al.*, 1977).

6.2.8.4 Triosephosphate Isomerase

In the glycolytic pathway, the action of aldolase on fructose-1,6-diphosphate produces dihydroxyacetone phosphate and glyceraldehyde

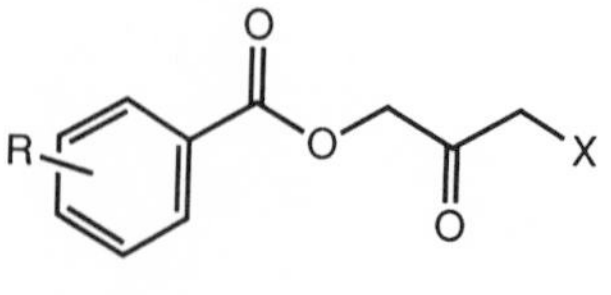

X = F, Cl, Br; R = H

X = Cl; R = para-NO$_2$

Figure 6-24. Nonphosphorylated esters of halogenated hydroxyacetone as potential antitumor agents (Pero *et al.*, 1977).

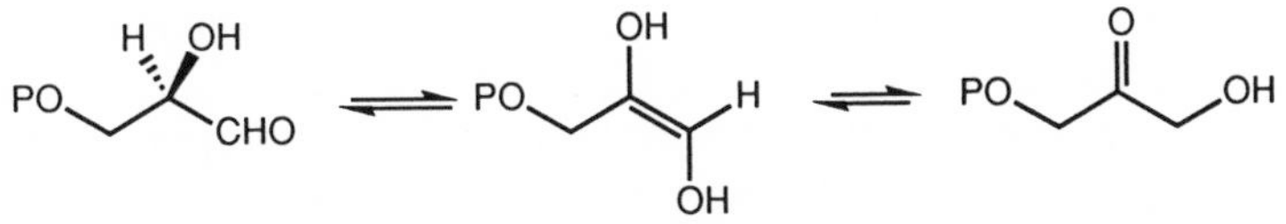

Figure 6-25. Triosephosphate isomerase-catalyzed interconversion, via enediol intermediate, of dihydroxyacetone phosphate and glyceraldehyde 3-phosphate, products of action of aldolase on fructose-1,6-diphosphate.

3-phosphate. Triosephosphate isomerase catalyzes the interconversion of these triosephosphates, and glyceraldehyde 3-phosphate is subsequently oxidized to 1,2-diphosphoglycerate. This reaction has been shown to proceed through an enediol intermediate (Fig. 6-25). Hartman (1971) measured the time-dependent inactivation of triosephosphate isomerase by chloro, bromo, and iodo analogues of dihydroxyacetone phosphate and obtained second-order rate constants of 2600, 2300, and 260 $M^{-1}s^{-1}$, respectively. Although Hartman was unable to detect measurable inactivation by the fluoro analogue, with modifications in experimental design, including the use of much higher concentrations, Silverman *et al.* (1975) obtained a second-order rate constant of 1.5 $M^{-1}s^{-1}$ for inactivation of the fluoro analogue. This inactivation process has been implicated in the toxic action of 3-FDG in locusts (see Section 6.2.5.1 above). The lower activity of the fluorinated compound is consistent with the sluggishness with which the $C-F$ bond participates in S_N2 processes.

6.2.8.5 Propanediol Dehydrase

Propanediol dehydrase is an adenosylcobalamin-dependent enzyme which catalyzes the rearrangement of the substrate propane-1,2-diol to

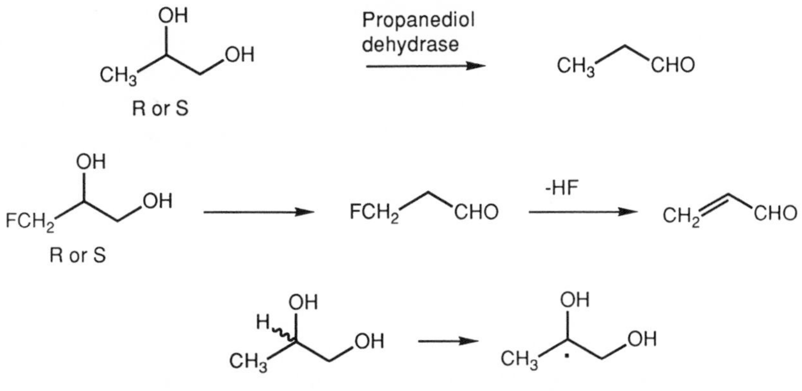

Figure 6-26. Fluoropropanediols as substrates for propanediol dehydrase (Eager *et al.*, 1975).

propionaldehyde. Despite a highly specific requirement for either (R)- or (S)-propanediol as substrate, both (R)- and (S)-3-fluoro-1,2-propanediol were found to be effective substrates, yielding β-fluoropropionaldehyde as the presumed initial product, which decomposes to acrolein as the isolable product (Fig. 6-26) (Eager $et\ al.$, 1975). While the K_m values were significantly higher for the fluoro analogues in each series, the results demonstrated that, in this case, fluorine is functioning as a replacement for hydrogen in the substrate. The fact that comparable k_{cat} values were derived for the four substrates was used as evidence that the critical carbon–hydrogen bond cleavage in the initial stages of the reaction probably proceeds by a radical mechanism (Fig. 6-26).

6.3 The Effect of Halogenation on the Participation of Carbohydrates in Biosynthetic Pathways

6.3.1 Incorporation of Halogenated Sugars into Polysaccharides

The biosynthesis of polysaccharides does not correspond to the reversal of polysaccharide breakdown. For example, whereas phosphorylase catalyzes the phosphorolytic cleavage of glucose units from the nonreducing end of glycogen to give glucose-1-phosphate, glycogen biosynthesis uses uridine diphosphate glucose (UDP-glucose; Fig. 6-27) as an activated form of glucose. Formation of nucleoside diphosphate esters is a prerequisite for incorporation of sugars into macromolecules.

Reports of incorporation of halogenated sugars into polysaccharides are quite limited. As part of an extensive research program on the biochemistry of fluorinated sugars, Agbanyo and Taylor (1986) recently reported the incorporation of 3-FDG into glycogen and the disaccharide, trehalose, of fat body and flight muscle of a species of locust (*Locusta migratoria*). In this instance, the inhibition of glycolysis by 3-FDG apparently leads to increased levels of glycogen, trehalose, and fructose synthesis, thus favoring analogue incorporation. The formation of "fluoroglycogen" was assumed to arise from action of glycogen synthase on UDP-3-FDG (Fig. 6-27).

6.3.2 Inhibition of Glycoprotein Biosynthesis by Halogenated Sugars

6.3.2.1 Glycoproteins—An Overview

Many proteins in eucaryotes contain covalently linked oligosaccharides as an integral structural feature. Protein glycosylation serves

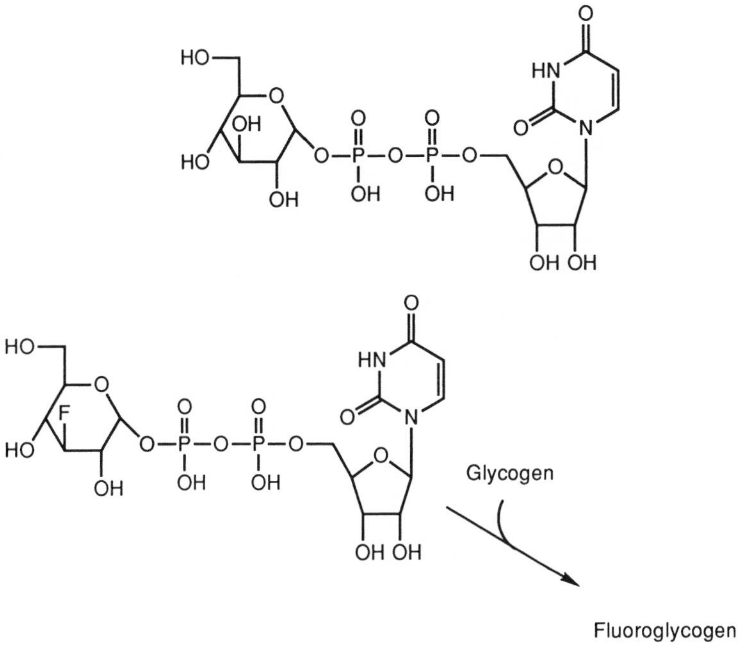

Figure 6-27. UDP-glucose and the formation of "fluoroglycogen" from UDP-3-FDG (Agbanyo and Taylor, 1986).

a complex variety of structural and functional roles. Extensive research has led to a current view that carbohydrates play specific and nonspecific roles that are secondary to the primary function of glycoproteins, but are nonetheless extremely relevant to the organism. Thus, glycoproteins are involved in such processes as compartmentalization, transport, conformational control, intermolecular association, and protection of the protein from its environment. Selective inhibitors of protein glycosylation, inhibitors which include halogenated sugar analogues, have been used effectively in research on the functions of glycoproteins, and more specific examples of these functions will be considered as this work is discussed below. It must be stressed that the results described below represent only a small fraction of the progress being realized through these approaches, and the original papers should be consulted to appreciate fully the complexity of the processes in which glycoproteins are involved and the skill with which sugar analogues are being used to understand these processes. For a thorough, but concise, review of the functions of protein glycosylation, a recent review by West (1986) is recommended.

A major class of soluble and membrane-bound glycoproteins contains oligosaccharides linked to asparagine through an N-glycosidic bond. The biosynthesis of asparagine-linked oligosaccharides involves complex processes which have evolved essentially intact from yeast to man. While there is wide diversity in the structures of asparagine-linked glycoproteins, the presence of an asparagine-linked pentasaccharide [$Man_3(GlcNAc)_2$] is ubiquitous, reflecting a unifying biosynthetic pattern. As shown schematically in Fig. 6-28, a common oligosaccharide precursor [$Glc_3Man_9(GlcNac)_2$] is synthesized by the stepwise addition of activated monosaccharides (derivatives of UDP, GDP, and the polyisoprenoid dolichol) to a carrier, dolichyl phosphate, situated on the endoplasmic reticulum membrane. The N-glycosidic linkage is formed by the transfer *en bloc* of the dolichol-linked oligosaccharide to asparagine with release of dolichyl pyrophosphate. Trimming of two of the three glucose residues

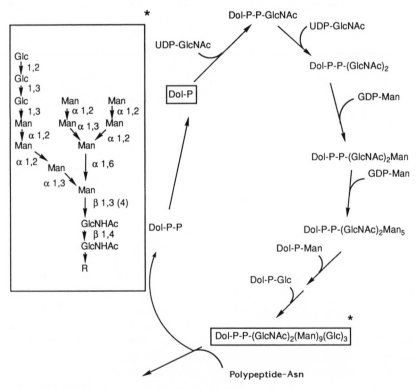

Figure 6-28. The dolichol pathway for the synthesis of glycoproteins.

rapidly follows. After transportation to the Golgi apparatus, the precursor glycoproteins are further processed by trimming and addition of new sugars to produce the high-mannose, hybrid, and complex asparagine-linked oligosaccharides, the three types which comprise this class of glycoprotein (Fig. 6-29). For more detailed discussions of these processes, several recent reviews are available including those by Kornfeld (1982), Kornfeld and Kornfeld (1985), and Schwarz and Datema (1982a).

The other major class of glycoprotein is formed by the covalent attachment of oligosaccharides to O-glycosidic linkages to serine and threonine. Unlike the N-linked glycoproteins, these are synthesized by the addition of sugar-nucleotide donors directly to the polypeptide, and subsequent trimming appears not to be important (West, 1986).

6.3.2.2 Inhibition of the Dolichol Pathway by Fluorinated Sugars

Compounds that interfere with the normal events in protein glycosylation have been used effectively to study the mechanism of biosynthesis of asparagine-linked glycoproteins, as well as to study the functions and

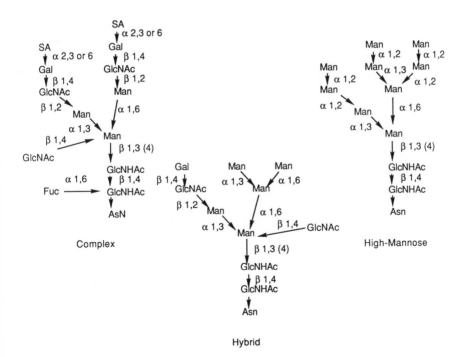

Figure 6-29. Three classes of glycoproteins.

importance of oligosaccharides present in these proteins [for reviews, see Schwarz and Datema (1982a, b)]. In addition, such compounds are often antiviral and antibacterial agents (Schwarz and Datema, 1980). Examination of sugar analogues as potential inhibitors of specific steps of the dolichol-dependent pathway presented itself as one obvious approach to the development of such mechanistic probes and medicinal agents. These analogues have the further potential of altering functional and structural parameters of the glycoprotein by becoming incorporated into the oligosaccharide.

2-Deoxyglucose, 2-deoxy-2-fluoroglucose, and 2-deoxy-2-fluoromannose (2-FDM) are all metabolized in yeast and chick embryo cells to their UDP and GDP derivatives (Fig. 6-30). Compared to 2-DG, the fluoro analogues were incorporated into macromolecules at a very low rate, although all had comparable inhibitory action on glycoprotein and polysaccharide formation. In spite of the different roles played by mannose and glucose in the dolichol pathway, these cells appeared not to distinguish

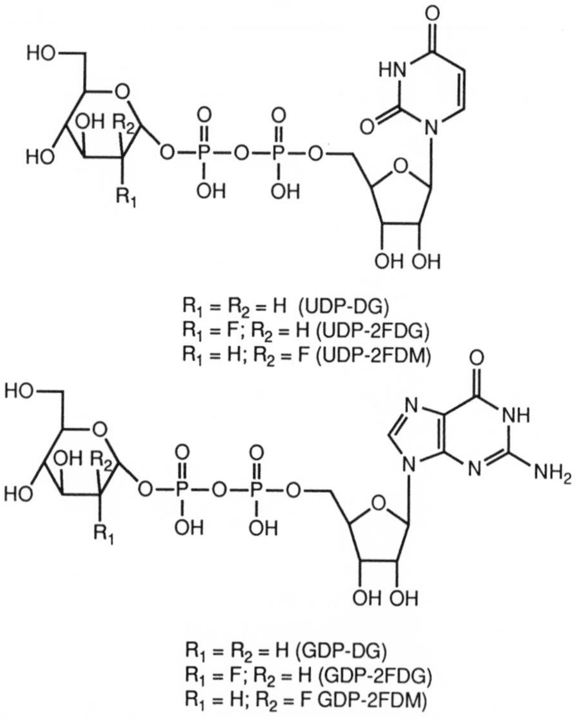

$R_1 = R_2 = H$ (UDP-DG)
$R_1 = F; R_2 = H$ (UDP-2FDG)
$R_1 = H; R_2 = F$ (UDP-2FDM)

$R_1 = R_2 = H$ (GDP-DG)
$R_1 = F; R_2 = H$ (GDP-2FDG)
$R_1 = H; R_2 = F$ GDP-2FDM)

Figure 6-30. UDP and GDP conjugates of 2-DG, 2-FDG, and 2-FDM.

between 2-FDM and 2-FDG (Schmidt *et al.*, 1978). 2-FDG subsequently was shown to inhibit the formation *in vivo* of dolichyl phosphate glucose (Glc-P-Dol) and dolichyl phosphate mannose (Man-P-Dol), but not to inhibit the transfer of these residues from the lipid carrier to the lipid-linked oligosaccharide. This inhibition caused the formation of smaller than normal lipid-linked oligosaccharides as biosynthetic intermediates (see Fig. 6-28). There was no evidence that the analogue was incorporated into the lipid-linked oligosaccharide (Datema *et al.*, 1980a).

Animal cells infected with enveloped viruses have been used extensively as model systems in the study of the biochemistry of protein-bound oligosaccharides. Using this technique, Datema *et al.* (1980b) were able to detect an alternative pathway for glycosylation of influenza virus protein in the presence of 2-FDG. This pathway has also been detected in a mutant, mouse lymphoma cell line incapable of synthesizing Man-P-Dol (Chapman *et al.*, 1980). Thus, while 2-FDG blocks formation of Man-P-Dol, the cell is not depleted of GDP-Man, and $Man_5(GlcNAc)_2$-PP-Dol is formed in 2-FDG-treated cells (see Fig. 6-28). This oligosaccharide, after attachment of three glucose residues, can be transferred to protein *in vivo*, with subsequent formation of complex and (truncated) high-mannose oligosaccharides (Kornfeld, 1982). Since both lipid-linked $Glc_3Man_5(GlcNAc)_2$ and $Glc_3Man_9(GlcNAc)_2$ can serve as precursors for protein-linked oligosaccharides, previous proposals that both complex and high-mannose oligosaccharides come from a common precursor have been questioned (Schwarz and Datema, 1982a). Thus, the possibility has been considered that the heptasaccharide [$Man_5(GlcNAc)_2$] represents a branching point in glycosylation, where attachment of glucose triggers transfer to protein, bypassing Man-P-Dol-mediated mannosylation and formation of the tetradecasaccharide (see Fig. 6-28). A possible regulatory role requiring the presence of two mannose donors in the dolichol pathway has also been considered (Chapman *et al.*, 1980).

The mechanism of inhibition of protein glycosylation by 2-FDM also has been studied in detail. Early steps in the dolichol pathway were inhibited in influenza virus-infected cells in the presence of 2-FDM and guanosine (the latter added to avoid inhibition of viral protein synthesis). GDP-2-FDM and UDP-2-FDM (see Fig. 6-30) were prepared as the likely active metabolites and used to study lipid-oligosaccharide bioassembly in cell-free systems. GDP-2-FDM blocked the glycosylation of $(GlcNAc)_2$-PP-Dol by GDP-Man and, in addition, trapped Dol-P as 2-FMan-P-Dol, causing a decrease in Man-P-Dol levels. UDP-2-FDG only inhibited Glc-P-Dol synthesis (Fig. 6-31). From these experiments, the conclusion was reached that inhibition by GDP-2-FDM of $Man(GlcNAc)_2$-PP-Dol formation from $(GlcNAc)_2$-PP-Dol and GDP-Man makes the largest con-

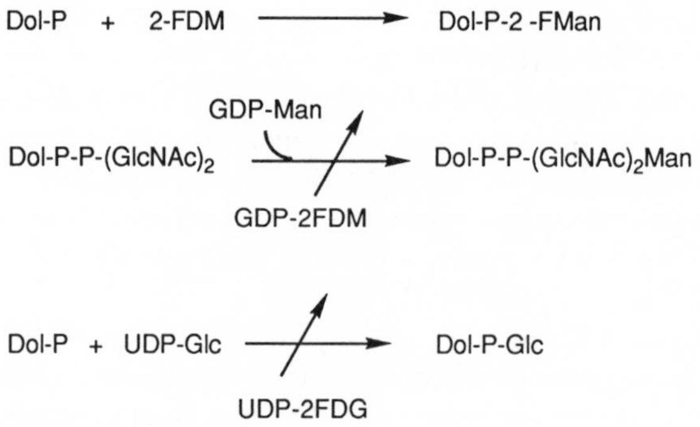

Dol-P   +   2-FDM   ⟶   Dol-P-2-FMan

GDP-Man

Dol-P-P-(GlcNAc)$_2$ ⟶   Dol-P-P-(GlcNAc)$_2$Man

GDP-2FDM

Dol-P   +   UDP-Glc   ⟶   Dol-P-Glc

UDP-2FDG

Figure 6-31. Mechanism for the inhibition of protein glycosylation by 2-FDM and 2-FDG (see text for details) (McDowell *et al.*, 1985).

tribution to the inhibition of oligosaccharide-lipid synthesis in 2-FDM-treated cells (McDowell *et al.*, 1985).

Grier and Rasmussen (1984) have shown that 4-deoxy-4-fluoromannose (4-FDM) inhibits the glycosylation of the viral G protein in vesicular stomatitis virus (VSV)-infected BHK-21 cells, as a consequence of a depletion of intracellular pools of Glc$_3$Man$_9$(GlcNAc)$_2$. In later studies *in vitro* to explore the mechanism of action of 4-FDM, GDP-4-FDM (Fig. 6-32) was shown to inhibit the addition of GDP-Man to Man(Glc-NAc)$_2$-PP-Dol (McDowell *et al.*, 1987).

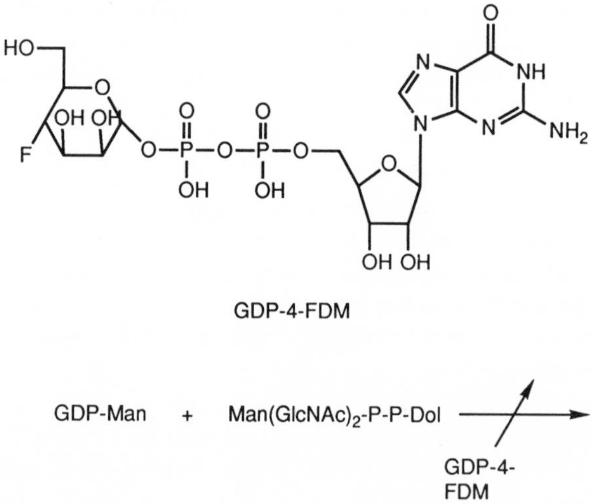

GDP-4-FDM

GDP-Man   +   Man(GlcNAc)$_2$-P-P-Dol   ⟶

GDP-4-FDM

Figure 6-32. Inhibition of protein glycosylation by GDP-4-FDM (see text for details) (McDowell *et al.*, 1987).

It should be noted that other inhibitors of protein glycosylation, including deoxysugars and a number of antibiotics, have received much attention. This material is beyond the scope of this chapter, and the reviews cited at the beginning of this section should be consulted for surveys of this work.

6.3.2.3 Inhibition by Halogenated Sugars of the Processing of Precursor Oligosaccharides to Complex Asparagine-Linked Oligosaccharides

In the preceding section, examples were given of studies wherein fluorinated sugars were used to interrupt early steps in the dolichol pathway. Halogenated sugar analogues have also been used to interdict the processing of the oligosaccharide–protein conjugate as a strategy to develop selectively acting inhibitors of membrane complex glycoprotein biosynthesis. The biochemical rationale for this approach has been discussed concisely in reviews by Sharma et al. (1988) and by Bernacki and Korytnyk (1982). A major impetus for this work comes from the potential of inhibitors of glycoprotein biosynthesis to be selectively toxic to cancer cells because of altered membrane properties found in oncogenically transformed cells, alterations that have been associated with cell-surface glycoproteins. Several structural units have been targeted. For example, since the sequence SA → Gal → GlcNAc (Fig. 6-33; see also Fig. 6-29) is most often present in the outer chains of complex-type glycoproteins— although numerous variations are found—analogues of SA have been studied (see below). [Sialic acids (SA) are amino sugars containing nine or more carbons (e.g., N-acetylneuraminic acid; Fig. 6-33) that are regular components of mucoproteins, mucopolysaccharides, and certain mucolipids.] As other examples, a GlcNAc linked $\beta 1 \rightarrow 4$ to Man and an L-fucose linked to the innermost GlcNAc are found in some complex types (Fig. 6-33). Analogues of these residues have been prepared in potential chemotherapeutic schemes based on their potential to interfere with, or regulate, the attachment of the natural relative.

Since hexosamines are precursors of a number of sugars found in cell-surface glycoproteins, a series of glucosamine, galactosamine, and mannosamine analogues have been synthesized and evaluated with respect to their abilities to inhibit tumor cell growth. Fondy et al. (1978) prepared haloacetamido analogues of 2-amino-2-deoxy-D-glucose and 2-amino-2-deoxy-D-galactose and O-acetylated derivatives thereof (Fig. 6-34). Increased toxicity was caused by O-acetylation of the N-bromo and N-chloro derivatives, and the degree of toxicity was comparable in each series. This suggested that these compounds act in vivo as nonspecific lipophilic alkylating agents. Toxicity of the N-fluoroacetamide derivatives was

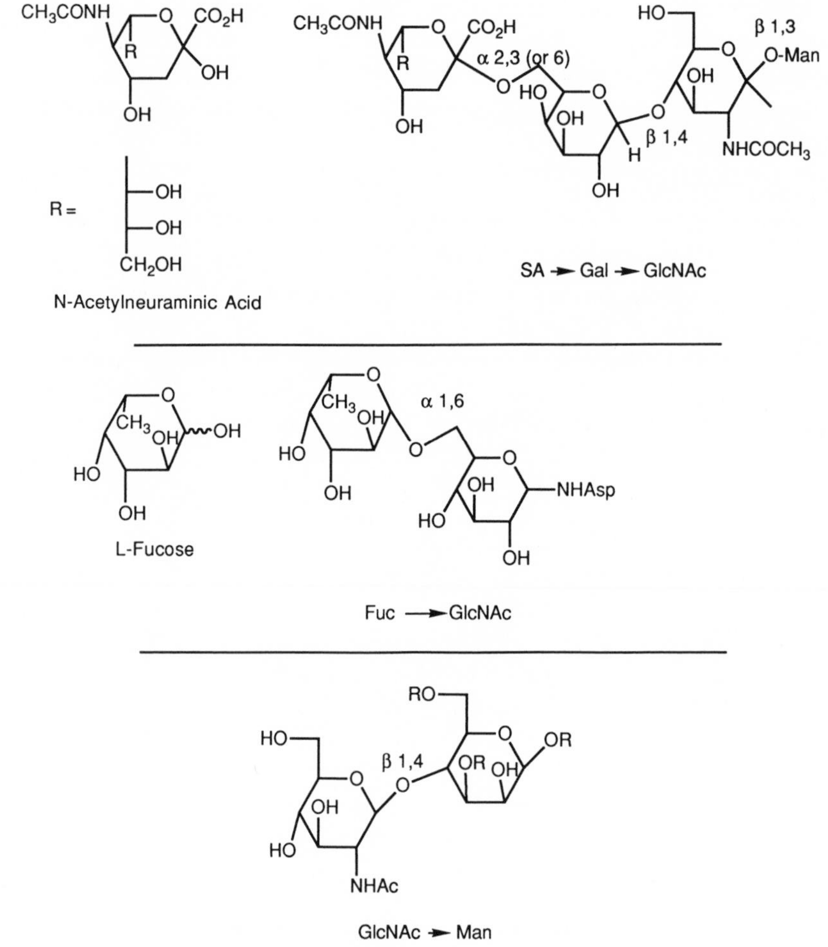

Figure 6-33. Structural units of glycoproteins targeted for selective inhibition as an approach to antitumor drugs: N-acetylneuraminic acid and the SA → Gal → GlcNAc sequence; L-fucose linked to the innermost GlcNAc; GlcNAc linked $\beta1 \rightarrow 4$ to Man.

unaffected by esterification, and the Glc derivative was severalfold more toxic than the Gal derivative. Specific uptake and metabolism to fluoroacetamide, followed by fluoroacetate formation, was considered as a possible mechanism of toxicity of 2-N-fluoroacetamido-2-deoxy-D-glucose. Tetra-O-acetyl-2-N-bromoacetamido analogues of glucose and galactose were subsequently found to be quite effective in the treatment of Ehrlich ascites tumor-bearing mice and were curative in some cases (Simon *et al.*,

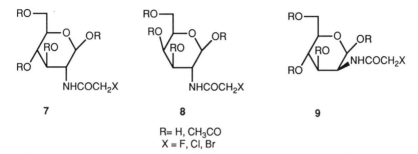

7   8   9

R= H, CH₃CO
X = F, Cl, Br

Figure 6-34. Haloacetamido analogues of 2-amino-2-deoxy-D-glucose (**7**), 2-amino-2-deoxy-D-galactose (**8**), and 2-amino-2-deoxy-D-mannose (**9**) and O-acetylated derivatives (Fondy *et al.*, 1978).

1979). Observed tumor selectivity may be related to increased sugar uptake in tumor cells. The chloro and fluoro derivatives, as well as nonacylated analogues, were inactive. Haloacetamido analogues of 2-amino-2-deoxy-D-mannose were found in a later study also to be effective and were significantly less toxic to animals at therapeutic doses (Fondy and Emlich, 1981). The tetra-O-acetylated chloroacetamido analogue as well as the bromoacetamido free sugar also gave high proportions of cures.

In a similar approach, Paul *et al.* (1980) prepared 1-N-substituted derivatives of 2-acetamido-2-deoxy-β-D-glucosamine, including the haloacetamido analogues (Cl, Br, I), and O-acetylated derivatives thereof (Fig. 6-35), as potential inhibitors of glycosidic bond formation between

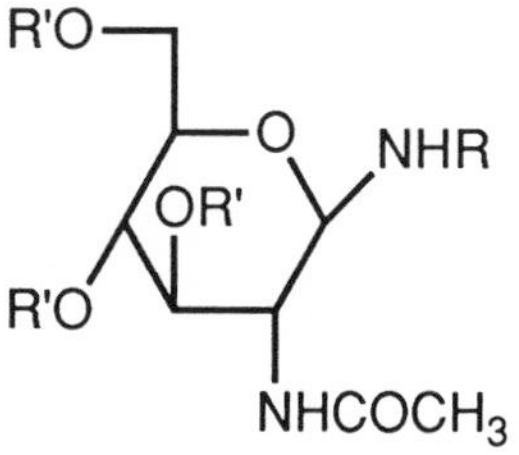

10

R = COCH₂Cl, R' = H or Ac
R = COCH₂Br, R' = H or Ac
R = COCH₂I, R' = H or Ac
R = COCF₃, R' = H or Ac
R = SO₂CF₃, R' = H or Ac

Figure 6-35. 1-N-Substituted analogues of 2-acetamido-2-deoxy-β-D-glucosamine and O-acetylated derivatives (Paul *et al.*, 1980).

asparagine and glucosamine. Both the *N*-bromo- and *N*-chloro-tri-*O*-acetyl derivatives had significant chemotherapeutic activity *in vivo* in mice.

Several analogues of aminohexoses, including 4- and 6-fluorinated D-glucosamine and D-galactosamine analogues and 6-fluorinated D-mannosamine analogues (**11**, **12**, and **13**, respectively; Fig. 6-36), were evaluated as potential inhibitors and/or modifiers of plasma membrane glycoproteins (Bernacki and Korytnyk, 1982, and references therein). In general, 6-substituted analogues such as 6-deoxy-6-fluoro-D-glucoseamine, 6-deoxy-6-fluoro-D-galactosamine, and 6-deoxy-6-fluoro-D-mannosamine (and their N-acetylated derivatives) were ineffective, possibly because of a requirement for phosphorylation at C-6 for biological activity. Per-O-acetylation of the analogues increased their activities considerably, presumably because of enhanced entry into the cell by simple diffusion. N-Acetyl per-O-acetylated analogues of glucosamine and galactosamine having fluorine at C-4 were active against L1210 murine leukemia cells (Paul *et al.*, 1980).

The attachment of sialic acid to galactose with an $\alpha(2 \rightarrow 6)$glycosidic linkage in cell-surface complex-type glycoproteins (see above, Fig. 6-33) has been shown in many cases to be associated with expression of the tumor cell phenotype (Morin *et al.*, 1983; Bernacki and Korytnyk, 1982; and references therein). For example, a correlation has been found between the amount of cell-surface sialic acid present on tumor cells and the ability to metastasize. Several sialic acid analogues, including 9-deoxy-9-fluoro-*N*-acetylneuraminic acid (9-F-NANA) (Fig. 6-37), have been synthesized as potential antitumor and/or antimetastatic agents. At $10^{-3}\,M$, 9-F-NANA

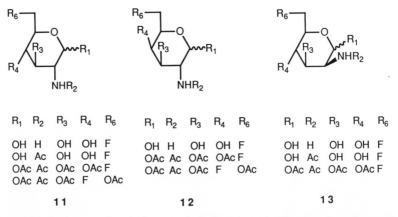

Figure 6-36. Analogues of aminohexoses, including 4- and 6-fluorinated D-glucosamine, D-galactosamine, and D-mannosamine analogues (Bernacki and Korytnyk, 1982, and references therein).

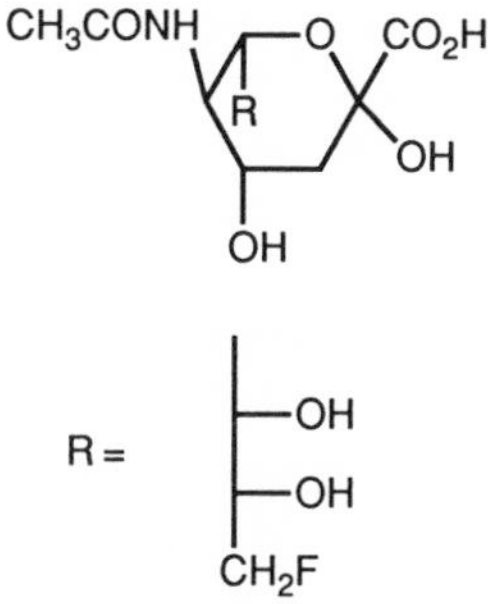

CH₃CONH ─O CO₂H
R
OH
OH

R =   ─OH
      ─OH
      CH₂F

9-Deoxy-9-fluoro-N-Acetylneuraminic Acid

(9-F-NANA)

Figure 6-37. 9-F-NANA, a potential antitumor agent (Bernacki and Korytnyk, 1982).

showed modest inhibition *in vitro* of L1210 cell growth (Bernacki and Korytnyk, 1982).

As a terminal sugar in complex-type glycoproteins, L-fucose presents an attractive target for alterations in membrane properties. Several halogenated L-fucose analogues (**14** and **15**; Fig. 6-38) were studied with respect to their ability to be incorporated into tumor cell glycoproteins and to interfere with the incorporation of the corresponding natural sugars (Sufrin *et al.*, 1980). At 10^{-3} M, fluoro-L-fucose (6-FFu) was metabolized to GDP-6-FFu and incorporated into the macromolecular fraction of leukemia L1210 cells and inhibited the incorporation of fucose. With increasing size of the halogen substituent at C-6, cytotoxicity increased and effects on fucose incorporation decreased. C-2-Halogenated-2-deoxy-L-fucose analogues also showed significant toxicity to both L1210 and SW613 cells, with effects decreasing with the size of the halogen atom. O-Acetylation was found again to increase cytotoxicity (Sufrin *et al.*, 1980).

X = F, Cl, Br, I   X = H, Cl, Br, I

14   **15**

Figure 6-38. Halogenated fucose analogues (Sufrin *et al.*, 1980).

2-Deoxy-2-fluoro-L-fucose (2-FDFu) specifically inhibited L-[^{14}C]fucose incorporation into cultured mouse fibroblasts cells with apparent incorporation of the analogue (Winterbourne *et al.*, 1979).

L1210 leukemia cells have an unusually active cell-surface sialyltransferase system and, for this reason, have provided a good model system for the study of analogue-induced alterations in cell-surface function. 6-Deoxy-6-fluoro-D-galactose (6-FDGal; Fig. 6-39) was investigated by Morin *et al.* (1983) as an analogue which, if incorporated into the oligosaccharide of L1210 cell-surface glycoprotein, would block formation of the $\alpha(2 \rightarrow 6)$glycosidic bond with sialic acid (see above, Fig. 6-33). In fact, [^{3}H]-6-FDGal was incorporated into L1210 cell acid-insoluble material to a six- to sevenfold greater extent than was [^{3}H]Gal. This was accompanied by a dose- and time-dependent decrease (up to 50%) in ectosialyltransferase-catalyzed transfer of *N*-acetylneuraminic acid (NANA) at the cell surface. This decrease in activity was attributed to a decrease in the number of suitable membrane-bound acceptors as a result of the presence of the C-6 fluorine in incorporated 6-FDGal. Nonetheless, there was no decrease in the total cellular sialic acid content nor in the amount of neuraminidase-releasable NANA from the cell surface. The attachment of sialic acid to galactose by an $\alpha(2 \rightarrow 3)$glycosidic linkage is known to be common in mammalian glycoproteins (Fig. 6-33), and the involvement of such a bond in other glycoproteins of the L1210 cell was suggested as an explanation for these results.

6.3.2.4 Fluoromannose Analogues and Functions of High-Mannose-Type Glycoprotein

Patients suffering from I-cell disease are characterized by inclusions of large quantities of undigested glycosaminoglycans and glycolipids in lysosomes of the connective tissues. This reflects the absence of the required acid hydrolases in these lysosomes. These hydrolases are present in large amounts in the blood and urine of these patients. Thus, the active enzymes are synthesized but are not transported properly to the lysosomes. Evidence from the study of this genetic disorder has led to the hypothesis

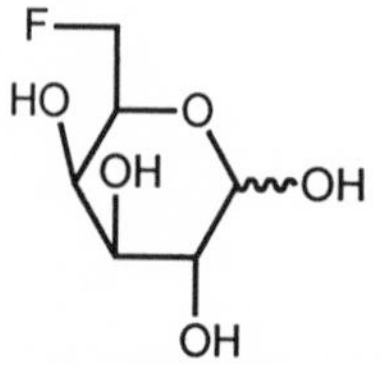

6-Deoxy-6-fluoro-D-galactose

Figure 6-39. 6-FDGal, an analogue designed to block formation of the $\alpha(2 \rightarrow 6)$glycosidic bond with sialic acid (Morin *et al.*, 1983).

that uptake of acid hydrolases into lysosomes is dependent on a common phosphomannosyl recognition marker [reviewed by Sly and Fischer (1982)]. Thus, in I-cell disease, active hydrolases are synthesized, but processing in the Golgi apparatus is incomplete due to the absence of mannose-6-phosphate (Man-6-P) as part of the high-mannose-type conjugate on the completed enzymes.

An enzyme required for the synthesis of the Man-6-P recognition marker is UDP-GlcNAc:lysosomal enzyme precursor N-acetylglucosamine-1-phosphotransferase (GlcNAc-P-transferase). Research to identify structural requirements for phosphorylation by this enzyme had established that a Man-α(1 → 2)-Man structural unit (Fig. 6-29) is recognized by the catalytic site, but the identity (terminal or penultimate) of the residue receiving the phosphate had remained unclear. To determine this, a 6-deoxy-6-fluoro-D-mannose residue was incorporated into the Man-α(1 → 2)-Man disaccharide as either the terminal or penultimate residue (16 and 17; Fig. 6-40). While Man-α(1 → 2)-6-FMan-α-methyl 17 was phosphorylated by GlcNAc-P-transferase from mouse fibroblasts at a comparable rate as was the unsubstituted disaccharide, enzyme activity was abolished by the presence of fluorine in the terminal residue. This establishes that phosphorylation of high-mannose-type residues by GlcNAc-P-transferase occurs at the terminal mannose (Madiyalakan et al., 1987).

The strategy of most of the above studies has been to develop agents which will block the assembly of dolichol-linked oligosaccharides or the

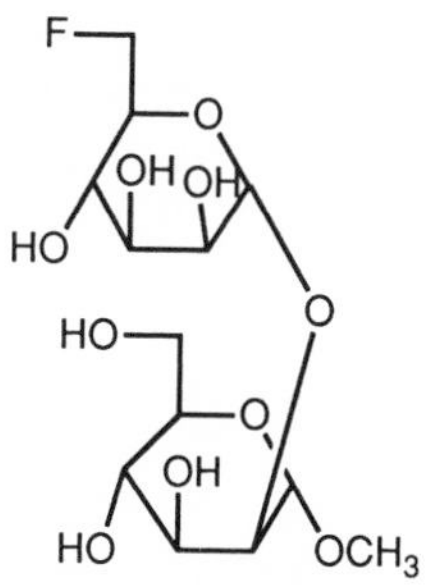

Methyl 2-O-(6-deoxy-6-fluoro-
α-D-mannopyranosyl)-α-D-manno-
pyranoside

16

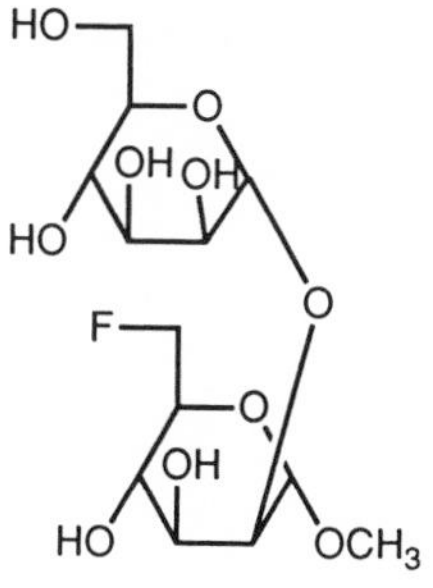

Methyl 2-O-(α-D-mannopyranosyl)-
6-deoxy-6-fluoro-α-D-manno-
pyranoside

17

Figure 6-40. Man-α(1 → 2)-6-FMan-α-methyl (17) is phosphorylated by GlcNAc-P-transferase while the structural isomer 6FMan-α(1 → 2)-6-Man-α-methyl (16) is not a substrate, demonstrating the requirement for a free terminal 6-OH group for activity (Madiyalakan et al., 1987).

subsequent attachment of critical sugar residues on an asparagine-linked precursor. The effects of bromoconduritol (Fig. 6-41) on complex- and high-mannose-type oligosaccharide biosynthesis were studied by Datema *et al.* (1982), based on the expectation that this α-glucosidase inhibitor would inhibit the trimming of glucose from $Glc_3Man_9(GlcNAc)_2$ (Fig. 6-28) required in glycoprotein synthesis. Treatment of influenza-infected cells with bromoconduritol led to inhibition of trimming of the innermost Glc residue from the precursor asparagine-linked oligosaccharide, as determined by the isolation of $GlcMan_9GlcNAc$, $GlcMan_8GlcNAc$, and $GlcMan_7NAc$ following pronase digestion and glucosamidase cleavage. This was accompanied by the inhibition of complex-type glycoprotein synthesis. There was no apparent effect on formation of the dolichol-linked oligosaccharide, transfer of oligosaccharide to protein, or synthesis of high-mannose-type glycoprotein. Thus, blockade of complex oligosaccharide synthesis appears to reflect the inhibition of glucose trimming.

6.3.2.5 Fluorinated Carbohydrates as Potential Inhibitors of Plasma Membrane O-Glycosylated Proteins

Most of the research to date has focused on the development of halogenated sugar analogues as antagonists for biosynthesis of asparagine-linked glycoproteins. Sharma *et al.* (1987, 1988) have recently prepared 2-acetamido-2,6-dideoxy-6-fluoro-D-galactose, the O-acetylated derivatives, and the N-deacetylated analogue (Fig. 6-42) as potential inhibitors of the attachment of *N*-acetyl-D-galactosamine to serine and threonine residues. Such O-linked moieties are present in many membrane glycoproteins. The fully acetylated analogue was active at $10^{-4}\ M$ as an inhibitor of L1210 leukemia cells in culture, while the N-deacetylated sugar was active at $10^{-3}\ M$.

6.3.2.6 Base-Halogenated Nucleotides and Glycoprotein Synthesis

As noted in Chapter 5 (Section 5.1.6), 5-fluorouracil is metabolized to fluorouridine nucleotide sugars (FUDP-sugars). While FUDP-sugars were

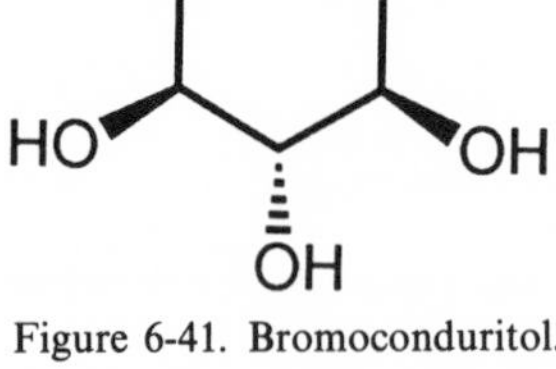

Figure 6-41. Bromoconduritol.

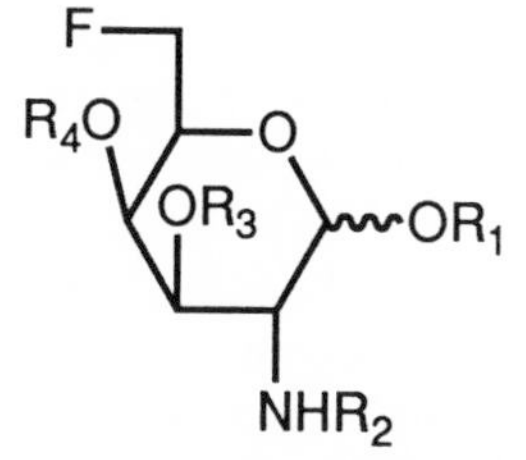

Figure 6-42. 2-Acetamido-2,6-dideoxy-6-fluoro-D-galactose, the O-acetylated derivatives, and the N-deacetylated analogue as potential inhibitors of the attachment of N-acetyl-D-galactosamine to serine and threonine residues (Sharma *et al.*, 1987, 1988).

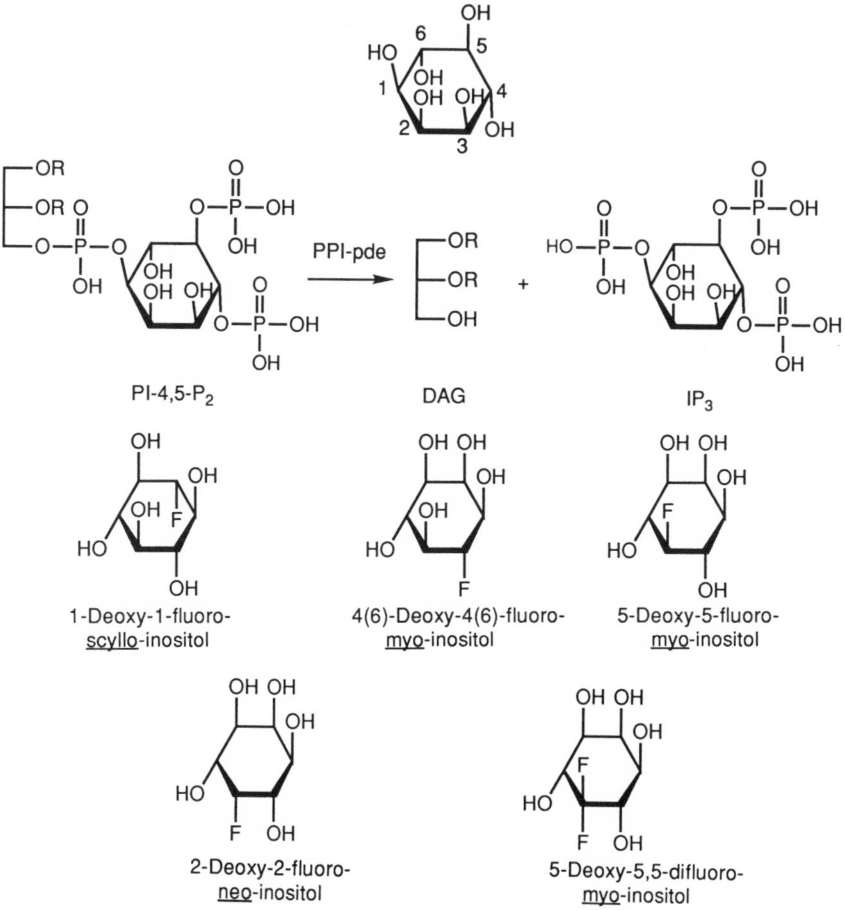

$R_1 = R_2 = R_3 = R_4 = H$
$R_1 = R_3 = R_4 = H; R_2 = Ac$
$R_1 = R_2 = R_3 = R_4 = Ac$

PI-4,5-P$_2$ PPI-pde   DAG   +   IP$_3$

1-Deoxy-1-fluoro-
scyllo-inositol

4(6)-Deoxy-4(6)-fluoro-
myo-inositol

5-Deoxy-5-fluoro-
myo-inositol

2-Deoxy-2-fluoro-
neo-inositol

5-Deoxy-5,5-difluoro-
myo-inositol

Figure 6-43. PI-4,5-P$_2$ hydrolysis and 1, 2-, 4(6)-, and 5-fluorinated analogues of inositol as potential substrates for phosphatidylinositol synthase (Moyer *et al.*, 1988).

found to behave comparably to UDP-sugars in several systems, this was found not to be the case with another potent antiviral drug, (E)-5-(2-bromovinyl)-2′-deoxyuridine (BVdU). In HSV-infected GMK cells, BVdU was found to cause an overall decrease in protein glycosylation as a result of inhibition of the synthesis of lipid-linked oligosaccharides. Processing of N-linked oligosaccharides also was inhibited, as was, to a small extent, the incorporation of glucosamine into O-linked oligosaccharides (Olofsson et al., 1985).

6.3.3 Incorporation of Fluorinated myo-Inositols into Phospholipids

The polyphosphoinositide phosphodiesterase (PPI-pde)-catalyzed hydrolysis of membrane-bound phosphatidylinositol-4,5-biphosphate (PI-4,5-P_2) provides the link between polyphosphoinositide turnover and several receptor-controlled cellular functions (see Chapter 2, Section 2.11, in Vol. 9A of this series). Moyer et al. (1988) have targeted the processes involved in PI-4,5-P_2 biosynthesis (Fig. 6-43) for drug design and have prepared 1, 2-, 4(6)-, and 5-fluorinated analogues of inositol (Fig. 6-43). Of these, 5-deoxy-5-fluoro-myo-inositol and 2-deoxy-2-fluoro-neo-inositol alone retained substrate activity for phosphatidylinositol synthase. The former was taken up into L1210 cells and incorporated into a lipid which was chromatographically similar, but not identical, to phosphatidylinositol.

6.4 Deoxyfluoro Carbohydrates as Tools for Probing Binding Subsites of Antisaccharide Antibodies

In previous sections, the use of deoxy- and deoxyfluorocarbohydrates to study transport mechanisms and enzyme specificities has been discussed. Glaudemans and co-workers have provided an example of the skillful extension of this strategy to the study of the structure of binding sites for monoclonal antisaccharide antibodies (Glaudemans and Kovac, 1988, and references therein). The importance of acquiring additional detailed information on the functioning of the immune response is seen in the devastating nature of immune deficiencies. Information from research on antibody–antigen binding can also be applied to the analysis of the nature of intermolecular interactions in general. To put this work into perspective, a brief discussion of the structure of immunoglobulins and of the origin and significance of monoclonal antibodies in the study of immunology follows. More thorough and concise discussions can be found in recent biochemistry textbooks (e.g., Stryer, 1988).

Specific antibodies present on the surface of B lymphocytes recognize and combine with antigens as part of an animal's multifaceted defense mechanism against the intrusion of foreign substances. While each lymphocyte has a preordained specificity for a given antigen, the immense variability of these antibodies provides protection against virtually any invading macromolecule, including proteins, polysaccharides, nucleic acids, or haptens attached to carriers. Binding of the antigen signals the differentiation of B lymphocytes to antibody-secreting plasma cells. Because a given antigen typically has several antigenic determinants, more than one type of antigen-specific B lymphocyte may be stimulated to differentiate, and a heterogeneous pool of antibodies normally is produced by the challenge of a foreign macromolecule. The problems in studying antigen–antibody binding mechanisms associated with this heterogeneity have been overcome by the increased availability of monoclonal antibodies, initially obtained by random selection from myeloma cell lines and now produced selectively by technology involving the fusion of a specific antibody-producing cell with a myeloma cell. In this latter procedure, the resulting *hybridoma* proliferates indefinitely, a property of the parent myeloma cell, and produces large amounts of the specific antibody characteristic of the other parent.

The characteristic Y-shaped or T-shaped immunoglobulin structure produced by association of two heavy (H) and two light (L) chains is now quite familiar (shown schematically in Fig. 6-44). Each H and each L chain has a variable and a constant region, and it is in the variable region that antibody specificity resides. The antibody-binding site is formed by amino acid residues of hypervariable portions of both H and L chains, residues that typically form a cleft between the H and L chains and that contribute electrostatic, hydrophobic, hydrogen bonding, and van der Waals type binding energy to the interaction.

The monoclonal myeloma immunoglobulin A, IgA J539, specific for $\beta(1 \rightarrow 6)$-D-galactopyrans (Fig. 6-45), has been the subject of detailed investigations of its binding mechanism. By strategic use of chemically altered ligand probes, including extensive use of deoxyfluoro analogues, the antibody has been shown to bind intercatenarily along the length of the linear $\beta(1 \rightarrow 6)$-D-galactopyranan antigen and to have contact with four sequential galactosyl residues, the four binding subsites have been mapped, the relative and absolute binding energies thereof have been determined, and the orientation of the polysaccharide in the binding region has been defined (Glaudemans, 1987; Glaudemans and Kovac, 1988, and references therein). In order to confirm the suspected contribution of hydrogen bonding to the binding energy and to provide a measure of the relative importance of each hydroxyl group in this interaction, the binding characteristics

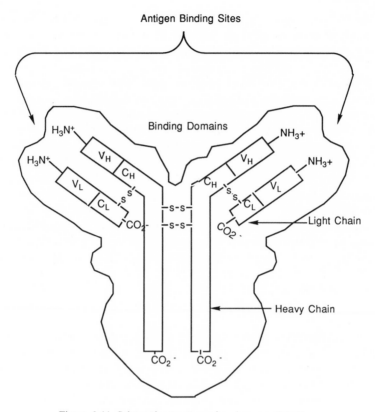

Figure 6-44. Schematic structure of an immunoglobulin.

of several deoxyfluorogalactosides were studied as determined by tryptophan fluorescence titration (Glaudemans *et al.*, 1984). The amino acid sequence of IgA J539 is known, and interpretation of the data was facilitated by the fact that antigen binding to the highest affinity site is accompanied by an increase in fluorescence of a putatively proximate tryptophan residue (Trp-33H), as shown by the ligand-induced fluorescence

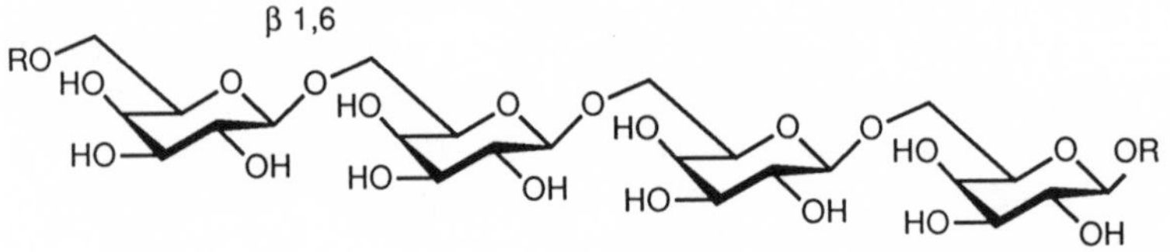

Figure 6-45. $\beta(1 \to 6)$-D-Galactopyrans   recognized   by   the   monoclonal   myeloma immunoglobulin A, IgA J539.

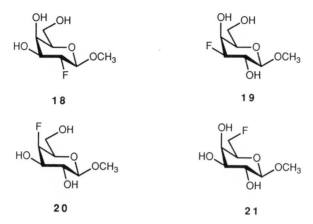

18   **1 9**

2 0   **2 1**

Figure 6-46. Deoxyfluorogalactosides used to probe the binding sites of IgA J539.

change (LIFC) accompanying the binding of a simple monogalactoside. A second tryptophan (Trp-91L) is proposed to be affected by binding at the next highest affinity site (LIFC is enhanced by binding of a digalactoside), a fact which has aided in defining the orientation of the reducing end of the galactan chain [for a complete discussion of the structural details of this immunoglobulin, see Glaudemans and Kovac (1988)]. Decreased binding of methyl-2-fluorogalactoside (**18**) and methyl-3-fluorogalactoside (**19**), but not of the 4-fluoro analogue **20** (Fig. 6-46), implicated the importance of the C-2 and C-3 hydroxyl groups as hydrogen bond *donors*, mandating

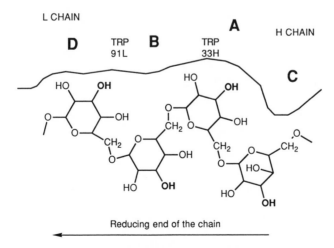

Reducing end of the chain

Figure 6-47. A model for the binding sites of IgA J539 (Glaudemans and Kovac, 1988).

orientation of the pyranose ring such that these hydroxyls can interact with the binding segment. Having defined the highest affinity site (A), binding affinities of fluorinated analogues of di-, tri-, and tetragalactosides were analyzed, following the principles that (1) site A requires the presence of C-2 and C-3 hydroxyl groups, (2) if possible, any oligogalactoside will occupy those sites that give the combined highest binding energy, (3) the reducing end will be oriented toward Trp-91L, and (4) binding energy will increase with increasing number of residues, up to four. By the strategic situation of fluorinated galactose residues, the oligogalactosides could be "walked" along the four contiguous binding sites, with predictable changes in binding affinities and fluorescence. Using this strategy, four binding sites were identified, designated A, B, C, and D in decreasing order of binding affinities, and a model for the binding sites of IgA J539 was constructed (Fig. 6-47) (Glaudemans and Kovac, 1988).

This same strategy was used to study seven other $\beta(1 \rightarrow 6)$galacto-pyaranan specific monoclonal antibodies, and the same K_a pattern was observed for subsites C, A, B, and D as found for IgA J539. This indicates that all have the same subsite arrangement. On the basis of comparison of amino acid sequence homologies of these immunoglobulins, other anticarbohydrate antibodies, and monoclonal antibodies of varied specificities, Glaudemans and Kovac (1988) have suggested that Asn-58H, Pro-94L, and Trp-33H define the subsite with the highest galactosyl binding affinity.

The dextran-specific myeloma monoclonal immunoglobulin IgA W3129 is precipitated by *branched* $\alpha(1 \rightarrow 6)$-linked dextrans, but not linear

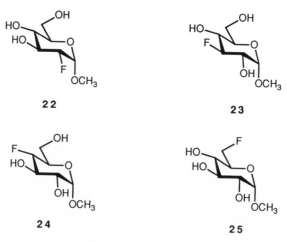

22   **23**

24   **25**

Figure 6-48. Fluorinated methyl α-D-glucosides used to probe the binding sites of the dextran-specific myeloma monoclonal immunoglobulin IgA W3129 (Glaudemans *et al.*, 1989).

ones, implicating the nonreducing terminus of the polysaccharide chain as an important binding determinant. This specificity for the chain-terminal glucosyl residues shown only by IgA W3129, the amino acid sequence of which is known, is unusual among characterized immunoglobulins and made it an attractive target for binding site studies using the above ligand analogue strategy. Accordingly, methyl α-D-glucosides and methyl α-D-glycosides of isomalto-oligosaccharides, in which selected OH groups were replaced by fluorine, were used to elucidate the hydrogen bonding requirements for binding and to locate the position of particular glucose moieties in the tetrose antigen determinant. Affinity constants were determined by ligand-induced changes in tryptophan fluorescence measurements (LIFC).

Methyl α-D-glucopyranoside binds to the immunoglobulin with a total free energy of binding equal to 62% of that of the maximally binding tetrasaccharide. However, methyl 6-deoxy-6-fluoro-α-D-glucopyranoside (**25**; Fig. 6-48) induced no measurable LIFC, revealing the requirement for a free 6-OH, presumably as a hydrogen bond donor, at the highest affinity subsite (A). Such a requirement readily explains the specificity of the antibody for branched dextrans, since only the nonreducing end(s) of the polysaccharide have a free 6-OH group. The 4-fluoro analogue, **24**, also showed lack of significant binding at subsite A, implicating the 4-OH as a second hydrogen bond donor. The 2-fluoro analogue, **22**, had comparable binding to that of the unsubstituted glucoside, while the 3-fluoro analogue, **23**, showed an affinity fourfold higher. This latter result may reflect an increased acidity of the 4-OH group as a consequence of the neighboring electronegative fluorine. The 3 position also was shown to be largely insensitive to bulky substituents, since 3-*O*-(6-deoxy-6-fluoro-*β*-D-glucopyranosyl)glucose (Fig. 6-49) was found to bind with 70% of maximal binding. This result suggests that the binding site is relatively exposed to the antigenic determinant of the dextran. The presence of additional binding sites could be confirmed by the observation of weak binding of 6-deoxy-6-fluoro analogues of di-, tri-, and tetraglucosides, showing that binding energy lost as a result of the absence of the 6-OH terminus can be partially compensated by an anchoring of the ligand by the additional sites. However, in these "anchored" ligands, the 6-fluorinated glucosyl terminus has a relatively constant K_a for its site (A), as determined by LIFC,

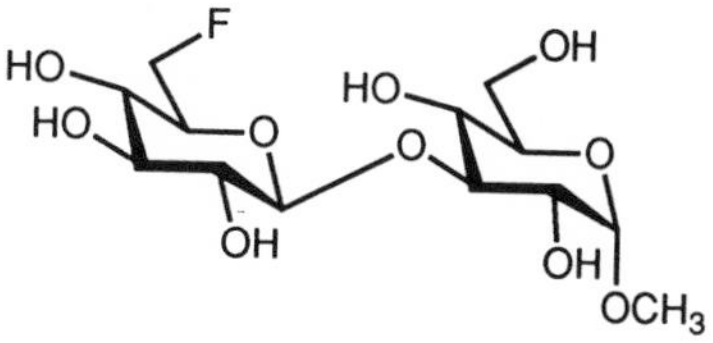

Figure 6-49. 3-*O*-(6-Deoxy-6-fluoro-*β*-D-glucopyranosyl)glucose, an example of a fluorinated glycoside used to study the binding sites of IgA W3129 (Glaudemans *et al.*, 1989).

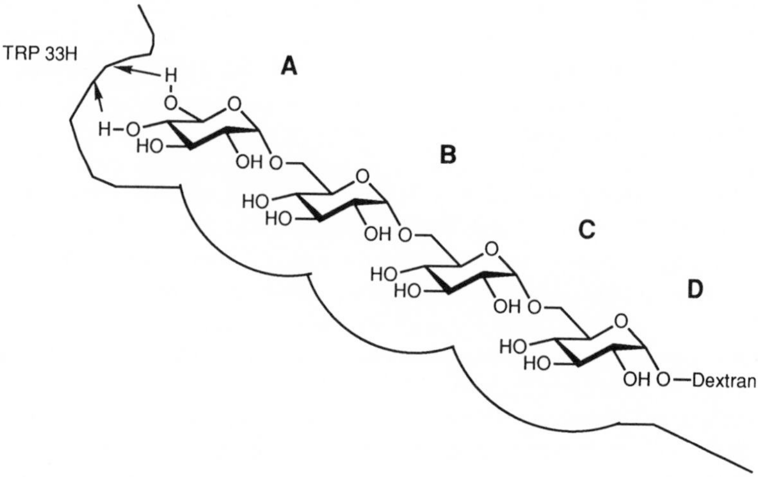

Figure 6-50. A model for the binding sites for dextran on IgA W3129 (Glaudemans *et al.*, 1989).

demonstrating a relatively free exchange independent of the anchor. From these and other data, a model for the binding sites for dextran in this immunoglobulin was constructed as shown in Fig. 6-50 (Glaudemans *et al.*, 1989).

6.5 Effects of Halogenation on the Sweetness of Sugars

An obvious physiological property of many sugars is their sweet taste. However, on a molar basis, natural sugars, including sucrose, are only moderately sweet compared to a number of artificial sweeteners. Indeed, elucidation of the molecular basis for sweetness has been quite challenging because of the wide variability of compounds that have a sweet taste [for a recent thorough review of this subject, see Lee (1987a)]. A major step in understanding the structural basis of sweetness came from the advancement of the "AH,B" theory of sweetness by Shallenberger and Acree (1967). A common feature of many sweet compounds is the presence of two electro-negative atoms, A and B, separated by a distance greater than 2.5 Å but less than 4 Å, where AH is a proton donor and B is a proton acceptor. Intermolecular interaction of AH and B with the receptor stimulates the sensation of sweetness (Fig. 6-51). According to this theory, gauche or staggered conformations of vicinal glycols of sugars constitute the saporous units. Subsequently, an AH,B,X tripartite glucophore was proposed (Fig.

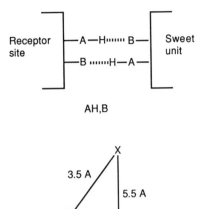

AH,B

AH,B,X

Figure 6-51. The "AH,B" theory of sweetness (Shallenberger and Acree, 1967) and the AH,B,X tripartite glucophore (Kier, 1972).

6-51), wherein the third (hydrophobic site) (X) was proposed to explain high sweetness of nonsugars such as certain D-amino acids (Kier, 1972).

A search for chemically modified sugars having enhanced sweetness was unsuccessful until the synthesis by Hough and Phadnis (1976) of 4,6-dichloro-4,6-dideoxy-α-D-galactopyranosyl 1,6-dichloro-1,6-dideoxy-β-D-fructofuranoside (1′,4,6,6′-*galacto*-sucrose) (**26**; Fig. 6-52), an analogue that was found to be several hundred times sweeter than sucrose. Subsequent examination of other deoxyhalosucrose analogues showed that halogen substitution on C-4, C-1′, and C-6′ is important for the enhancement, while a free 6-OH increases the sweetness of the halogenated analogues. Thus, substitution of chlorine at C-4, C-1′, and C-6′ positions

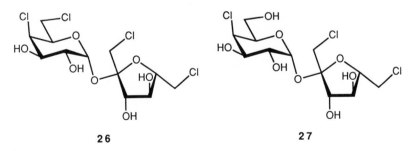

26   **27**

Figure 6-52. Intensely sweet sugars: 1′,4,6,6′-tetrachloro-*galacto*-sucrose (**26**) (Hough and Phadnis, 1976) and 1′,4,6′-trichloro-*galacto*-sucrose (**27**) (Lee, 1987a, and references therein).

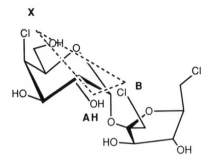

Figure 6-53. The extreme sweetness of trichlorogalactosucrose based on an optimal configuration of an AH,B,X tripartite glucophore (Hough and Khan, 1978).

of sucrose resulted in a 2000-fold increase in sweetness (**27**; Fig. 6-52). Other deoxyhalosucrose analogues having sweetness several thousand times that of sucrose were found (Lee, 1987a, and references therein). This enhanced sweetness appears unique to sucrose derivatives, and, in fact, other deoxyhalo sugars tested were extremely bitter (Lee, 1987a, and references therein). Hough and Khan (1978) rationalized the extreme sweetness of trichlorosucrose and trichlorogalactosucrose on the basis of an optimal configuration of an AH,B,X tripartite glucophore, an example of which is shown in Fig. 6-53. Mathlouthi *et al.* (1986) have proposed that the enhanced hydrophobicity of the CH_2Cl group plays the additional role of stabilizing the AH $\cdots$ B hydrogen bond between the sugar and the receptor. Lee (1987b) recently has extended the series of intensely sweet chlorodeoxy sucrose analogues with the first successful chlorination at C-4′ of sucrose. Further analyses of structure–sweetness relationships of this growing family of supersweet sugar analogues will clearly be important in defining more clearly the mechanism of glucophore–receptor interactions.

6.6 Conclusion

As in other chapters, fluorinated analogues have received the most attention in this review. The use of fluorinated carbohydrates in the study of cell-surface glycoproteins has been one topic stressed, since this is a particularly active area of current research, with many potential medicinal applications. Other topics, including phenomena involving recognition of carbohydrates by such macromolecular systems as enzyme active sites, transport mechanisms, and immunoglobulins are likewise areas of much current activity. Fluorine-18-labeled carbohydrates are becoming more readily available with new synthetic methodology, and broader applications in PET scanning are certain to result. All of these areas have benefited immensely from new synthetic methodologies.

References

Agbanyo, M., and Taylor, N. F., 1986. Incorporation of 3-deoxy-3-fluoro-D-glucose into glycogen and trehalose in fat body and flight muscle in *Locusta migratoria, Biosci. Rep.* 6:309–316.

Barnett, J. E. G., 1972. Fluorine as a substituent for oxygen in biological systems: Examples in mammalian membrane transport and glycosidase action, in *Ciba Foundation Symposium: Carbon–Fluorine Compounds: Chemistry, Biochemistry, and Biological Activities,* Associated Scientific Publishers, New York, pp. 95–115.

Barnett, J. E. G., Holman, G. D., and Munday, K. A., 1973. Structural requirements for binding to the sugar-transport system of the human erythrocyte, *Biochem. J.* 131:211–221.

Barnett, J. E. G., Holman, G. D., Chalkley, R. A., and Munday, K. A., 1975. Evidence for two asymmetric conformational states in the human erythrocyte sugar-transport system, *Biochem. J.* 145:417–429.

Bernacki, R. J., and Korytnyk, W., 1982. Development of membrane sugar and nucleotide sugar analogues as potential inhibitors or modifiers of cellular glycoproteins, in *The Glycoproteins,* Vol. IV, *Glycoproteins, Glycolipids, and Proteoglycans, Part B* (M. I. Horowitz, ed.), Academic Press, New York, pp. 245–263.

Bessel, E. M., and Thomas, P., 1973a. The effect of substitution at C-2 of D-glucose 6-phosphate on the rate of dehydrogenation by glucose 6-phosphate dehydrogenation by glucose 6-phosphate dehydrogenase (from yeast and from rat liver), *Biochem. J.* 131:83–89.

Bessel, E. M., and Thomas, P., 1973b. The deoxyfluoro-D-glucopyranose 6-phosphates and their effect on yeast glucose phosphate isomerase, *Biochem. J.* 131:77–82.

Bessel, E. M., Foster, A. B., and Westwood, J. H., 1972. The use of deoxyfluoro-D-glucopyranoses and related compounds in a study of yeast hexokinase specificity, *Biochem. J.* 128:199–204.

Bigham, E. C., Gragg, C. E., Hall, W. R., Kelsey, J. E., Mallory, W. R., Richardson, D. C., Benedict, C., and Ray, P. H., 1984. Inhibition of arabinose 5-phosphate isomerase. An approach to the inhibition of bacterial lipopolysaccharide biosynthesis, *J. Med. Chem.* 27:717–726.

Briley, P. A., Eisenthal, R., and Harrison, R., 1975. Fluorine as a hydroxy analogue, *Biochem. J.* 145:501–507.

Card, P. J., 1985. Synthesis of fluorinated carbohydrates, *J. Carbohydr. Res.* 4:451–487.

Card, P. J., and Hitz, W. D., 1984. Synthesis of 1′-deoxy-1′-fluorosucrose via sucrose synthetase mediated coupling of 1-deoxy-1-fluorofructose with uridine diphosphate glucose, *J. Am. Chem. Soc.* 106:5348–5350.

Card, P. J., Hitz, W. D., and Ripp, K. G., 1986. Chemoenzymatic syntheses of fructose-modified sucroses via multienzyme systems. Some topographical aspects of the binding of sucrose to a sucrose carrier protein. *J. Am. Chem. Soc.* 108:158–161.

Chapman, A., Fujimoto, K., and Kornfeld, S., 1980. The primary glycosylation defect in class E thy-1-negative mutant mouse lymphoma cells is an inability to synthesize dolichol-P-mannose, *J. Biol. Chem.* 255:4441–4446.

Chiba, S., Brewer, C. F., Okada, G., Matsui, H., and Hehre, E. J., 1988. Stereochemical studies of D-glucal hydration by α-glucosidases and exo-α-glucanases: Indications of plastic and conserved phases in catalysis by glycosylases, *Biochemistry* 27:1564–1569.

Datema, R., Schwarz, R. T., and Jankowski, A. W., 1980a. Fluoroglucose-inhibition of protein glycosylation *in vivo.* Inhibition of mannose and glucose incorporation into lipid-linked oligosaccharides, *Eur. J. Biochem.* 109:331–341.

Datema, R., Schwarz, R. T., and Winkler, J., 1980b. Glycosylation of influenza virus proteins in the presence of fluoroglucose occurs via a different pathway, *Eur. J. Biochem.* 110:355–361.

Datema, R., Romero, P. A., Legler, G., and Schwarz, R. T., 1982. Inhibition of formation of complex oligosaccharides by the glucosidase inhibitor bromoconduritol, *Proc. Natl. Acad. Sci. USA* 79:6787–6791.

Dick, A. P., and Harik, S. I., 1986. Distribution of the glucose transporter in the mammalian brain, *J. Neurochem.* 46:1406–1411.

Eager, R. G., Jr., Bachovchin, W. W., and Richards, J. H., 1975. Mechanism of action of adenosylcobolamin: 3-Fluoro-1,2-propanediol as substrate for propanediol dehydrase—mechanistic implications, *Biochemistry* 14:5523–5528.

Eisenthal, R., Harrison, R., Lloyd, W. J., and Taylor, N. F., 1972. Activity of fluoro and deoxy analogues of glycerol as substrates and inhibitors of glycerol kinase, *Biochem. J.* 130:199–205.

Fondy, T. P., and Emlich, C. A., 1981. Haloacetamido analogues of 2-amino-2-deoxy-D-mannose. Syntheses and effects on tumor-bearing mice, *J. Med. Chem.* 24:848–852.

Fondy, T. P., Changes, G. S., and Reza, M. J., 1970. Synthesis of 1-halo analogues of DL-glycerol 3-phosphate and their effects on glycerol phosphate dehydrogenase, *Biochemistry* 9:3272–3280.

Fondy, T. P., Roberts, S. B., Tsiftsoglou, A. S., and Sartorelli, A. C., 1978. Haloacetamido analogues of 2-amino-2-deoxy-D-glucose and 2-amino-2-deoxy-D-galactose. Syntheses and effects on the Friend murine erythroleukemia, *J. Med. Chem.* 21:1222–1225.

Fondy, T. P., Pero, R. W., Karker, K. L., Ghanges, G. S., and Batzold, F. H., 1974. Synthesis of L-1-deoxyfluoroglycerol and its 3-phosphate ester. Effects of the L and D enantiomers in BDF₁ mice, *J. Med. Chem.* 17:697–702.

Fowler, J. S., and Wolf, A. P., 1986. 2-Deoxy-2-[^{18}F]fluoro-D-glucose for metabolic studies: Current status, *Appl. Radiat. Isot.* 37:663–668.

Genghof, D. S., Brewer, C. F., and Hehre, E. J., 1978. Preparation and use of α-maltosyl fluoride as a substrate by beta amylase, *Carbohydr. Res.* 61:291–299.

Glaudemans, C. P. J., 1987. Seven structurally different murine monoclonal galactan-specific antibodies show identity in their galactosyl-binding subsite arrangements, *Mol. Immunol.* 24:371–377.

Glaudemans, C. P. J., and Kovac, P., 1988. Deoxyfluoro carbohydrates as probes of binding sites of monoclonal antisaccharide antibodies, in *Fluorinated Carbohydrates, Chemical and Biological Aspects* (N. F. Taylor, ed.), ACS Symposium Series, No. 374, American Chemical Society, Washington, D.C., pp. 78–108.

Glaudemans, C. P. J., Kovac, P., and Rasmussen, K., 1984. Mapping of subsites in the combining area of monoclonal anti-galactan immunoglobulin A J539, *Biochemistry* 23:6732–6736.

Glaudemans, C. P. J., Kovac, P., and Rao, A. S., 1989. The subsites of monoclonal anti-dextran W3129, *Carbohydr. Res.* 190:267–277.

Grier, T. J., and Rasmussen, J. R., 1984. 4-Deoxy-4-fluoro-D-mannose inhibits the glycosylation of the G protein of vesicular stomatitis virus, *J. Biol. Chem.* 259:1027–1030.

Halama, J. R., Gatley, J. S., DeGrado, T. R., Bernstein, D. R., Ng, C. K., and Holden, J. E., 1984. Validation of 3-deoxy-3-fluoro-D-glucose as a glucose transport analogue in rat heart, *Am. J. Physiol.* 247:H754–H759.

Halton, D. M., Taylor, N. F., and Lopes, D. P., 1980. The uptake of 3-deoxy-3-fluoro-D-glucose by synaptosomes from rat brain cortex, *J. Neurosci. Res.* 5:241–252.

Hartman, F. C., 1970. Haloacetol phosphates. Potential active-site reagents for aldolase, triose phosphate isomerase, and glycerol phosphate dehydrogenase. II. Inactivation of aldolase, *Biochemistry* 9:1783–1791.

Hartman, F. C., 1971. Haloacetol phosphates. Characterization of the active site of rabbit muscle triosphosphate isomerase, *Biochemistry* 10:146–154.

Hehre, E. J., Okada, G., and Genghof, D. S., 1969. Configurational specificity: Unappreciated key to understanding enzymic reversions and *de novo* glycosidic bond synthesis. 1. Reversal of hydrolysis by α, β, and glucoamylases with donors of correct anomeric form, *Arch. Biochem. Biophys.* 135:75–89.

Hehre, E. J., Brewer, C. F., and Genghof, D. S., 1979. Scope and mechanism of carbohydrase action. Hydrolytic and nonhydrolytic actions of β-amylase on α- and β-maltosyl fluoride, *J. Biol. Chem.* 254:5942–5950.

Hehre, E. J., Sawai, T., Brewer, C. F., Nakano, M., and Kanda, T., 1982. Trehalase: Stereocomplementary hydrolytic and glucosyl transfer reactions with α- and β-D-glucosyl fluoride, *Biochemistry* 21:3090–3097.

Hitz, W. D., 1988. Sucrose transport in plants using monofluorinated sucroses and glucosides, in *Fluorinated Carbohydrates, Chemical and Biological Aspects* (N. F. Taylor, ed.), ACS Symposium Series, No. 374, American Chemical Society, Washington, D.C., pp. 138–155.

Hitz, W. D., Card, P. J., and Ripp, K. G., 1986. Substrate recognition by a sucrose transporting protein, *J. Biol. Chem.* 261:11986–11991.

Hough, L., and Khan, R., 1978. Intensification of sweetness, *Trends Biochem. Sci.* 3:61–63.

Hough, L., and Phadnis, S. P., 1976. Enhancement in the sweetness of sucrose, *Nature* 263:800.

Jeffery, J., and Jörnvall, H., 1983. Enzyme relationships in a sorbitol pathway that bypasses glycolysis and pentose phosphates in glucose metabolism, *Proc. Natl. Acad. Sci. USA* 80:901–905.

Kanazawa, Y., Momozono, Y., Ishikawa, M., Yamada, T., Yamane, H., Haradahira, T., Maeda, M., and Kojima, M., 1986. Metabolic pathway of 2-deoxy-2-fluoro-D-glucose studied by F-19 NMR, *Life Sci.* 39:737–742.

Kasumi, T., Brewer, C. F., Reese, E. T., and Hehre, E. J., 1986. Catalytic versatility of trehalase: Synthesis of α-D-glucopyranosyl α-D-glucopyranoside from β-D-glucosyl fluoride and α-D-xylose, *Carbohydr. Res.* 146:39–49.

Kasumi, T., Tsumuraya, Y., Brewer, C. F., Kersters-Hilderson, H., Claeyssens, M., and Hehre, E. J., 1987. Catalytic versatility of *Bacillus pumilus* β-xylosidase: Glycosyl transfer and hydrolysis promoted with α- and β-D-xylosyl fluoride, *Biochemistry* 26:3010–3016.

Kent, P. W., 1972. Synthesis and reactivity of fluorocarbohydrates, in *Ciba Foundation Symposium: Carbon–Fluorine Compounds: Chemistry, Biochemistry, and Biological Activities,* Associated Scientific Publishers, New York, pp. 169–213.

Kent, P. W., 1988. Retrospect and prospect, in *Fluorinated Carbohydrates, Chemical and Biological Aspects* (N. F. Taylor, ed.), ACS Symposium Series, No. 374, American Chemical Society, Washington, D.C., pp. 1–12.

Kier, L. B., 1972. A molecular theory of sweet taste, *J. Pharm. Sci.* 61:1394–1397.

Kitahata, S., Brewer, C. F., Genghof, D. S., Sawai, T., and Hehre, E. J., 1981. Scope and mechanism of carbohydrase action. Stereocomplementary hydrolytic and glucosyl-transferring actions of glucoamylase and glucodextranase with α- and β-D-glucosyl fluoride, *J. Biol. Chem.* 256:6017–6026.

Kornfeld, S., 1982. Oligosaccharide processing during glycoprotein biosynthesis, in *The Glycoproteins*, Vol. III, *Glycoproteins, Glycolipids, and Proteoglycans, Part A* (M. I. Horowitz, ed.), Academic Press, New York, pp. 3–23.

Kornfeld, R., and Kornfeld, S., 1985. Assembly of asparagine-linked oligosaccharide, *Annu. Rev. Biochem.* 54:631–664.

Kwee, I. L., Nakada, T., and Card, P. J., 1987. Noninvasive demonstration of *in vivo* 3-fluoro-3-deoxy-D-glucose metabolism in rat brain by ^{19}F nuclear magnetic resonance spectroscopy: Suitable probe for monitoring cerebral aldose reductase activity, *J. Neurochem.* 49:428–433.

Lee, C.-K., 1987a. Chemistry and biochemistry of sweetness, in *Advances in Carbohydrates Chemistry and Biochemistry*, Vol. 45 (R. S. Tipson and D. Horton, eds.), Academic Press, San Diego, pp. 199–351.

Lee, C.-K., 1987b. Synthesis of an intensely sweet chlorodeoxysucrose. Mechanism of 4′-chlorination of sucrose by sulfuryl chloride, *Carbohydr. Res.* 162:53–63.

Madiyalakan, R., Jain, R. K., and Matta, K. L., 1987. Phosphorylation of the α1,2-linked mannosylsaccharide by *N*-acetylglucosamine-1-phosphotransferase from fibroblasts occurs at the terminal mannose, *Biochem. Biophys. Res. Commun.* 142:354–358.

Mathlouthi, M., Seuvre, A.-M., and Birch, G. G., 1986. Relationship between the structure and the properties of carbohydrates in aqueous solutions: Sweetness of chlorinated sugars, *Carbohydr. Res.* 152:47–61.

McDowell, W., Datema, R., Romero, P. A., and Schwarz, R. T., 1985. Mechanism of inhibition of protein glycosylation by the antiviral sugar analogue 2-deoxy-2-fluoro-D-mannose: Inhibition of synthesis of Man(GlcNAc)$_2$-PP-Dol by the guanosine diphosphate ester, *Biochemistry* 24:8145–8152.

McDowell, W., Grier, T. J., Rasmussen, J. R., and Schwarz, R. T., 1987. The role of C-4-substituted mannose analogues in protein glycosylation, *Biochem. J.* 248:523–531.

Miles, R. J., and Pirit, S. J., 1973. Inhibition by 3-deoxy-3-fluoro-D-glucose of the utilization of lactose and other carbon sources by *Escherichia coli*, *J. Gen. Microbiol.* 76:305–318.

Morin, M. J., Porter, C. W., Petrie, C. R., III, Korytnyk, W., and Bernacki, R. J., 1983. Effects of a membrane sugar analogue, 6-deoxy-6-fluoro-D-galactose, on the L1210 leukemic cell ectosialyltransferase system, *Biochem. Pharmacol.* 32:553–561.

Moyer, J. D., Reizes, O., Malinowski, N., Jiang, C., and Baker, D., 1988. Fluorinated analogues of *myo*-inositols as biological probes of phosphatidylinositol metabolism, in *Fluorinated Carbohydrates, Chemical and Biological Aspects* (N. F. Taylor, ed.), ACS Symposium Series, No. 374, American Chemical Society, Washington, D.C., pp. 43–58.

Nakada, T., and Kwee, I. L., 1986. *In vivo* metabolism of 2-fluoro-2-deoxy-D-glucose in the aldose reductase pathway in rat brain demonstrated by ^{19}F NMR spectroscopy, *Biochem. Arch.* 2:52–61.

Nakada, T., Kwee, I. L., and Conboy, C. B., 1986. Noninvasive *in vivo* demonstration of 2-fluoro-2-deoxy-D-glucose metabolism beyond the hexokinase reaction in rat brain by ^{19}F nuclear magnetic resonance spectroscopy, *J. Neurochem.* 46:198–201.

Olofsson, S., Lundstroem, M., and Datema, R., 1985. The antiherpes drug (*E*)-5-(2-bromovinyl)-2′-deoxyuridine (BVDU) interferes with formation of N-linked and O-linked oligosaccharides of the herpes simplex virus type I glycoprotein C, *Virology* 147:201–205.

Paul, B., Bernacki, R. J., and Korytnyk, W., 1980. Synthesis and biological activity of some 1-N-substituted 2-acetamido-2-deoxy-β-D-glycopyranosylamine derivatives and related analogues, *Carbohydr. Res.* 80:99–115.

Penglis, A. A. E., 1981. Fluorinated carbohydrates, *Adv. Carbohydr. Res.* 38:195–285.

Pero, R. W., Babiarz-Tracy, P., and Fondy, T. P., 1977. 3-Fluoro-1-hydroxypropan-2-one (fluorohydroxyacetone) and some esters. Syntheses and effects in BDF$_1$ mice, *J. Med. Chem.* 20:644–647.

Phelps, M. E., and Mazziotta, J. C., 1985. Positron emission tomography: Human brain function and biochemistry, *Science* 228:799–809.

Phelps, M. E., Huang, S. C., Hoffman, E. J., Selin, C., Sokoloff, L., and Kuhl, D. E., 1979. Tomographic measurement of local cerebral glucose metabolic rate in humans with (F-18)2-fluoro-2-deoxy-D-glucose: Validation of method, *Ann. Neurol.* 6:371–388.

Rees, W. D., and Holman, G. D., 1981. Hydrogen bonding requirements for the insulin-sensitive sugar transport system of rat adipocytes, *Biochim. Biophys. Acta* 646:251–260.

Riley, G. J., and Taylor, N. F., 1973. The interaction of 3-deoxy-3-fluoro-D-glucose with the hexose-transport system of human erythrocytes, *Biochem. J.* 135:773–777.

Romaschin, A., and Taylor, N. F., 1981. The *in vivo* effects of 3-deoxy-3-fluoro-D-glucose metabolism on respiration in *Locusta migratoria, Can. J. Biochem.* 59:262–268.

Romaschin, A., Taylor, N. F., Smith, D. A., and Lopes, D., 1977. The metabolism of 3-deoxy-3-fluoro-D-glucose by *Locusta migratoria* and *Schistocerca gregaria, Can. J. Biochem.* 55:369–375.

Schelbert, H. R., and Phelps, M. E., 1984. Positron computer tomography for the *in vivo* assessment of regional myocardial function, *J. Mol. Cell. Cardiol.* 16:683–693.

Schmidt, M. F. G., Biely, P., Krátkyz, and Schwarz, R. T., 1978. Metabolism of 2-deoxy-2-fluoro-D-[^{3}H]glucose and 2-deoxy-2-fluoro-[^{3}H]mannose in yeast and chick-embryo cells, *Eur. J. Biochem.* 87:55–68.

Schwarz, R. T., and Datema, R., 1980. Inhibitors of protein glycosylation, *Trends Biochem. Sci.* 5:65–67.

Schwarz, R. T., and Datema, R., 1982a. The lipid pathway of protein glycosylation and its inhibitors: The biological significance of protein-bound carbohydrates, *Adv. Carbohydr. Chem. Biochem.* 40:287–379.

Schwarz, R. T., and Datema, R., 1982b. Inhibition of lipid-dependent glycosylation, in *The Glycoproteins*, Vol. III, *Glycoproteins, Glycolipids, and Proteoglycans, Part A* (M. I. Horowitz, ed.), Academic Press, New York, pp. 47–79.

Semenza, G., Kessler, M., Hosang, M., Weber, J., and Schmidt, U., 1984. Biochemistry of the Na$^+$, D-glucose cotransporter of the small-intestinal brushborder membrane. The state of the art in 1984, *Biochim. Biophys. Acta* 779:343–379.

Shallenberger, R. S., and Acree, T. E., 1967. Molecular theory of sweet taste, *Nature* 216:480–482.

Sharma, M., Potti, G. G., Simmons, O. D., and Korytnyk, W., 1987. Fluorinated carbohydrates as potential plasma membrane modifiers and inhibitors. Synthesis of 2-acetamido-2,6-dideoxy-6-fluoro-D-galactose, *Carbohydr. Res.* 163:41–51.

Sharma, M., Bernacki, R. J., and Korytnyk, W., 1988. Fluorinated derivatives of cell-surface carbohydrates as potential chemotherapeutic agents, in *Fluorinated Carbohydrates, Chemical and Biological Aspects* (N. F. Taylor, ed.), ACS Symposium Series, No. 374, American Chemical Society, Washington, D.C., pp. 191–206.

Silverman, J. B., Babiarz, P. S., Mahajan, K. P., Buschek, J., and Fondy, T. P., 1975. 1-Halo analogues of dihydroxyacetone 3-phosphate. The effects of the fluoro analogue on cytosolic glycerol-3-phosphate dehydrogenase and triosephosphate isomerase, *Biochemistry* 14:2252–2258.

Simon, P., Burlingham, W. J., Conklin, R., and Fondy, T. P., 1979. *N*-Bromoacetyl-β-D-glucosamine tetra-*O*-acetate and *N*-bromoacetyl-β-D-galactosamine tetra-*O*-acetate as chemotherapeutic agents with immunopotentiating effects in Ehrlich ascites tumor-bearing mice, *Cancer Res.* 39:3897–3902.

Sly, W. S., and Fischer, H. D., 1982. The phosphomannosyl recognition system for intracellular and intercellular transport of lysosomal enzymes, *J. Cell. Biochem.* 18:67–85.

Smith, H., Anderson, J. D., Keppie, J., Kent, P. W., and Timmis, G. M., 1965. The inhibition of growth of brucellas *in vitro* and *in vivo* by analogues of erythritol, *J. Gen. Microbiol.* 38:101–108.

Sokoloff, L., Reivich, M., Kennedy, C., Des Rosiers, M. H., Patlak, C. S., Pettigrew, K. D., Sakurada, O., and Shinohara, M., 1977. The [^{14}C]deoxyglucose method for the measurement of local cerebral glucose utilization: Theory, procedure, and normal values in the conscious and anesthetized albino rat, *J. Neurochem.* 28:897–916.

Stein, W. D., 1986. *Transport and Diffusion across Cell Membranes*, Academic Press, Orlando, Florida, pp. 231–361.

Street, I. P., Armstrong, C. R., and Withers, S. G., 1986. Hydrogen bonding and specificity. Fluorodeoxy sugars as probes of hydrogen bonding in the glycogen phosphorylase–glucose complex, *Biochemistry* 25:6021–6027.

Stryer, L., 1988. *Biochemistry*, 3rd ed., W. H. Freeman and Company, New York, pp. 890–920.

Sufrin, J. R., Bernacki, R. J., Morin, M. J., and Korytnyk, W., 1980. Halogenated L-fucose and D-galactose analogues: Synthesis and metabolic effects, *J. Med. Chem.* 23:143–149.

Taylor, N. F., 1972. The metabolism and enzymology of fluorocarbohydrates and related compounds, in *Ciba Foundation Symposium: Carbon–Fluorine Compounds: Chemistry, Biochemistry, and Biological Activities*, Associated Scientific Publishers, New York, pp. 215–238.

Taylor, N. F. (ed.), 1988. *Fluorinated Carbohydrates, Chemical and Biological Aspects*, ACS Symposium Series, No. 374, American Chemical Society, Washington, D.C.

Taylor, N. F., and Louie, L.-Y., 1977. Some biochemical effects of 4-deoxy-4-fluoro-D-glucose on *Escherichia coli*, *Can. J. Biochem.* 55:911–915.

Taylor, N. F., Romaschin, A., and Smith, D., 1976. Metabolic and transport studies with deoxyfluoro-monosaccharides, in *Biochemistry Involving Carbon–Fluorine Bonds* (R. Filler, ed.), American Chemical Society, Washington, D.C., pp. 99–116.

Taylor, N. F., Sbrissa, C., Squire, S. T., D'Amore, T., and McIntosh, J. M., 1988. Metabolic and enzymatic studies with deoxyfluoro carbohydrates, in *Fluorinated Carbohydrates, Chemical and Biological Aspects* (N. F. Taylor, ed.), ACS Symposium Series, No. 374, American Chemical Society, Washington, D.C., pp. 109–137.

Tong, M. K., and Ganem, B., 1988. A potent new class of active-site-directed glycosidase inactivators, *J. Am. Chem. Soc.* 110:312–313.

West, C. M., 1986. Current ideas on the significance of protein glycosylation, *Mol. Cell. Biochem.* 72:3–20.

Wheeler, T. J., and Hinkle, P. C., 1985. The glucose transporter of mammalian cells, *Annu. Rev. Physiol.* 47:503–517.

Wheeler, T. J., and Whelan, J. D., 1988. Infinite-cis kinetics support the carrier model for erythrocyte glucose transport, *Biochemistry* 27:1441–1450.

Winterbourne, D. J., Butchard, C. G., and Kent, P. W., 1979. 2-Deoxy-2-fluoro-L-fucose and its effect on L-[1-^{14}C]fucose utilization in mammalian cells, *Biochem. Biophys. Res. Commun.* 87:989–992.

Withers, S. G., Street, I. P., Bird, P., and Dolphin, D. H., 1987. 2-Deoxy-2-fluoroglucosides: A new class of mechanism-based glucosidase inhibitors, *J. Am. Chem. Soc.* 109: 7530–7531.

Withers, S. G., Street, I. P., and Percival, M. D., 1988. Fluorinated carbohydrates as probes of enzyme specificity and mechanism, *Fluorinated Carbohydrates, Chemical and Biological Aspects* (N. F. Taylor, ed.), ACS Symposium Series, No. 374, American Chemical Society, Washington, D.C., pp. 59–77.

Biochemistry of Halogenated Amino Acids

7

7.1 Introduction

When a biologically active analogue is processed to metabolites which themselves are biologically active—for example, the lethal synthesis of fluorocitrate from fluoroacetate—a combination of effects may be seen, which can complicate interpretation of biological behavior. The situation becomes even more complex when the analogue can be incorporated into macromolecular systems essential to the functioning of the organism—the diversities of biological actions of fluorouracil and bromodeoxyuridine are notable examples. This latter situation definitely pertains in the study of the biological effects of halogenated amino acids, since these analogues can function as inhibitors of specific enzymes, as substrates for incorporation into enzymes and other proteins, and as precursors of other critical biomolecules, such as aminergic neurotransmitters. Fraudulent enzymes, inactive regulatory proteins, and conformationally altered structural proteins are examples of possible consequences of analogue incorporation into protein. While this situation has complicated interpretation of biochemical observations, it also enhances the value of these analogues by virtue of their potential application to the study of a broad spectrum of cellular mechanisms.

Halogenated amino acids have been available for biochemical studies for decades, and consequently there exists a vast body of literature describing this research. Several reviews of this work are available, and these will be cited where appropriate. Primarily because of the favorable steric factors discussed in Chapter 1, fluorinated analogues have received the most attention. While this review will consider only halogenated amino acids, research with other amino acid analogues has also been important in many of the biochemical areas discussed.

7.1.1 A Brief Summary of Early Research

Exploitation of intermediates made available by newly developed procedures for the introduction of fluorine into aromatic rings led in 1932 to the synthesis of 2-, 3-, and 4-fluorophenylalanine (FPhe) (Fig. 7-1) (Weygand and Oettmeier, 1970, and references therein). Several reports in the 1950s described the effects of 4-FPhe on bacterial and yeast growth and protein synthesis (Marquis, 1970, and references therein). The substitution of 4-FPhe for phenylalanine in protein by *Escherichia coli* was documented by Munier and Cohen (1959), and later reports followed which explored this phenomenon in depth. In an example of an early study with mammalian systems, Westhead and Boyer (1961) measured the extent of incorporation of 4-FPhe into muscle aldolase and glyceraldehyde-3-phosphate dehydrogenase of rabbits maintained on a 4-FPhe-containing diet. Extensive studies on analogues of other amino acids also were initiated [reviewed by Richmond (1962) and Fowden (1972)], and the following decades experienced an explosion of activity in several related chemical and biological disciplines.

7.1.2 Overview of Cellular Effects of Amino Acid Analogues

The effects of analogues of amino acids on cell function are a sum of the effects on transport, feedback inhibition of the biosynthesis of endogeneous amino acids, repression of amino acid biosynthetic enzymes, inhibition of specific enzymes required for amino acid metabolism and catabolism, and the several effects which can result from the incorporation of the analogue into protein. In this chapter, topics to be covered include (1) the effects of halogenated amino acids on amino acid biosynthesis, (2) interactions of halogenated amino acids with catabolic enzymes, (3) the use of halogenated amino acids as mechanism-based irreversible enzyme inhibitors, (4) effects of halogenation on amino acid transport and storage,

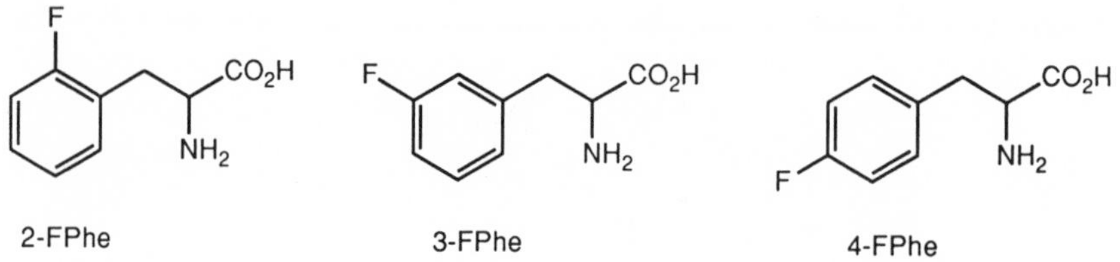

2-FPhe   3-FPhe   4-FPhe

Figure 7-1. 2-, 3-, and 4-Fluorophenylalanine.

and (5) the incorporation of halogenated amino acids into protein. The biochemistry of thyroxine and related iodinated amino acids is the subject of Chapter 6 in Vol. 9A of this series.

7.2 Effects of Halogenated Amino Acids on Amino Acid Biosynthesis

7.2.1 Mechanisms of Analogue Inhibition of Amino Acid Biosynthesis

Typical effects of amino acid analogues on microorganisms include alterations in growth patterns (log phase to linear growth, for example), toxicity, and the development of strains that are resistant to these effects. Consideration of the analysis presented by Marquis (1970) of the effects of fluoroamino acids on specific cell functions (examples of topics discussed include effects on cell division, motility, capsule formation, enzyme activity and regulation, transport systems, and ribosome synthesis) reveals the complexity of these actions (see also Singh and Sinha, 1979, and references therein).

The biosynthesis of amino acids is controlled by end-product feedback inhibition and the repression of synthesis of biosynthetic enzymes. The ability of analogues to function as allosteric feedback inhibitors and as repressors of amino acid biosynthetic enzymes often contributes significantly to their toxicity. Thus, biosynthesis of the natural amino acid is blocked by the presence of the analogue, yet the analogue is an ineffective replacement with respect to other biological functions. As noted above, a common observation is the development of mutants which are resistant to the toxic effects of amino acid analogues. Mechanisms by which mutants establish resistance include the production of biosynthetic enzymes that are no longer subject to feedback inhibition, alteration of the repressibility of biosynthetic enzymes, and a blockade in the biosynthesis of fraudulent polypeptides through alteration in the specificity of the aminoacyl-tRNA required for activation of the amino acid and its analogue (see below) or, more rarely, a restriction of uptake of the analogue (Bussey and Umbarger, 1970). In general, an understanding of the mechanism of resistance provides valuable information on normal cell functions, and, accordingly, the selection of resistant mutants has been, and is, an important part of microbiology.

7.2.2 Study of Essential Amino Acid Biosynthesis with Mutants Resistant to Halogenated Amino Acids

The 10 (nonessential) amino acids which can be synthesized by humans (Ala, Asn, Asp, Cys, Gln, Glu, Gly, Pro, Ser, Tyr) are made by simple biochemical transformations. The 10 (essential) amino acids (Arg, His, Ile, Leu, Lys, Met, Phe, Thr, Trp, Val) required in the human diet are synthesized by nonhuman organisms by multistep processes from simple metabolic intermediates (Stryer, 1988, pp. 575–598; White *et al.*, 1973). The first step committed to a specific amino acid generally is subject to feed-back inhibition by that amino acid. Other enzymes may also be sensitive to end-product inhibition. The genes encoding these enzymes typically make up a coordinated unit of genetic expression, termed the *operon*, which, through the action of repressors, activators, and attenuator sequences, regulates the enzyme synthesis [see Stryer (1988, pp. 799–822) for a thorough discussion of the regulation of gene expression in procaryotes]. An extensive literature exists on the use of analogue-resistant mutants in

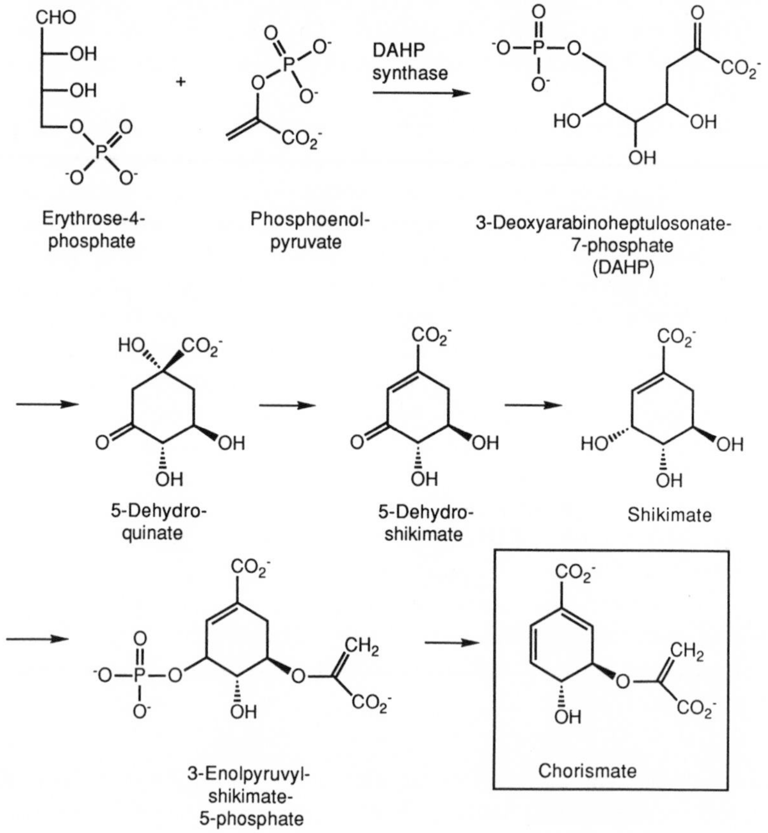

Figure 7-2. Biosynthesis of chorismate, the common precursor of phenylalanine, tyrosine, and tryptophan.

the study of the several aspects of these complex mechanisms. Examples are given below which illustrate some of these strategies using halogenated amino acid analogues.

7.2.2.1 Analogues of Phenylalanine and Tyrosine

In the initial steps in the biosyntheses of Phe, Tyr, and Trp in microorganisms such as *E. coli*, 3-deoxy-D-arabinoheptulosonic acid 7-phosphate (DAHP) is synthesized from erythrose 4-phosphate and phosphoenol-

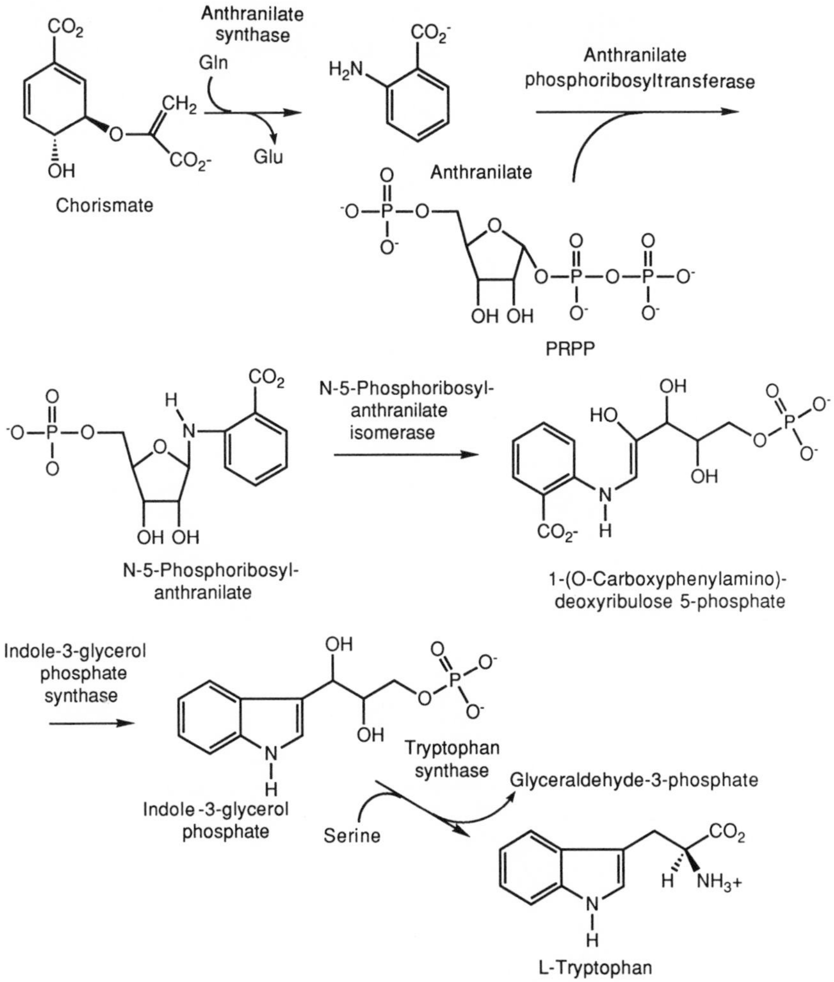

Figure 7-3. Biosynthesis of tryptophan from chorismate.

pyruvate. DAHP is converted in several steps to chorismate, the common intermediate for Phe, Tyr, and Trp (Fig. 7-2). Branching occurs at this point to give Trp through the intermediate anthranilate (Fig. 7-3) and Phe and Tyr through prephenate (Fig. 7-4). The first enzyme of this sequence, DAHP synthase, is subject to allosteric feedback inhibition of end products. Other key enzymes in the sequence are chorismate mutase-P:prephenate dehydratase, a bifunctional enzyme (P) which converts chorismate to Phe through phenylpyruvate (PPY), and chorismate mutase-T:prephenate dehydrogenase, a bifunctional enzyme (T) which channels chorismate to Tyr through p-hydroxyphenylpyruvate (PPA).

Mutants of bacteria resistant to 4-FPhe have been isolated that have altered amino acid transport (Brown, 1970), altered phenylalanine tRNA synthase (unable to activate 4-FPhe, but capable of activating Phe and 2- and 3-FPhe) (Fangman and Neidhardt, 1964), altered feedback inhibition of DAHP (Gollub et al., 1973), altered regulation of the Phe operon (Gollub et al., 1973; Im and Pittard, 1971), and constitutive expression

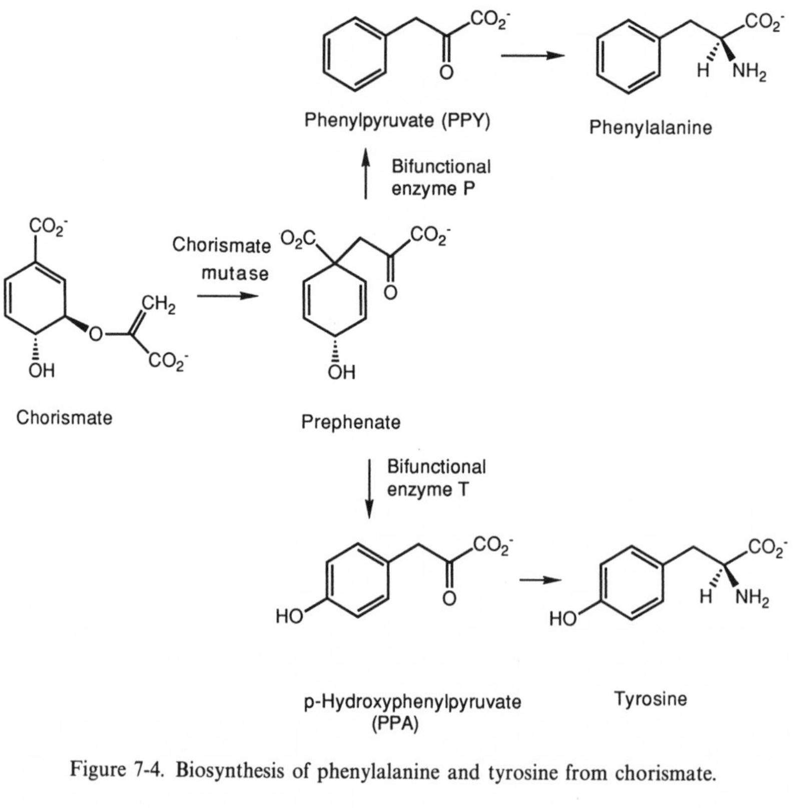

Figure 7-4. Biosynthesis of phenylalanine and tyrosine from chorismate.

of the *PhA* operon (Borg-Olivier *et al.*, 1987). The regulation of the channeling of chorismate to Phe and to Tyr (Fig. 7-4) is highly complex. An example of the fine-tuning that can come into play is seen in the bacterium *Pseudomonas aeruginosa*. This organism has especially complex gene-enzyme patterns for the biosynthesis of aromatic amino acids. The presence of multiple flow routes to end products has complicated the isolation of regulatory mutants. With fructose as the sole source of carbon to render Phe growth limiting, two classes of mutants resistant to 3- or 4-FPhe were isolated. The first lacked sensitivity to Phe feedback inhibition of prephenate dehydratase and was a dramatic excreter of Phe. Tyr inhibited Phe excretion by one-third, as a result of early-pathway feedback inhibition of DAHP synthase. A second class of mutant had a Tyr-insensitive DAHP-synthase-Tyr isozyme. The chorismate molecules produced by the deregulation of DAHP-synthase-Tyr were able to escape through an unregulated, previously hidden pathway to Phe, by way of arogenate (Fig. 7-5) (Fiske *et al.*, 1983).

Tyr is a potent inhibitor of prephenate dehydrogenase, a component of the bifunctional enzyme T (hydroxypyruvate synthase) required in the channeling of chorismate to Tyr (see Fig. 7-4). Christopherson (1985) has examined the effects of a variety of structural analogues of Tyr and Phe, including halogenated derivatives (Fig. 7-6), on the mutase and dehydrogenase activities of hydroxypyruvate synthase. 3-FTyr was found to be a more potent inhibitor of the dehydrogenase activity than Tyr and 3-ITyr and 3-ClTyr were less effective, while 3-FPhe, 4-FPhe, and 4-ClPhe had no effect. These and other data from this study support the existence of spatial overlap of the mutase and dehydrogenase sites of this bifunctional enzyme.

7.2.2.2 Tryptophan

Similar to the behavior of the two bifunctional proteins which compete for chorismate at the branching point in the biosynthesis of Phe and Tyr, the competition of anthranilate synthase, an enzyme committed to Trp biosynthesis, and chorismate mutase for chorismate is also subject to end-product regulation. Fluorinated analogues of Trp have been used to compare substrate and inhibitor specificities of the two enzymes in the yeast *Candida maltose* (Bode *et al.*, 1985).

Halogenated analogues of Trp have also been used effectively to study feedback inhibition by Trp of anthranilate synthase, anthranilate phosphoribosyltransferase, and Trp synthase and the repression by Trp of the synthesis of Trp biosynthetic enzymes (Matsui *et al.*, 1987, and references therein). Repression of the Trp operon mRNA synthesis is

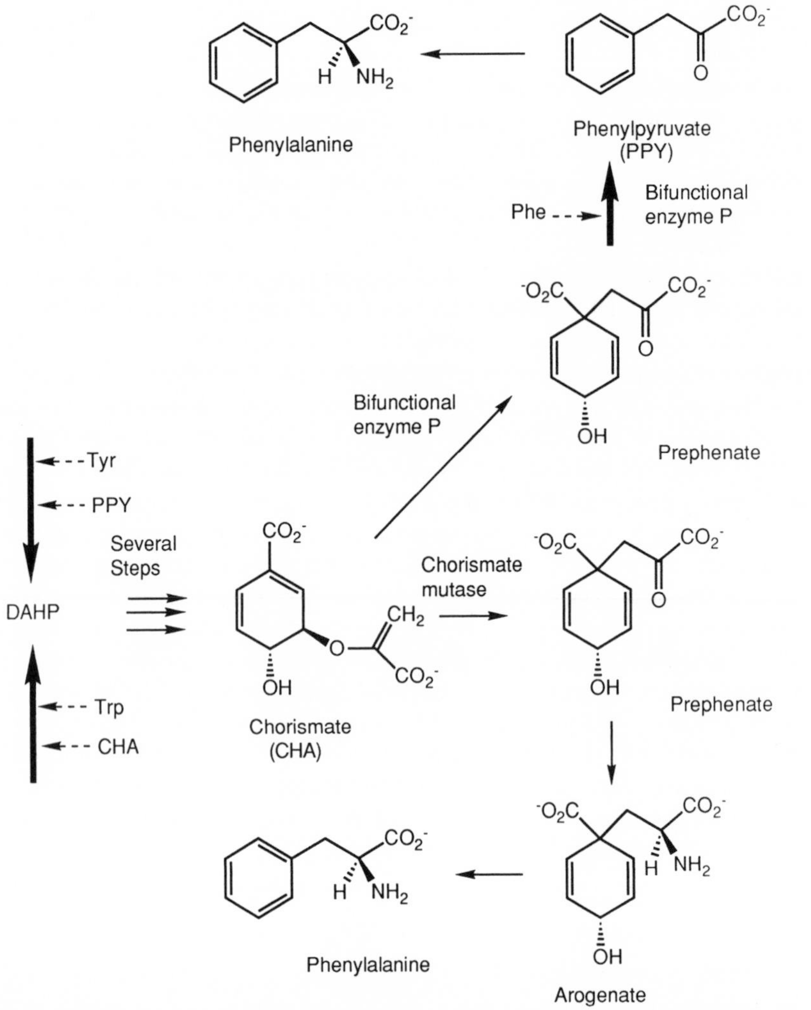

Figure 7-5. Allosteric control points along the biosynthetic pathway leading to phenylalanine in *P. aeruginosa*. The heavy arrows indicate control points mediated by effector molecules (dashed arrows). Despite inhibition of the bifunctional enzyme P-catalyzed conversion of PPA to PPY, phenylalanine was still produced, revealing a previously hidden route through arogenate. See text for further details (Fiske *et al.*, 1983).

mediated by association of Trp with an aporepressor to form the functional repressor complex. Repression of gene product synthesis by 5-F- and 6-FTrp (Fig. 7-7) was comparable to the repression elicited by Trp and confirmed the abilities of these analogues to form a functional complex with the repressor (Pauley *et al.*, 1978).

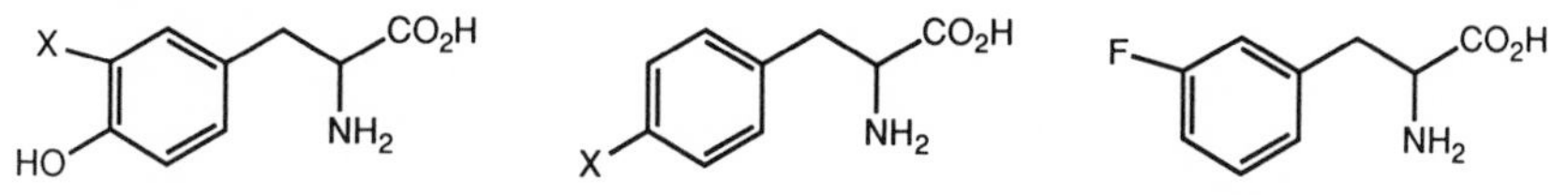

X = F; 3-FTyr
X = Cl; 3-ClTyr
X = I; 3-ITyr

X = Cl; 4-ClPhe
X = F; 4-FPhe

3-FPhe

Figure 7-6. Examples of halogenated analogues of tyrosine and phenylalanine.

Matsui *et al.* (1987) have investigated the molecular basis for the insensitivity to end-product feedback inhibition of anthranilate synthase and the enhanced expression of Trp genes in a 5-FTrp-resistant mutant of *Brevibacterium lactofermentum*. A single-base change (adenine to cytosine) in the gene coding anthranilate synthase resulted in a Ser-38 to Arg-38 substitution. Furthermore, the amino acid sequence Leu-35 to Ser-38 (Leu-Leu-Glu-Ser) was conserved in the corresponding regions of anthranilate synthase from *E. coli* and *B. subtilis*, suggesting that this region is essential for allosteric regulation of anthranilate synthase. The mechanism by which the Ser to Arg substitution blocks feedback inhibition has not been determined.

A frequent consequence of the development of resistance by mutants to the toxic effects of an amino acid analogue is their overproduction of the natural amino acid, and advantage of this is taken in the development of fermentation processes for the production of amino acids. For example, a mutant strain of *Pseudomonas hydrogenothermophila* was cultivated which could produce Trp efficiently, using carbon dioxide, or carbon dioxide and pyruvate, as the sole carbon source (Igarashi *et al.*, 1982).

Mechanistic details of the interaction of Trp and its analogues with tryptophan synthase have been the subject of recent NMR measurements. This enzyme, which catalyzes the synthesis of Trp from indole-3-glycerol phosphate and serine (Fig. 7-3), consists of an $\alpha_2\beta_2$ complex. The cleavage of indole-3-glycerol phosphate occurs at the active site of the α subunit

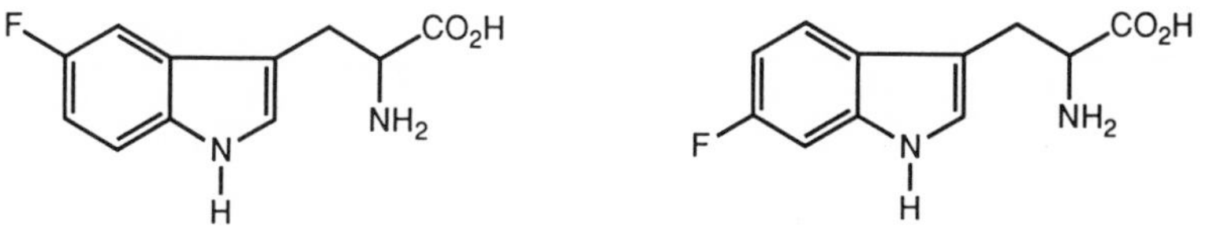

5-F-Trp   6-F-Trp

Figure 7-7. 5- and 6-Fluorotryptophan.

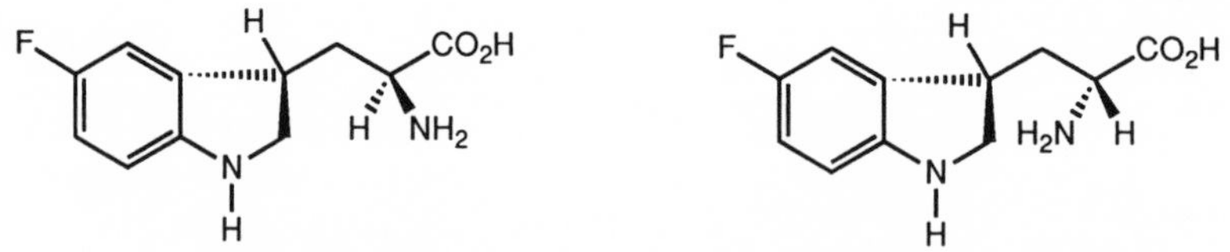

(3*S*)-2,3-Dihydro-5-F-L-Trp   (3*S*)-2,3-Dihydro-5-F-D-Trp

Figure 7-8. 2,3-Dihydro analogues of 5-fluorotryptophan used to explore the mechanism of tryptophan synthase using ^{19}F-NMR and absorption spectroscopy (Miles *et al.*, 1986).

while the β subunit catalyzes the pyridoxal-dependent addition of serine to the 3 position of the indole ring (Miles *et al.*, 1986, and references therein). Miles *et al.* (1986) used ^{19}F-NMR and absorption spectroscopy to study the binding of D- and L-5-FTrp, of D- and L-Trp, and of the 3*S* and 3*R* isomers of 2,3-dihydro-D- and 2,3-dihydro-L-5-FTrp (Fig. 7-8). Trp synthase tightly bound the 3*S* diastereomer of 2,3-dihydro-D- and 2,3-dihydro-L-5-FTrp but bound D-5-FTrp more tightly than L-5-FTrp. An unexpected result was the observation of a slow interconversion of the D and L isomers of Trp, 5-FTrp, and (3*S*)-2,3-dihydro-5-FTrp. Since these isomerizations are 10^3–10^5 times slower than the β-replacement and β-elimination reactions catalyzed by Trp synthase, they have no biochemical significance *in vivo*, but the results did provide useful information concerning the active site of the enzyme.

7.2.2.3 Histidine

The ring-halogenated L-His analogues 2- and 4-FHis, 2- and 4-ClHis, 2- and 4-BrHis, and 2- and 4-IHis (Fig. 7-9) have been screened for bacteriostatic activity. Of these, L-2-FHis alone was active, and completely inhibited growth of wild-type and a His auxotroph of *E. coli* at 10^{-5} M. Incorporation of L-2-FHis into bacterial protein presumably contributes to toxicity (see below). Antimetabolic activity may also stem

2 X = F, Cl, Br, I   4 X = F, Cl, Br, I

Figure 7-9. Halogenated analogues of histidine.

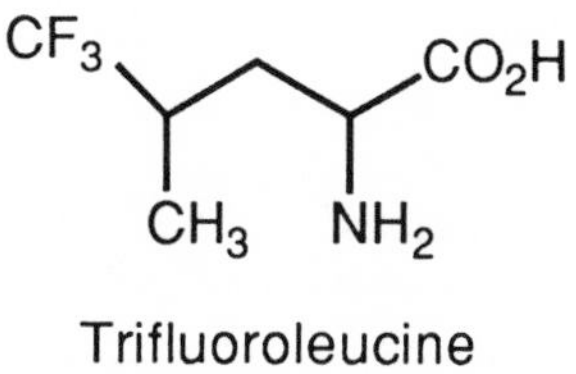

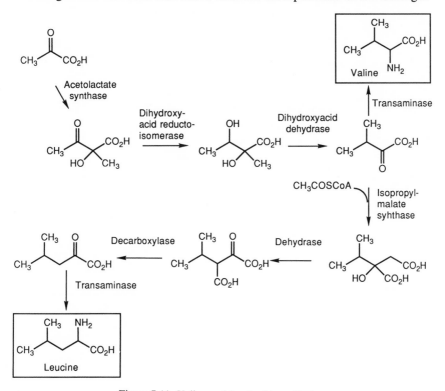

Figure 7-10. Trifluoroleucine, a toxic leucine analogue.   Trifluoroleucine

from the action of L-2-FHis as a potent inhibitor of the first committed enzyme for His biosynthesis, ATP:phosphoribosyltransferase (K_i for L-His $= 4 \times 10^{-5} M$; K_i for L-2-FHis $= 7 \times 10^{-7} M$) (S. Naghshineh, K. L. Kirk, and L. A. Cohen, unpublished results).

7.2.2.4 Effects of Trifluoroleucine (TFLeu) on the Biosynthesis of Leucine, Isoleucine, and Valine

The Leu-reversible toxicity of trifluoroleucine (TFLeu) (Fig. 7-10) to microorganisms has been associated with the incorporation of the analogue

Figure 7-11. Valine and leucine biosynthesis.

into protein (*E. coli*) (Rennert and Anker, 1963) and with the action of the analogue as a false-feedback inhibitor of α-isopropylmalate synthase, the enzyme catalyzing the first committed step in Leu biosynthesis (Fig. 7-11) (*Salmonella typhimurium*) (Calvo *et al.*, 1969). Three classes of mutants of *S. typhimurium* resistant to the toxicity of TFleu were isolated. These mutants overproduced Leu, Val, and Ile because of defective regulation of branched-chain amino acid synthesis [Figs. 7-11 (Leu and Val) and 7-12 (Ile)]. Mutants were of three classes, having (1) resistance to feedback inhibition of α-isopropylmalate synthase, (2) high levels of Leu biosynthetic enzymes, or (3) high levels of Leu, Val, and Ile biosynthetic enzymes (Calvo *et al.*, 1969). Although the initial precursors for the biosynthesis of Ile and of Leu and Val are different, these complex pathways are interrelated, as is apparent from their comparable sensitivity to TFLeu. The reactions involved indeed differ only in the presence of an additional methylene group in the Ile scheme, and the routes share common enzymes. For regulation of Leu biosynthetic enzymes, only Leu is required. In contrast, both end products, Ile and Val, are required, as well as Leu and pantothenic acid, in the multivalent repression of the enzymes that are required for the synthesis of Ile and Val.

Freundlich and Trela (1969) made the surprising observation that, although TFLeu could replace Leu as a corepressor of Ile–Val biosynthetic enzymes in *S. typhimurium*, it had no effect on Leu enzymes. Since activation of Leu by Leu tRNA synthase is required for it to function as a repressor, this anomalous result was interpreted as indicating that a dif-

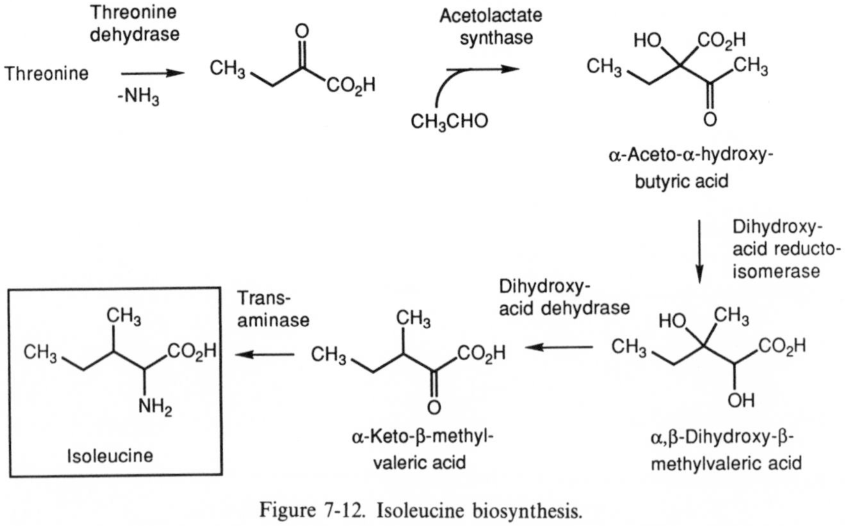

Figure 7-12. Isoleucine biosynthesis.

ferent Leu tRNA species is required for repression of each of the systems. Thus, TFLeu apparently is able to attach to some, but not all, of the Leu tRNA species present.

7.3 Halogenated Aromatic Amino Acids and Amino Acid-Processing Enzymes

The behavior of halogenated amino acids as substrates or inhibitors of amino acid-processing enzymes will be covered in the next three sections. In this first section, enzymes for the processing of aromatic amino acids will be discussed. The following section (Section 7.4) will treat selected topics concerning the enzymatic behavior of aliphatic amino acids. The development of active-site-directed inhibitors of pyridoxal phosphate-dependent enzymes will be reviewed in Section 7.5. The aminoacyl-tRNA synthases which activate amino acids prior to their incorporation into protein will be discussed in Section 7.7.1.

7.3.1 Halogenated Analogues of Phenylalanine

7.3.1.1 Hydroxylation, Dehalogenation, and the "NIH Shift"

The critical role of Phe as a biosynthetic precursor of ring-hydroxylated amino acids, and thus of phenolic biogenic amines, prompted early interest in behavior of ring-halogenated Phe analogues toward hydroxylating enzymes. The demonstration that 4-FPhe is rapidly defluorinated *in vivo* (Weissman and Koe, 1967, and references therein) was followed by experiments *in vitro* which implicated phenylalanine hydroxylase as the enzyme responsible for this (Kaufman, 1961). Three types of monooxygenases have been found to catalyze defluorination of aryl fluorides, including a pteridine-linked phenylalanine hydroxylase [for a concise discussion, see Walsh (1983)]. Using purified rat liver extracts, Kaufman (1961) found that 4-FPhe is converted *in vitro* to tyrosine by this enzyme system, but only about one-sixth as efficiently as is Phe. In the reaction, an equivalent of fluoride is released. The requirement for oxygen, tetrahydropteridine, and NADPH indicated that a simple hydrolytic cleavage of fluorine was not occurring. Based on the utilization of between three and four moles of reducing agent for each mole of product, Kaufman postulated the existence of an intermediate containing both molecular oxygen-derived oxygen and fluorine (Fig. 7-13). Regeneration of 4-FPhe by a NADPH-dependent reductive removal of oxygen would result in the

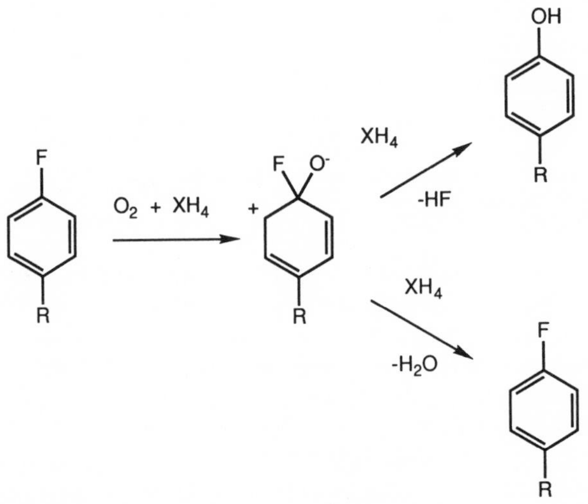

Figure 7-13. The intermediate containing both molecular oxygen-derived oxygen and fluorine proposed by Kaufman (1961) to rationalize the low efficiency of the conversion of *p*-fluorophenylalanine to tyrosine.

higher than predicted NADPH consumption. A result which was noted briefly in Kaufman's preliminary report, but assumed greater significance subsequently, was the fact that formation of Tyr from 4-ClPhe was accompanied by an even higher NADPH consumption for each mole of tyrosine formed (5–8-fold).

This early observation of Kaufman, important in its demonstration of facile enzymatic cleavage of the C—F bond, had further ramifications. A recurring observation in the hydroxylation of aromatic substrates by oxygenase enzymes had been that *para* hydroxylation of an aromatic substrate was accompanied by migration of radioactive label from the *para* to the *meta* position. Having noted Kaufman's observation on the hydroxylation of 4-ClPhe, Guroff *et al.* (1966) reinvestigated this reaction to see if the excess NADPH was used to form additional, undetected hydroxylated product. The major product formed by the action of bacterial phenylalanine hydroxylase on 4-ClPhe was in fact 3-ClTyr, with Tyr formation representing only approximately 10% of the product. In analogous experiments, 4-BrPhe produced 3-BrTyr, but, significantly, no 3-FTyr was produced from 4-FPhe (Fig. 7-14). The mechanism of uncoupling of NADPH oxidation from the hydroxylation of 4-FPhe has been discussed more recently by Kaufman and Fisher (1974) in terms of a competing reduction of oxygen to hydrogen peroxide.

The formation of 3-ClTyr from 4-ClPhe was one of the pieces of

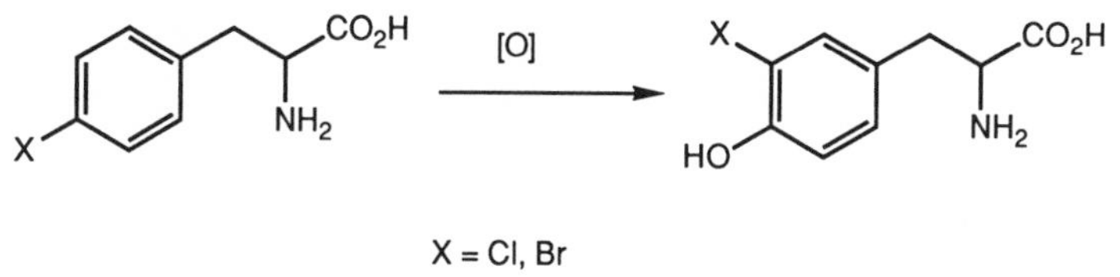

X = Cl, Br

Figure 7-14. Migration of halogen during monooxygenase-catalyzed oxidation of *p*-chloro-and *p*-bromophenylalanine (Guroff *et al.*, 1966).

evidence which led to the elucidation of the mechanism of migration of a *para* substituent to the *meta* position during enzymatic hydroxylation of aromatic substrates, a rearrangement termed the "NIH shift" (Guroff *et al.*, 1967). The key intermediate in this process is an arene oxide formed by insertion of molecular oxygen into the carbon–carbon double bond (Fig. 7-15). As illustrated for 4-ClPhe, migration and loss of chlorine produces Tyr, while migration and retention of chlorine produces 3-ClTyr. To a lesser extent, migration of OH occurs to give 4-chloro-3-hydroxyphenylalanine. This arene oxide mechanism has been extended to other monooxygenase systems, such as the liver microsomal cytochrome P-450 heme monooxygenase, and has facilitated greatly the understanding of such important processes as aromatic polycyclic hydrocarbon carcinogenesis (see Chapter 9 in Vol. 9A of this series).

3-FPhe is also a substrate for hepatic phenylalanine hydroxylase. The product, 3-FTyr, is acutely toxic to animals by virtue of its further metabolism to fluorocitrate through the shikimate pathway (Koe and Weissman, 1967) (see below, Section 7.3.1.2). In fact, the absence of acute toxicity of 4-FPhe is further evidence that the NIH shift does not occur during *in vivo* hydroxylation of this isomer.

The potential lethality of 3-FPhe to cells has been exploited in a prodrug approach to fungal chemotherapy. 3-FPhe is a weak inhibitor of fungal growth, presumably as a result of poor transport into the cell. Simple di- and tripeptides containing 3-FPhe, however, have potent activity against several fungal organisms (Kingsbury *et al.*, 1983). This activity is attributed to the ability of specific peptide transport mechanisms of the microorganisms to carry the peptides into the cell, where 3-FPhe is released by intracellular peptidases.

2-FPhe also is hydroxylated by phenylalanine hydroxylase to 2-FTyr, a product that, unlike 3-FTyr, is not acutely toxic. Thus, even though 2-FTyr may be degraded by the same metabolic sequence as 3-FTyr, fluoroacetate is not the ultimate product (Weissman and Koe, 1967).

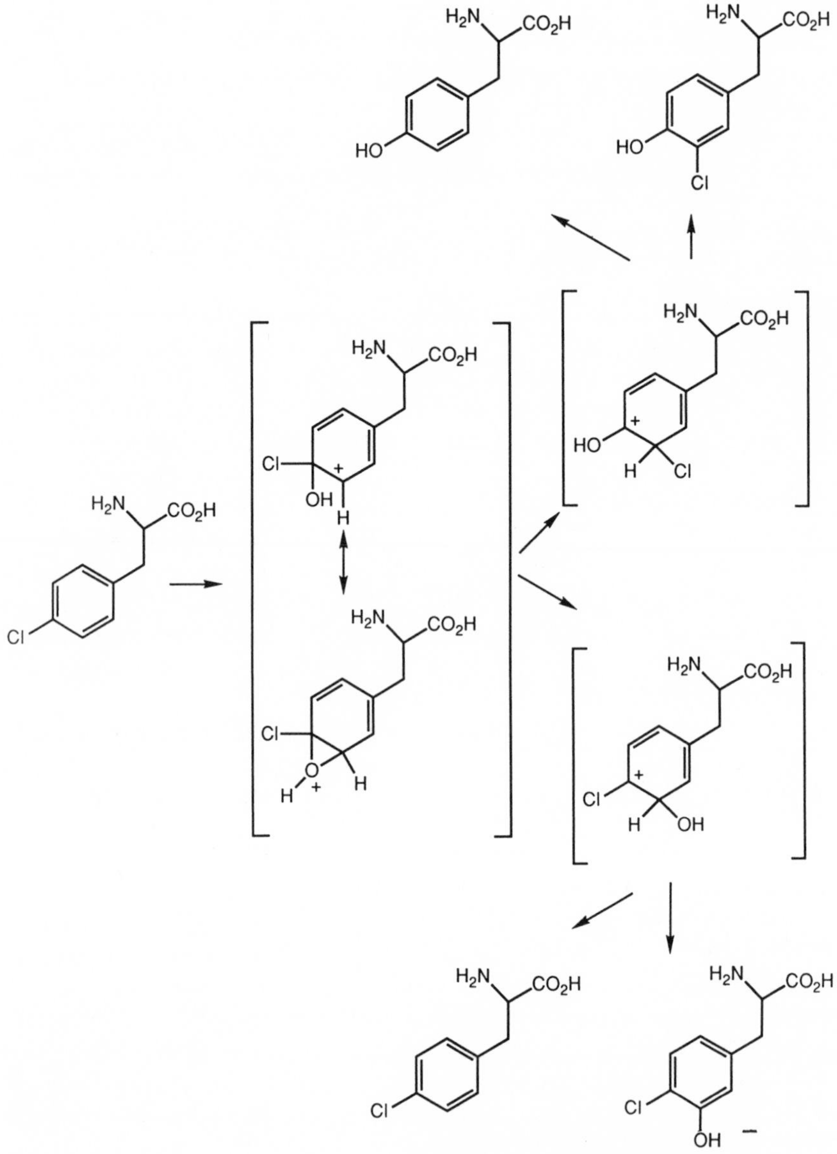

Figure 7-15. Mechanism of the "NIH shift" (Guroff *et al.*, 1967).

7.3.1.2 Inhibition of Phenylalanine and Tryptophan Hydroxylases by 4-ClPhe

Tryptophan hydroxylase and tyrosine hydroxylase catalyze the formation of 5-hydroxytryptophan (5-HOtrp) and 3,4-dihydroxyphenylalanine (DOPA), respectively, in the initial and rate-limiting steps of the synthesis of 5-hydroxytryptamine (serotonin, 5HT) and catecholamines (dopamine, norepinephrine, epinephrine). Selective modulation of these reactions *in vivo* is an attractive approach to the modification of levels of these important neurochemicals. Several ring-halogenated aromatic amino acids are effective inhibitors of these hydroxylase enzymes, both *in vitro* and *in vivo* (Miwa *et al.*, 1987, and references therein; McGeer *et al.*, 1968). For example, 4-ClPhe acts as a competitive inhibitor of both phenylalanine and tryptophan hydroxylases *in vitro* and produces selective irreversible inactivation *in vivo* of both enzymes (Gál and Whitacre, 1982, and references therein). The discovery by Koe and Weissman (1966a) that 4-ClPhe specifically depletes brain serotonin was particularly significant and had immediate far-reaching pharmacological implications. As a tryptophan hydroxylase inhibitor, this analogue is still used extensively in a wide range of psychopharmacological experiments based on *in vivo* serotonin depletion (Vorhees *et al.*, 1981, and references therein; see also, e.g., Pappius *et al.*, 1988). Because 4-ClPhe produces increased levels of Phe *in vivo*, it has been used as a phenylalanine hydroxylase inhibitor to develop models for phenylketonuria (Vorhees *et al.*, 1981).

The irreversible, selective, and long-lasting inhibition of phenylalanine and tryptophan hydroxylase *in vivo* by 4-ClPhe was attributed initially to the substitution of the analogue for Phe in the two enzymes at positions critical to enzyme activity (Gál, 1972). A revision of this interpretation was mandated by new evidence which revealed that, in the doubly labeled analogue [2-^{14}C,4-^{36}Cl]*p*-chlorophenylalanine, radioactive carbon, but not radioactive chlorine, was incorporated. Thus, the loss of chlorine from this analogue indicates that a substitution of Phe by Tyr more likely is occurring. Selectivity of action would result from substitution at the active sites of phenylalanine and tryptophan hydroxylases, but in noncritical sites in other enzymes (Gál and Whitacre, 1982). Regardless of mechanistic details, 4-ClPhe has been used effectively in an enormous number of studies as a selective inhibitor of these two hydroxylase enzymes.

7.3.2 Halogenated Analogues of Tyrosine

The metabolism of 3-FTyr was investigated in detail by Weissman and Koe (1967) in an attempt to determine the mechanism for the extreme toxicity of this analogue. 3-FTyr had been shown to lead to convulsions in mice characteristic of fluoroacetate poisoning, and it is, in fact, more toxic to mice than is fluoroacetate. Elevation of tissue citrate concentrations by lethal doses of 3-FTyr further emphasized the parallel toxic manifestations of 3-FTyr and fluoroacetate. These and other data confirmed that 3-FTyr is metabolized by Tyr metabolic enzymes to fluoroacetate, which is converted to fluorocitrate by the citric acid cycle [reviewed by Smith (1970)].

7.3.3 Halogenated Analogues of Tryptophan

Metabolic studies have shown that the C-5 carbon of Trp is converted to C-2 of acetic acid, suggesting that 5-FTrp could, by a parallel metabolic route, produce fluoroacetate. Koe and Weissman (1966b) studied the effects of 5-FTrp on citrate levels in mice and concluded that, despite the relatively low toxicity of 5-FTrp, fluoroacetate is a product of one of the available metabolic pathways.

In comparison to 4-ClPhe, various halo- and methyl-substituted tryptophans are as good or better inhibitors of tryptophan hydroxylase (McGeer et al., 1968). The 6-substituted analogues are more potent than 5-substituted analogues as inhibitors of tryptophan hydroxylase, while the opposite is true for inhibition of tyrosine hydroxylase. The only analogue tested in vivo, 6-FTrp, caused a greater drop in 5HT levels than did 4-ClPhe. The duration of this action, however, was much shorter (ca. 24 h) than the duration of action of 4-ClPhe (several days). 6-FTrp has been used subsequently as an effective inhibitor of serotonin biosynthesis (e.g., Peters, 1971). Miwa et al. (1987) have recently developed a procedure for the determination of 5HT turnover in vivo in rat brain based on the complete inhibition of tryptophan hydroxylase by 6-FTrp. The absence of any significant effects on 5HT release and on tyrosine hydroxylase activity increases the reliability of this procedure.

7.3.4 Halogenated Analogues of 3,4-Dihydroxyphenylalanine (DOPA)

3,4-Dihydroxyphenylalanine (DOPA), the product of tyrosine hydroxylase-catalyzed hydroxylation of Tyr, is the immediate precursor of catecholamines, which function as neurotransmitters and hormones. Thus, DOPA decarboxylase catalyzes the decarboxylation of DOPA to the neurotransmitter dopamine (DA), which in turn is hydroxylated by dopamine β-hydroxylase to norepinephrine (NE) (Fig. 7-16). NE functions

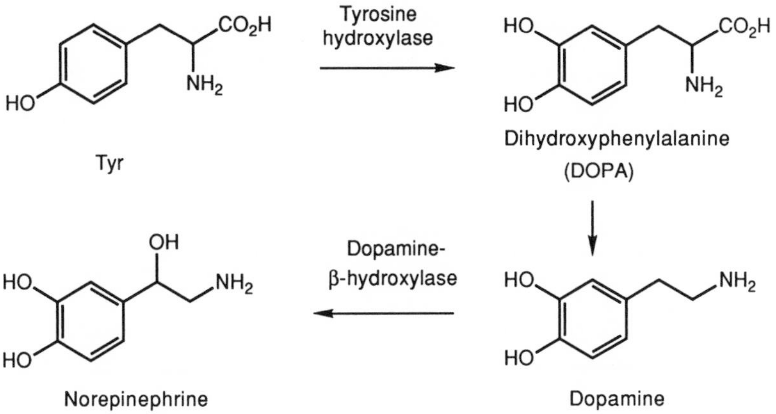

Figure 7-16. Biosynthesis of biogenic amines from tyrosine.

as a neurotransmitter for the sympathetic nervous system and in the central nervous system (CNS). NE is also a precursor of the neurotransmitter and hormone epinephrine (EPI). There has been much research done on the effects of ring fuorination on the biological behavior of catecholamines, a topic to be discussed in Chapter 8. As DOPA is the precursor of these amines, fluorinated analogues of DOPA have also received much attention.

A major impetus to the study of ring-fluorinated analogues of DOPA (Fig. 7-17) has been the interest in the development of PET scanning agents capable of quantitating central dopaminergic activity, particularly as a diagnostic tool for Parkinson's disease. In this approach, an appropriately labeled ring-^{18}F-fluorinated DOPA, after crossing the blood–brain barrier, serves as a biological precursor of the corresponding ^{18}F-fluorinated DA. The latter is taken up and stored in the central dopaminergic neurons, and regions of dopaminergic activity are quantitated by PET scanning (Fig. 7-18). In initial studies, Garnett et al. (1978) prepared [^{18}F]-5-FDOPA and studied its applicability for the in vivo measurement of storage and metabolism of intracerebral DA. While

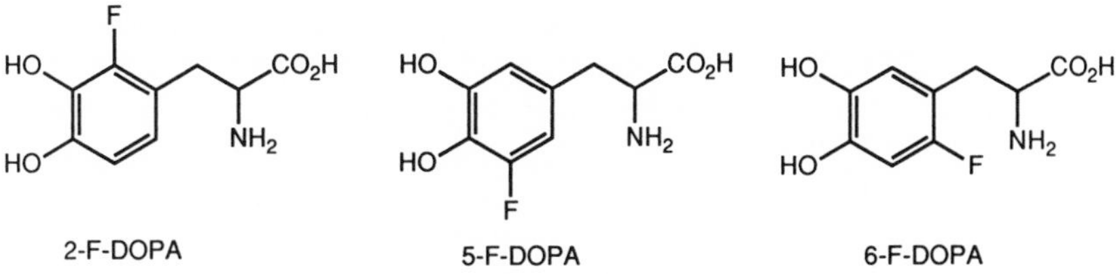

Figure 7-17. Fluorinated analogues of DOPA.

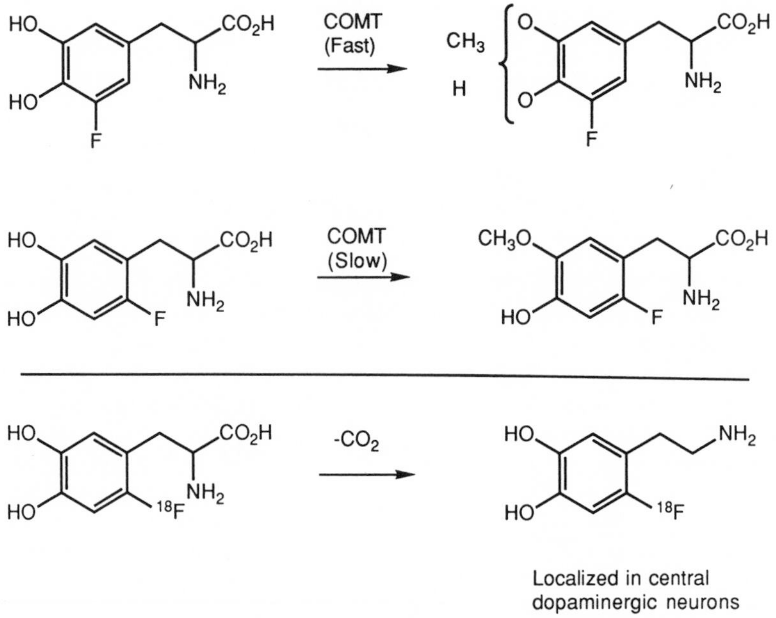

Figure 7-18. Contrasting behavior of 5- and 6-FDOPA toward catechol O-methyltransferase (COMT) (Creveling and Kirk, 1985); formation of 6-[^{18}F]fluorodopamine, an effective PET scanning agent for central dopaminergic function (Garnett *et al.*, 1983).

metabolic and PET studies confirmed the intracerebral formation of 5-FDA, rapid O-methylation catalyzed by catechol O-methyltransferase (COMT) has complicated the development of 5-FDOPA as a PET scanning agent (Firnau *et al.*, 1981). Since this rapid rate of O-methylation was attributed to the greater acidity of the hydroxyl *ortho* to fluorine, 6-FDOPA was investigated with the expectation that this isomer would not be subject to rapid O-methylation. Indeed, [^{18}F]-6-FDOPA has been shown to be an effective PET scanning agent for the localization of dopaminergic pathways in the human brain and for the quantitation of dopamine turnover (Fig. 7-18) (Garnett *et al.*, 1983). Several biochemical and metabolic studies, necessary to provide a foundation for interpretation of data obtained from PET scans, have confirmed that [^{18}F]-6-FDA is an important brain metabolite derived from [^{18}F]-6-FDOPA (e.g., Cumming *et al.*, 1987, and references therein). Of particular interest with respect to the biological half-life of 6-FDOPA and 6-FDA is the fact that fluorine situated at the 6 position of DOPA (and the related catecholamines; see Chapter 8) actually retards COMT-catalyzed O-methylation (Fig. 7-18) (Creveling and Kirk, 1985). Several examples have been reported of applications of $[^{18}F]$-6-FDOPA for the *in vivo* determination of dopaminergic activity under a variety of physiological conditions—for example, following 1-methyl-4-phenyl-1,2,3,6-tetrahydropyridine (MPTP) induced parkinsonism in primates (Chiueh *et al.*, 1986).

7.3.5 Halogenated Analogues of Histidine

Histidine ammonia-lyase effects an α, β-elimination of ammonia from L-histidine to give *trans*-urocanic acid in the first step of catabolism of His (Fig. 7-19). 4-FHis (Fig. 7-20) has a higher affinity for this enzyme ($K_i = 1.25$ mM) than does His ($K_m = 2.7$ mM), but a much lower V_{max}, and thus is an effective reversible inhibitor. 2-FHis (Fig. 7-20) has low affinity ($K_i = 170$ mM) as well as a low V_{max}. A reinvestigation of the mechanism of this reaction using 4-FHis and β-deuterated 4-FHis revealed that the rate-limiting step is the almost concerted loss of a β hydrogen and the amino group (Klee *et al.*, 1975).

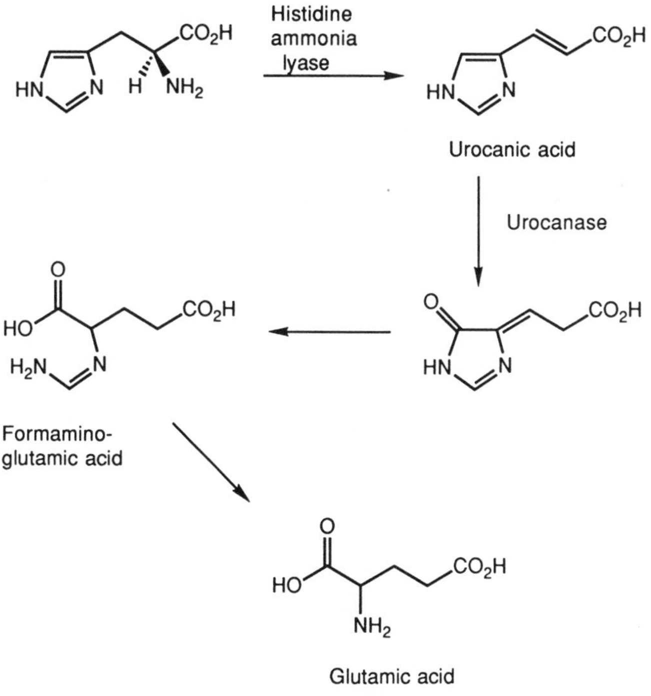

Figure 7-19. Catabolism of histidine.

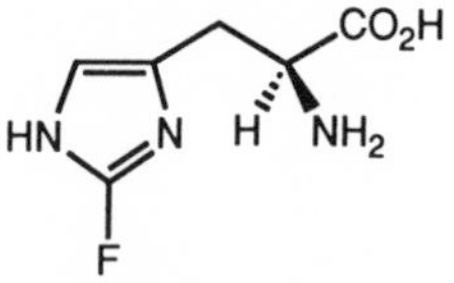

2-F-L-His 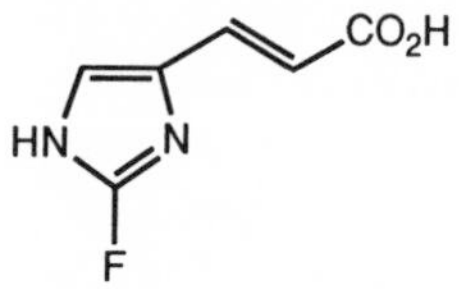   2-Fluorourocanic acid

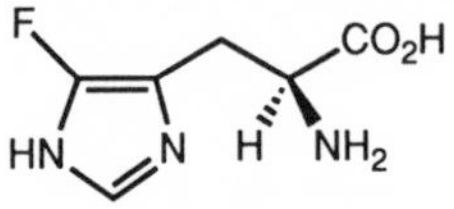

4-F-L-His 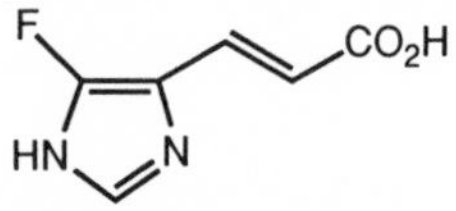   4-Fluorourocanic acid

Figure 7-20. Fluorinated histidine and urocanic acid analogues.

The next enzyme in the catabolic sequence for His, urocanase, catalyzes the hydration rearrangement of urocanic acid, the first step in the degradation of the imidazole ring to glutamic acid (Fig. 7-19). 2-Fluorourocanic acid (Fig. 7-20), the product of deamination of 2-FHis, is a potent inhibitor of urocanase, with 1000-fold higher affinity for the enzyme than urocanic acid, the natural substrate. Hydration is 100-fold slower than for urocanic acid. 4-Fluorourocanic acid (Fig. 7-20) acts neither as a substrate nor as an inhibitor of this enzyme, another example of the disparate behaviors of these two isomers (Klee *et al.*, 1977).

7.4 Behavior of Halogenated Aliphatic Amino Acids toward Amino Acid-Processing Enzymes

7.4.1 Nonfunctionalized Aliphatic Amino Acids

Compared to the aromatic amino acids discussed above, which are subject to many enzymatic conversions targeted to the aromatic ring, simple aliphatic amino acids, such as Ala, Val, and Leu, have relatively limited metabolic choices available. Included are pyridoxal phosphate (PLP)-dependent decarboxylation and transamination reactions that have been the target for mechanism-based enzyme inhibitors, a topic covered separately in Section 7.5.

7.4.2 Aliphatic Amino Monocarboxylic Acids Possessing an Additional Functional Group

S-Trifluoromethyl-L-homocysteine (TFMet) (Fig. 7-21) competes effectively with methionine (Met) in a number of biological processes. For example, TFMet represses the enzymes of the methionine biosynthetic pathway, is esterified by Met-tRNA synthase, and is incorporated into protein *in vivo* and *in vitro* (Colombani *et al.*, 1975). TFMet also competitively inhibits S-adenosylmethionine (SAM) synthase, a key enzyme in the "active methyl" cycle (Chou and Talalay, 1966). While such behavior likely contributes to the high toxicity of TFMet, an additional mechanism for toxicity has been suggested by Alston and Bright (1983). TFMet is converted stoichiometrically to 2-oxobutanoate and fluoride by rat liver γ-cystathionase (homoserine dehydratase). The highly reactive acylating agent carbonothioic difluoride, proposed as an obligatory intermediate in this process, may be toxic by functioning as a cross-linking agent (Fig. 7-21).

7.4.3 Amino Dicarboxylic Acids

Aspartic acid (Asp) is required for the biosynthesis of both purines and pyrimidines. Thus, Asp donates one of the nitrogens present in the six-membered ring of inosine and is also the source of the amino group required for the conversion of inosine to adenosine (Stryer, 1988, pp. 603–604). In addition, Asp is required for pyrimidine biosynthesis, wherein the first committed step is the formation of N-carbamoylaspartate

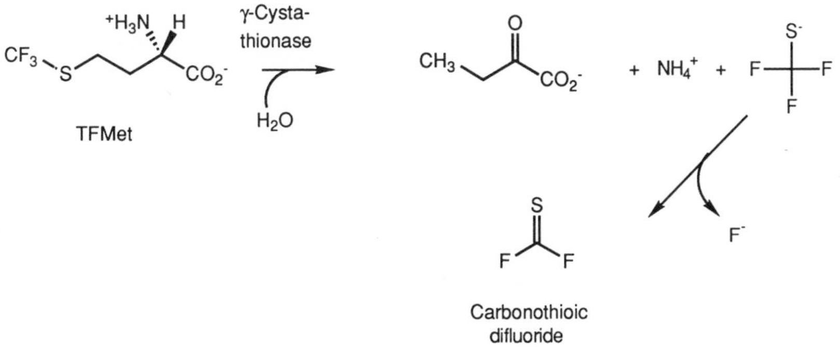

Figure 7-21. Formation of carbonothioic difluoride by the action of γ-cystathionase on S-trifluoromethyl-L-homocysteine (TFMet).

from Asp and carbamoyl phosphate (Stryer, 1988, pp. 607–608). While these important roles of Asp in nucleotide biosynthesis made *threo*- and *erythro*-FAsp attractive synthetic targets—having the potential, for example, to block adenosine biosynthesis (Wanner *et al.*, 1980) or to be precursors of 5-fluorouracil (Stern *et al.*, 1982)—early synthetic progress was impeded by instability of these analogues. Kollonitsch *et al.* (1979) reported the synthesis of *erythro*- and *threo*-3-F-D,L-Asp without configurational assignments while Wanner *et al.* (1980) reported the syntheses of *erythro*- and *threo*-3-F-D,L-Asp, established the configuration of the *erythro* isomer, and converted it to *erythro*-3-F-D,L-Asn (Fig. 7-22). In the systems studied, no antileukemic activity was found for the *erythro* analogues. Stern *et al.* (1982) independently prepared *erythro*- and *threo*-3-F-D,L-Asp (Fig. 7-22) and the corresponding asparagines and found significant selective toxicity to tumor cells in the *threo* series. This activity appears not to be related to nucleotide biosynthesis, but rather to incorporation into protein (see Section 7.7.3 for a further discussion).

Glutamic acid (Glu) and glutamine (Gln) also have critical biochemical functions apart from their presence in proteins. Thus, Gln is an amino-group donor in many enzymatic amination reactions, Glu is present as polyglutamic acid in the key coenzyme tetrahydrofolate, and Glu is involved in neurotransmission as an excitatory amino acid, while its decarboxylatin product, γ-aminobutyric acid (GABA), is a neuroinhibitory amino acid. All of these functions have been the subject of analogue studies.

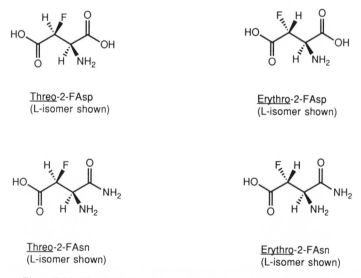

Threo-2-FAsp
(L-isomer shown)

Erythro-2-FAsp
(L-isomer shown)

Threo-2-FAsn
(L-isomer shown)

Erythro-2-FAsn
(L-isomer shown)

Figure 7-22. Fluorinated analogues of aspartic acid and of asparagine.

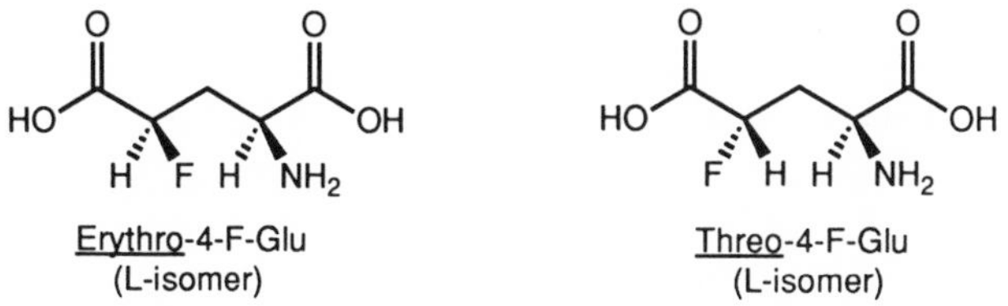

Figure 7-23. Fluorinated analogues of glutamic acid.

Glutamine synthase, a key enzyme in nitrogen metabolism, catalyzes the synthesis of glutamine from L-glutamate and ammonia. Both *threo-* and *erythro*-4-fluoro-D,L-glutamate (4-FGlu) (Fig. 7-23) are noncompetitive inhibitors of glutamine synthase. This inhibition has been proposed as a possible explanation for the antitumor and antiviral activities of these analogues (Firsova *et al.*, 1986a, b).

Glutamate decarboxylase is involved with the regulation of concentrations of Glu and GABA. Both *threo-* and *erythro*-3-F-L-Glu are substrates for glutamate decarboxylase but follow different reaction pathways. The

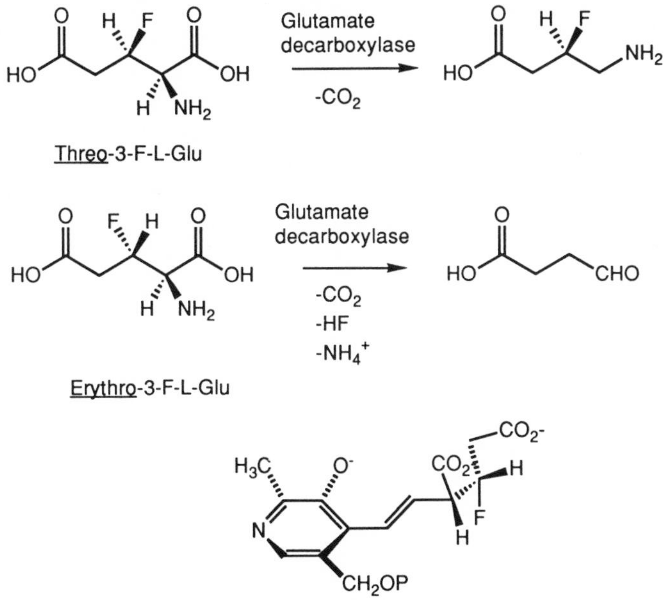

Figure 7-24. Contrasting behavior of *threo-* and *erythro*-3-F-L-Glu toward glutamate decarboxylase. From this, a model for the conformation of the Glu Schiff base at the active site was proposed, based on the assumption that the carboxylate and fluorine substituents in *erythro*-3-F-Glu should be in a *trans* antiparallel configuration in order for elimination to occur (Vidal-Cros *et al.*, 1985).

Folylpolyglutamate
synthase

H$_4$PteGlu   +   ATP   +   L-Glu$_n$ $\xrightarrow{\hspace{3cm}}$ H$_4$PteGlu$_{n+1}$ +   ADP   +   P$_i$

H$_4$PteGlu   =   Tetrahydrofolic Acid

Figure 7-25. Folylpolyglutamate synthase-catalyzed formation of pteroyl-L-glutamyl γ-peptides, the predominant form of intracellular folate coenzymes.

threo isomer forms optically active 4-amino-3-fluorobutyrate whereas the *erythro* isomer loses fluoride during the reaction, giving succinic semialdehyde after hydrolysis of the unstable intermediate enamine (Fig. 7-24). Based on the assumptions that the C−F bond broken must be perpendicular to the planar conjugated Schiff base system and that elimination is anti, the conformation of the 3-F-L-Glu–PLP Schiff base at the active site shown in Fig. 7-24 was constructed. Since elimination of a β-leaving group during PLP-dependent decarboxylations can lead to irreversible inactivation (Section 7.5), *erythro*-3-F-L-Glu had the potential, unrealized, to be a mechanism-based inhibitor (Vidal-Cros *et al.*, 1985).

Folylpolyglutamate synthase catalyzes the formation of pteroyl-L-glutamyl γ-peptides, the predominant form of intracellular folate coenzymes (Fig. 7-25). With either tetrahydrofolic acid (H$_4$PteGlu) or methotrexate (MTX, 4-NH$_2$-10-CH$_3$PteGlu) as substrates, *threo*-4-FGlu is incorporated into folate products by this enzyme with the formation of predominantly one product. The product from MTX was identified as 4-NH$_2$-10-CH$_3$PteGlu-*threo*-4-FGlu. This analogue thus serves as a substrate for folylpolyglutamate synthase and as a chain-terminating inhibitor of polyglutamate synthesis (McGuire and Coward, 1985). This result has added significance because the narrow amino acid substrate specificity of this enzyme had thwarted previous attempts to substitute amino acid analogues for L-Glu in this reaction. Subsequent studies showed that 4-FGlu had no effect on the glutamylation of MTX by hepatic cells *in vitro*. Toxic effects and inhibition of accumulation of MTX polyglutamates appeared to be related to effects of the analogue on cell membrane integrity (Galivan *et al.*, 1985).

7.5 Halogenated Amino Acids as Mechanism-Based Enzyme Inactivators

7.5.1 Introduction

The essential macromolecular component of bacterial cell walls, the peptidoglycan, consists of linear polysaccharide chains cross-linked by short peptides (Stryer, 1988, pp. 195–197). The disruption of peptidoglycan synthesis has been a basic strategy for the design of antibacterial reagents. In particular, the unique constituents of peptidoglycan—D-Ala, D-Glu, and muramic acid—have been recognized for decades as offering chemists attractive targets for drug design (Park, 1958). Included among the enzymes involved in peptidoglycan biosynthesis which have been the subjects of such research are alanine racemase, D-Ala:D-Ala ligase, and D-amino acid transaminase. The D-Ala analogues cycloserine and O-carbamoyl-D-serine (Fig. 7-26) are early examples of antibacterial agents that interfere with incorporation of D-Ala into the bacterial cell wall.

During the last two decades, much research has focused on the use of halogenated alanine analogues as potential antibacterial agents. Efficient binding of β-chloroamino acids to transaminases and decarboxylases had been shown to lead to β-elimination and, in some cases, to irreversible inactivation of the enzymes. For example, Tate et al. (1969) reported that an L-aspartate β-decarboxylase-catalyzed conversion of β-chloroalanine (β-ClAla) to pyruvate, chloride, and ammonia was accompanied by a slow, irreversible active-site labeling and inactivation of the enzyme. Nucleophilic displacement of chloride from the β-chloroalanyl-PLP Schiff base by a lysine residue near or at the active site was proposed as the mechanism of inactivation. β-ClAla also inhibits threonine deaminase and transaminase. Threonine-reversible inhibition of the growth of Salmonella typhimurium by β-ClAla thus was attributed to interference with branched-chain amino acid biosynthesis (Arfin and Koziell, 1971).

Manning et al. (1974, and references therein) found that inhibition of growth of several bacterial cell lines by the D and L isomers of β-ClAla was blocked by D-Ala, or D-Ala-D-Ala, but not by L-Ala. Furthermore, intracellular levels of D-Ala were greatly reduced by β-Cl-D-Ala treatment. These results implicated an inhibition by β-Cl-D-Ala of the PLP-dependent alanine racemase required for the synthesis of D-Ala as a mechanism of the antibacterial properties of these analogues.

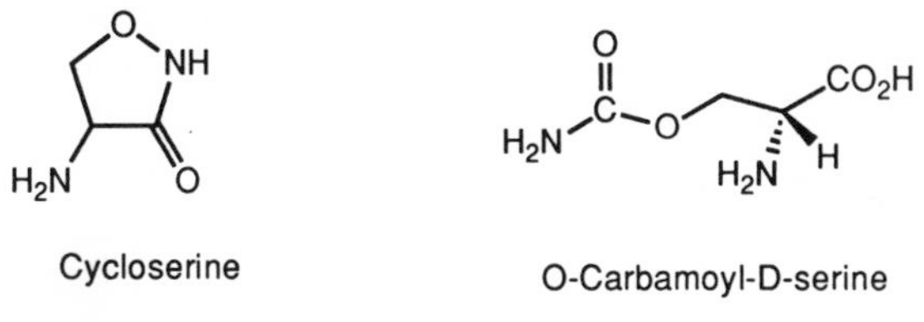

Cycloserine   O-Carbamoyl-D-serine

Figure 7-26. D-Ala analogues, cycloserine and O-carbamoyl-D-serine—early examples of antibacterial agents that inhibit the incorporation of D-Ala into the bacterial cell wall.

In 1970, Kollonitsch and his colleagues at Merck exploited a new photofluorination procedure to explore a new rational approach to drug design, synthesized β-F-D-Ala, and found this analogue to have potent *in vivo* antibacterial activity (Kollonitsch *et al.*, 1973). No detailed mechanistic proposals were advanced in the initial report, but elimination of HF from β-F-D-Ala and subsequent inhibition of PLP-dependent transaminases and racemases were suggested as explanations for the antibacterial properties of this analogue (Manning *et al.*, 1974). Recognition of the possibility that β-F-D-Ala could be acting as a suicide substrate for alanine racemase prompted early and intense interest in the further exploitation of this inhibitor concept with other PLP-linked enzymes (Maycock *et al.*, 1980). This line of research subsequently produced promising new antibacterial agents, as well as many other classes of medicinal and pharmacological agents. Several excellent and extensive reviews of this subject are available (e.g., Palfreyman *et al.*, 1987; Neuhaus and Hammes, 1981; Kollonitsch, 1982; Walsh, 1982, 1983, 1984). In this section, the rationale of this approach will be reviewed briefly, the scope of the applications will be summarized, and recent developments will be discussed.

7.5.2 PLP-Dependent Enzymes: The Mechanism upon Which Inactivation Is Based

PLP-dependent enzymes catalyze a wide range of reactions of amino acids, including decarboxylations, deaminations, racemizations, and aldol cleavages. PLP is initially attached to the enzyme by a Schiff base linkage to an ε-amino group of an active-site lysine. In the presence of substrate, a transimination occurs to produce a PLP–substrate Schiff base complex, a structure which can stabilize carbanion equivalents resulting from breaking of one of the three bonds to the α carbon (Fig. 7-27) (for a further discussion, see Stryer, 1988, pp. 496–499). The common feature of the enzymatic reactions targeted for inactivation by β-halo-Ala and related analogues is the potential for participation of the carbanion equivalent in the extrusion of halide ion or other leaving groups to produce a reactive enamine intermediate. Irreversible inactivation is a result of binding between the substrate (or substrate–cofactor complex, see below) and the enzyme. It is obvious, of course, that this approach is not confined to halogenated amino acids—indeed, several other classes of mechanism-based PLP-dependent enzyme inhibitors, for example, cyclopropyl and acetylenic amines, are based on similar rationales.

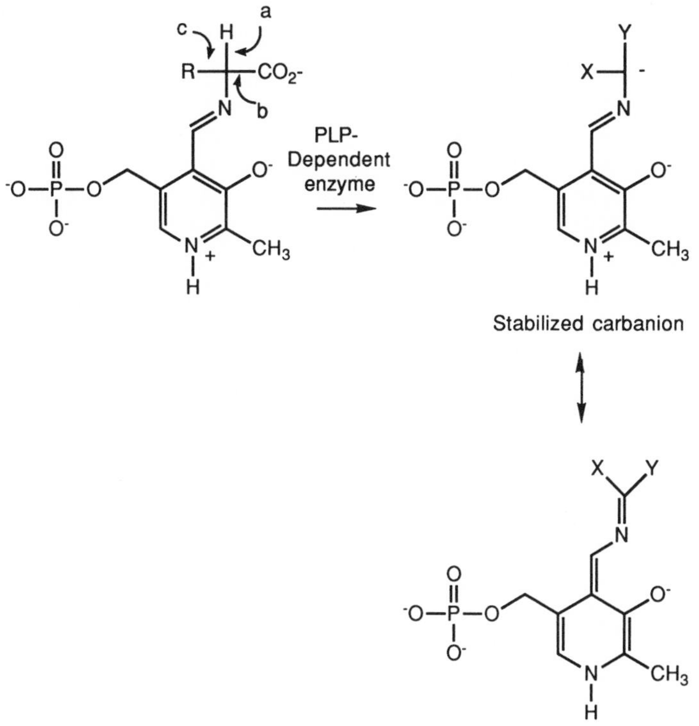

Figure 7-27. Reactivity of the PLP–substrate Schiff base complex. Enzyme-catalyzed reactions mediated by this complex include those catalyzed by transaminases (breaking bond a), those catalyzed by decarboxylases (breaking bond b), and aldol reactions (breaking bond c).

7.5.3 3-Haloalanines: Inhibitors of Alanine Racemase and Related Enzymes

7.5.3.1 Mechanism of Inactivation of Alanine Racemase by β-Substituted Alanines

Wang and Walsh (1978) compared the behavior of a number of irreversible inhibitors of *E. coli* alanine racemase and found several lines of evidence that pointed to the formation of a common inactivating species derived from these inhibitors. Purified enzyme catalyzes the formation of pyruvate from D- and L-β-chloroalanine and D- and L-β-fluoroalanine following a time course paralleling the inactivation of the enzyme, with the D isomers showing a higher rate of turnover. β-Fluoroalanine enantiomers were more reactive than the β-chloro analogues. In no case was a

β-halopyruvate formed (Fig. 7-28). Despite differences in affinities for the enzyme, and different rates of reaction, the partition ratio (the number of pyruvate molecules formed for each enzyme molecule inactivated) was essentially identical for all four inhibitors (between 790 and 870). Since this partition ratio was independent of the nature of the leaving group, the chirality of the substrate, and the V_{max} of the substrate, a common inactivating species was proposed. The mechanism suggested (Fig. 7-29) invokes an enamino acid–pyridoxal phosphate complex formed by elimination of HX from the Schiff base-bound inhibitor. As the partitioning intermediate, this can either partition to pyruvate and ammonia (the "harmless pathway") or, functioning as an electrophilic center, trap the lysine residue invariably present at the active site of PLP-dependent enzymes. Other inhibitors, including D- and L-O-carbamoylserine and D- and L-O-acetylserine, had partition ratios comparable to those of the β-haloalanines, implicating this same intermediate. It should be noted that this type of intermediate is also produced during certain normal enzyme-catalyzed β-eliminations without leading to inactivation (Section 7.5.3.3).

This "elimination–Michael addition" mechanism was widely accepted and extended to mechanism-based inactivators of a broad range of PLP-dependent enzymes (see below). However, recent research by Metzler and co-workers on the PLP-dependent enzyme glutamate decarboxylase has produced information which has required a reevaluation of this scheme (Likos et al., 1982). Irreversible inactivation of glutamate decarboxylase by L-serine-O-sulfate, followed by treatment of the inactivated species with base, released a low-molecular-weight yellow product. This was shown to be identical to the product formed by an aldol reaction between pyruvate and pyridoxal phosphate, followed by dehydration, a compound previously synthesized by Schnackerz et al. (1979) and now termed the Schnackerz adduct (Fig. 7-30). To account for this product, Metzler and co-workers proposed the release of α-aminoacrylate from the enamino acid–PLP com-

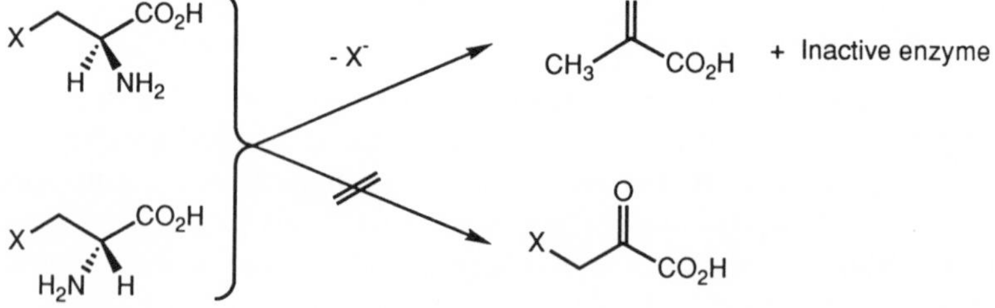

Figure 7-28. D- and L-β-Haloalanines as inhibitors of E. coli alanine racemase (Wang and Walsh, 1978).

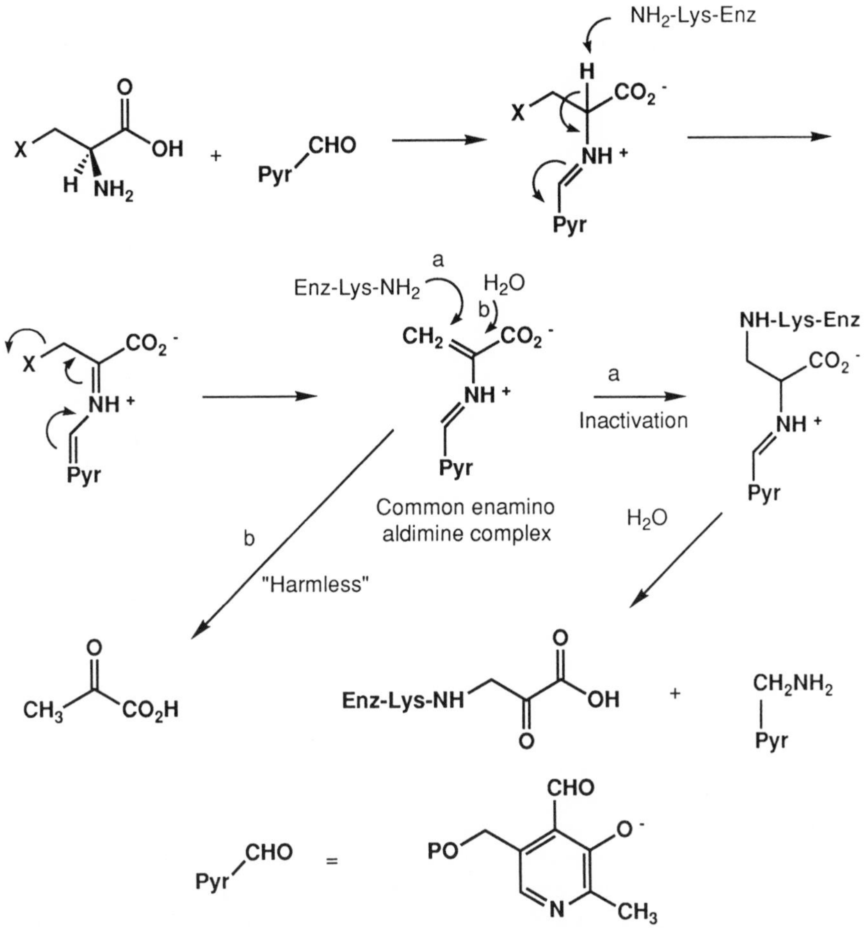

Figure 7-29. The initial "elimination–Michael addition" mechanism proposed for inhibition based on an enamino acid–pyridoxal phosphate complex formed by elimination of HX from the Schiff base-bound inhibitor (Wang and Walsh, 1979).

plex by transimination. Inactivation occurs by *nucleophilic* attack of this enamine on the internal Schiff base of pyridoxal phosphate and lysine. The Schnackerz adduct is formed by base cleavage of the lysine residue and hydrolysis (Fig. 7-30). The authors stressed in this and an accompanying report that this "reversed polarity" mechanism might be more common than expected, particularly since the inactivated enzyme is often trapped by borohydride reduction, a procedure which would render differentiation between this and the original Walsh mechanism difficult (Likos *et al.*, 1982; Ueno *et al.*, 1982).

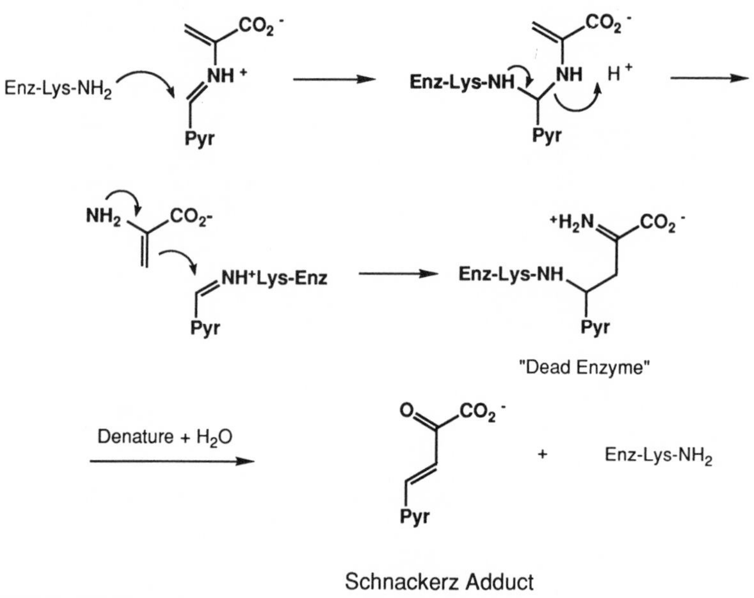

Schnackerz Adduct

Figure 7-30. Revised mechanism invoking the release of α-aminoacrylate from the enamino acid–PLP complex by transimination. Inactivation occurs by *nucleophilic* attack of this enamine on the internal Schiff base of pyridoxal phosphate and lysine (Likos *et al.*, 1982).

Walsh and co-workers subsequently studied the mechanism of inactivation of purified broad-specificity amino acid racemase from *Pseudomonas striata* and of cloned narrow-specificity alanine racemase from *Salmonella typhimurium* (Roise *et al.*, 1984; Badet *et al.*, 1984). In both systems, they confirmed that, while their initially proposed enamino acid–PLP complex is the first common species formed from a number of inhibitors (D- and L-3-ClAla, D- and L-3-FAla, O-acetyl-D-serine), the ultimate inactivator is aminoacrylate. There are clear implications for drug design in these results. As Walsh and co-workers noted, a consequence of this mechanism is that the efficiency of inactivation may be largely independent of active-site structure and local concentration of nucleophilic amino acid side chains. The probability of attack of aminoacrylate on ε-PLP-lysine would be controlled by rotational isomerization and rates of diffusion of aminoacrylate away from the active site (Roise *et al.*, 1984).

7.5.3.2 PLP-Linked Enzymes Inhibited by 3-Haloalanines

In addition to alanine racemase, 3-haloalanines interact with, and inhibit, transaminases and decarboxylases. For example, both 3-Cl- and

3-F-D,L-Ala inactivate glutamate-pyruvate transaminase and glutamate-oxaloacetate transaminase (Silverman and Abeles, 1976). Similarly, PLP-dependent alanine aminotransferase catalyzes the α, β-elimination of HCl from 3-ClAla, releasing pyruvate and ammonia. This reaction is accompanied by irreversible inactivation and covalent labeling of the enzyme (Morino et al., 1979). 3-ClAla inactivates L-aspartate β-decarboxylase and has been used as an affinity label for this enzyme (Relyea et al., 1974).

7.5.3.3 Polyhalogenated Alanine Analogues as Inhibitors of Alanine Racemase and Related Enzymes

During the course of β-elimination reactions catalyzed by PLP-linked enzymes, an enamine–PLP complex must be formed. Yet, in general, these reactions proceed without inactivation by substrate. To see if such a process could be subject to inactivation if the rate of addition of enzyme-bound nucleophile were enhanced, Silverman and Abeles (1976) examined the behavior of trifluoro-D,L-Ala and dichloro-D,L-Ala toward several PLP-dependent enzymes that catalyze β-eliminations, including β-cystathionase, γ-cystathionase, tryptophanase, D-serine dehydratase, tryptophan synthase (β_2 subunit and $\alpha_2\beta_2$ complex), cystathionine synthase, and threonine synthase. All of these enzymes, with the exception of cystathionine synthase, catalyze elimination reactions from Ser, 3-ClAla, and 3-FAla to give pyruvate, with no observable inactivation. In contrast, enzyme-catalyzed elimination from trifluoroalanine and dichloroalanine led to irreversible inactivation in most cases. This indicates that, with these enzymes which catalyze eliminations from their natural substrates, increased electronegativity of the β carbon indeed is required for addition of enzyme-bound nucleophile.

An increased rate of attack of enzyme-bound nucleophile on the enamine–PLP complex derived from alanine racemase could be expected to produce a lower partition ratio, and a more efficient antibacterial agent. To examine this issue, Wang and Walsh (1981) examined β, β-difluoroalanine and β, β, β-trifluoroalanine as suicide substrates for E. coli alanine racemase. Both D- and L-difluoroalanine were good substrates for enzyme-catalyzed loss of HF and fluoropyruvate formation (Fig. 7-31). However, neither was an effective inhibitor, giving partition ratios of 5000 and 2600, respectively, for less than complete inactivation. This was attributed to the loss of a second HF molecule from the product of Michael addition of the enzyme nucleophile and the enamino acid–PLP complex. The resulting enamine would be susceptible to hydrolytic cleavage, releasing active enzyme (Fig. 7-31). In contrast, trifluoro-D,L-alanine was found to be a very poor substrate for processing. However, the relationship between the amount of

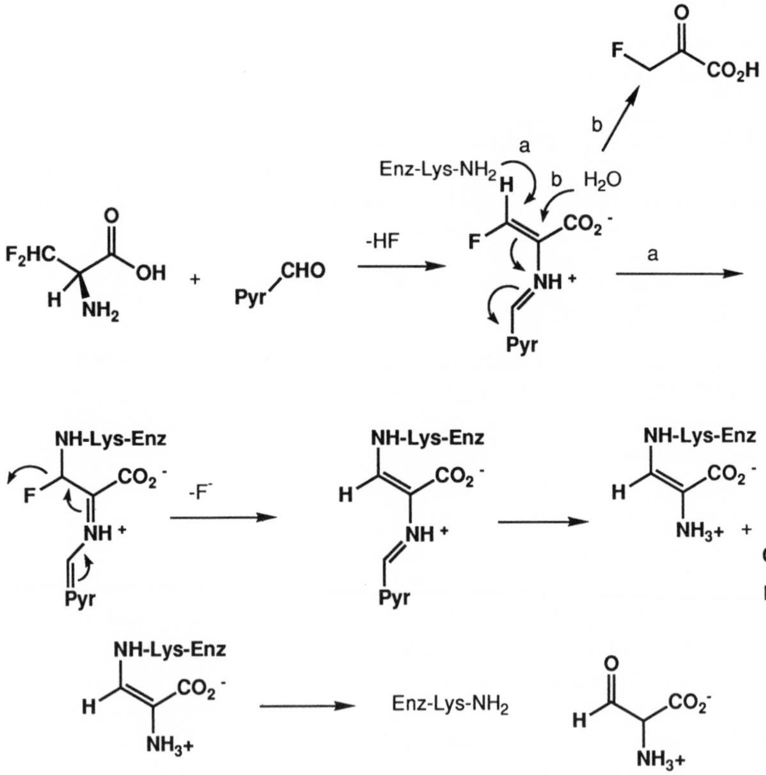

Figure 7-31. Difluoroalanine as a substrate for alanine racemase. Loss of a second HF molecule from the product of Michael addition of the enzyme nucleophile and the enamino acid–PLP complex (path a) produces an unstable enamine that releases active enzyme (Wang and Walsh, 1981).

fluoride released and the degree of inactivation shows that this analogue is an extremely effective inhibitor, with a partition ratio of < 10. In this case, loss of two additional HF molecules from the Michael product would produce the stable acyl enzyme adduct (Fig. 7-32). A similar structure had been characterized earlier by Silverman and Abeles (1977) following inactivation of γ-cystathionase by trifluoroalanine. Their identification of enzyme-bound glycine derived from trifluoroalanine by decarboxylation of the aminomalonyl-type intermediate (Fig. 7-32) implies that, in this case, inactivation has proceeded by addition of the enzyme-bound nucleophile to the β carbon of the inhibitor. The behavior of polyhaloalanines as racemase inhibitors is discussed further in the review by Walsh (1983).

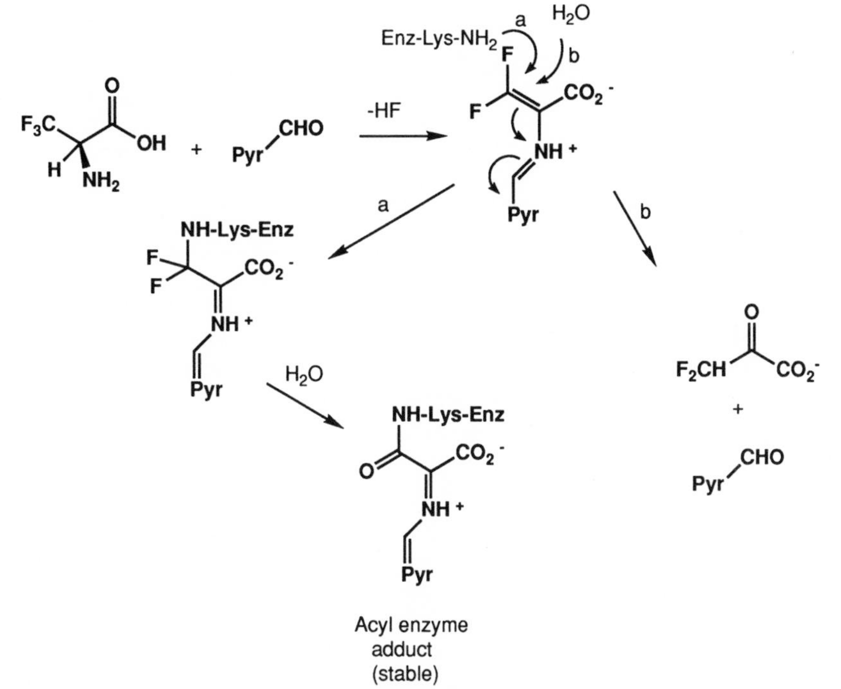

Figure 7-32. Trifluoroalanine as an inhibitor of alanine racemase. Loss of two additional HF molecules from the Michael product (path a) produces the stable acyl enzyme adduct (Wang and Walsh, 1981).

7.5.3.4 New Inhibitors of Alanine Racemase Based on Halogenated Analogues

In their review, Neuhaus and Hammes (1981) have described a wide range of analogues of Ala that have been developed for inhibition of bacterial cell wall biosynthesis. Examples of more recently developed inhibitors of alanine racemase will be considered in this section.

Clinical applications of alanine analogues, including β-halo-Ala, as antibiotics have been complicated by problems of transport and host toxicity. For example, trifluoroalanine is an effective inhibitor of alanine racemase but is a poor antibacterial agent, likely as a consequence of poor transport caused by a low amine pK_a (5.8) (Walsh, 1983). Also, 3-Cl-D-Ala is a substrate for renal D-amino acid oxidase, producing the potentially toxic β-chloropyruvate. Incorporation of antibacterial amino acids into transportable peptides has been exploited effectively to circumvent problems associated with poor transport, as well as to decrease potential

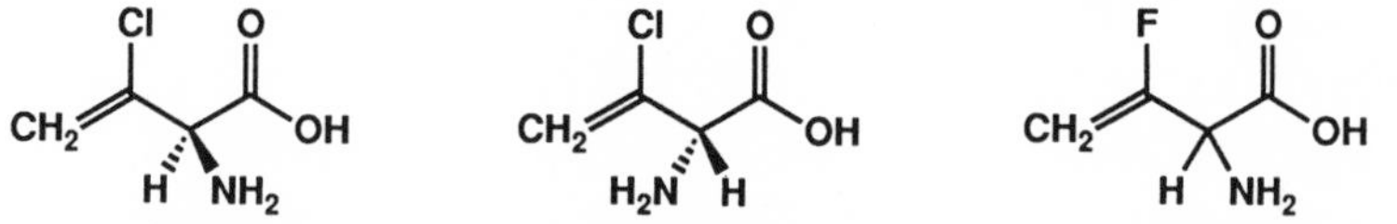

L-Chlorovinylglycine   D-Chlorovinylglycine   D,L-Fluorovinylglycine

Figure 7-33. 3-Halovinylglycines, mechanism-based alanine racemase inhibitors (Thornberry et al., 1987).

toxicity by providing a mechanism for targeted drug delivery [reviewed by Payne (1986)]. Accordingly, a series of peptides containing 3-Cl-D-Ala (and the related mechanism-based inhibitor propargylglycine) were synthesized. Several di- and tripeptides, including 3-Cl-L-Ala-3-Cl-L-Ala, displayed potent, broad-spectrum activity against both gram-positive and gram-negative species (Cheung et al., 1983, 1986).

3-Halovinylglycines (Fig. 7-33) represent a new class of mechanism-based alanine racemase inhibitors. D-Chlorovinylglycine, L-chlorovinyl-glycine, and D,L-fluorovinylglycine are all effective irreversible inactivators of E. coli alanine racemase, with partition ratios nearing unity for D-chloro-vinylglycine; that is, inactivation occurs with nearly every catalytic event. Denaturation of the enzyme–inhibitor complex after inactivation released free pyridoxal phosphate. This is in contrast to the formation of the Schnackerz adduct during inactivation by β-haloalanines, indicating that the two classes of inhibitors operate by different mechanisms (Thornberry et al., 1987). Despite extremely efficient enzyme inhibition, D-chlorovinyl-glycine had only modest antibacterial activity, and this behavior was attributed to inefficient transport. Consistent with this assumption, several L-norvalyl peptides containing halovinylglycines were found to have increased antibacterial activity (Patchett et al., 1988).

The phosphonic analogue of alanine, (1-aminoethyl)phosphonic acid (Ala-P) (Fig. 7-34), functions as a potent inhibitor of alanine racemase by forming a tight enzyme–Ala-P complex ($t_{1/2} = 25$ days) (Copié et al., 1988, and references therein). When incorporated into a peptide, it is an effective antibiotic agent. The β-chloro, β,β-dichloro, and β,β,β-trichloro analogues

Ala-P   1   2   3

Figure 7-34. (1-Aminoethyl)phosphonic acid (Ala-P) and β-chloro-, β,β-dichloro-, and β,β,β-trichloro analogues of Ala-P, inhibitors of alanine racemase (Vo-Quang et al., 1986).

of Ala-P (Fig. 7-34; structures **1**, **2**, and **3**, respectively) were synthesized in an attempt to enhance the inactivation of alanine racemase by Ala-P by incorporation of a suicide substrate structural moiety. The monochloro and dichloro analogues were strong inhibitors of alanine racemase but did not behave as suicide substrates. The poor antibacterial activity observed was attributed to poor transport, a problem that might be circumvented by the peptide-transport strategy (Vo-Quang *et al.*, 1986).

7.5.4 Halogenated Analogues as Inhibitors of Amino Acid Decarboxylases

The importance of the amine products of amino acid decarboxylases is clear. The aminergic neurotransmitters (dopamine and norepinephrine, 5HT, histamine, and γ-aminobutyric acid), in addition to serving as central neurotransmitters, are involved in the control of a host of important biological functions, such as regulation of blood pressure, maintenance of smooth muscle tone, and the allergic response. Accordingly, drugs designed to regulate selectively these amine levels have many clinical applications and are the targets of continuing active research in medicinal chemistry. Another biologically important amine, putrescine, the product of decarboxylation of ornithine, is a precursor of polyamines, increased levels of which have been associated with rapid cell division. The amino acid decarboxylases, required for the biosynthesis of these amines, are PLP linked.

7.5.4.1 Inhibition of Biogenic Amine Formation

Many α-fluoromethyl-α-amino acids have been synthesized and have been shown to be selective and potent irreversible inhibitors of amino acid decarboxylases. For example, at Merck, Kollonitsch *et al.* (1978) prepared (R,S)-α-fluoromethylglutamic acid (**4**), (R,S)-α-fluoromethylornithine (**5**), (R)- and (S)-α-fluoromethyl-DOPA (**6** and **7**), (S)-α-fluoromethyltyrosine (**8**), and (R,S)-α-fluoromethylhistidine (**9**) (Fig. 7-35). The selectivity of inhibition is illustrated by the potent irreversible inhibition of DOPA decarboxylase (also known as aromatic amino acid decarboxylase) by (S)-α-fluoromethyl-DOPA, a compound that has no effect on histidine decarboxylase. In contrast, (R,S)-α-fluoromethylhistidine irreversibly inhibits mammalian histidine decarboxylase (but not the bacterial enzyme) but has no effect on DOPA decarboxylase. α-Fluoromethyl-DOPA, but not the less effective inhibitor α,α-difluoromethyl-DOPA, synthesized independently by the Merrel group in Strasbourg, increased DOPA levels *in vivo* (Jung *et al.*, 1979). The inactivation of DOPA decarboxylase by α-fluoromethyl-DOPA was studied in detail by Maycock *et al.* (1980), and accumulated data were

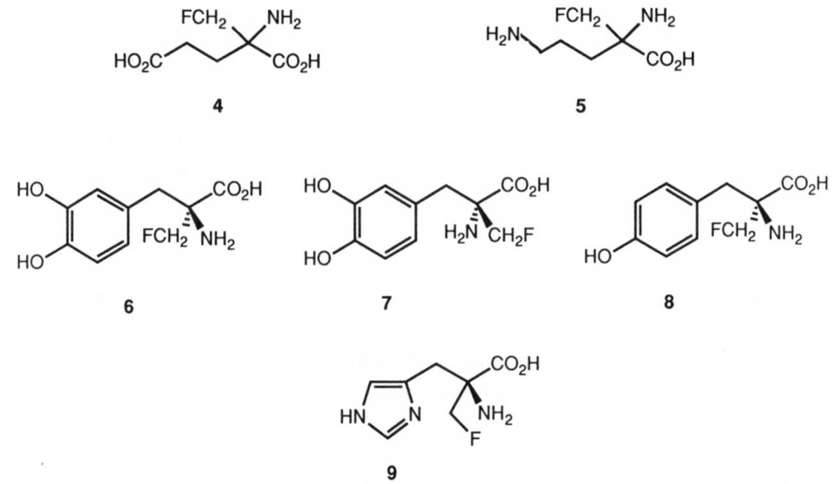

Figure 7-35. Examples of α-fluoromethyl-α-amino acids that function as inhibitors of amino acid decarboxylases.

consistent with mechanism-based inactivation. Kollonitsch *et al.* (1978) reported that α-(*R,S*)-fluoromethylglutamic acid was only a poor inhibitor of mammalian glutamate decarboxylase, but Kuo and Rando (1981) found it to be a potent inhibitor of the bacterial enzyme. Their detailed mechanistic studies supported mechanism-based inactivation. Certain α-chloromethyl analogues also were decarboxylase inhibitors but were less stable in solution. This observation illustrates a particular advantage of the incorporation of fluorine as the trigger for revealing latent reactivity, in that the C−F bond, unlike other carbon–halogen bonds, is quite resistant to normal nucleophile displacement reactions.

The examples cited above are analogues of the normal enzyme substrates. Based on the principle of microscopic reversibility, product analogues also can serve as mechanism-based inhibitors. As an illustration of product inhibition, (*R*)- and (*S*)-α-fluoromethylhistamine (**10** and **11**)

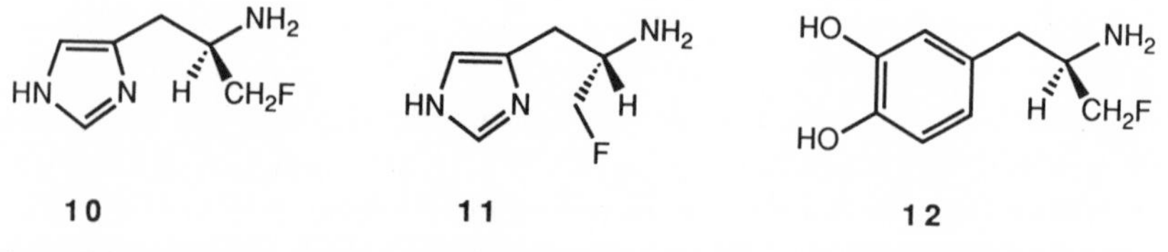

Figure 7-36. Examples of α-fluoromethylamines that function as product analogue inhibitors of amino acid decarboxylases.

and (R)-α-fluoromethyldopamine (12) (Fig. 7-36) irreversibly inhibit mammalian histidine decarboxylase and DOPA decarboxylase, respectively [Kollonitsch *et al.*, 1978; reviewed by Kollonitsch (1982)].

7.5.4.2 Inhibition of Polyamine Biosynthesis

The polyamines putrescine, spermidine, and spermine (Fig. 7-37), present in all living cells, have important roles in such cellular processes as replication and differentiation. An important strategy in the study of the functions of these amines has been to investigate the pharmacological consequences of their depletion. In mammalian cells (but not in bacterial cells), ornithine decarboxylase (ODC)-catalyzed decarboxylation of ornithine (Fig. 7-37) provides the sole source of putrescine, the biological precursor of spermidine and spermine.

While ODC inhibition represented an obvious method to lower polyamine levels, initial attempts to accomplish this with such competitive ODC inhibitors as α-methylornithine were disappointing (Sjoerdsma, 1981). In recognition of the potential of mechanism-based inhibitors to abolish activity more effectively, Metcalf *et al.* (1978) synthesized α-difluoromethylornithine (13) (DFMO), α-chloromethylornithine (14), and α-cyanomethylornithine (15) as substrate analogues of ornithine and 5-hexyne-1,4-diamine (16) and *trans*-hex-2-en-5-yne-1,4-diamine (17) as product analogues (Fig. 7-38). All functioned as irreversible inactivators of ODC,

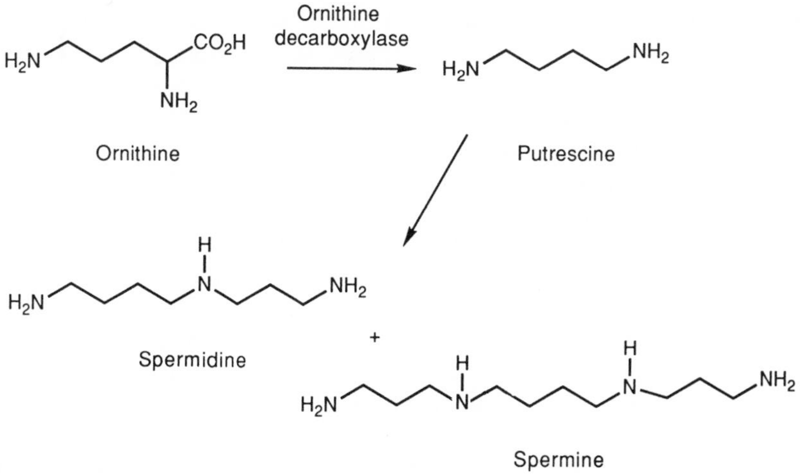

Figure 7-37. Biosynthesis of the polyamines putrescine, spermidine, and spermine from ornithine.

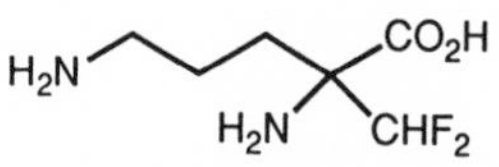

Figure 7-38. Examples of mechanism-based inhibitors of ornithine decarboxylase.

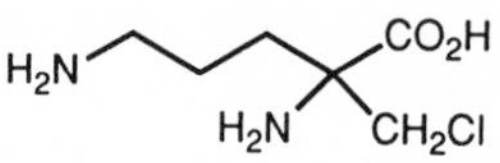

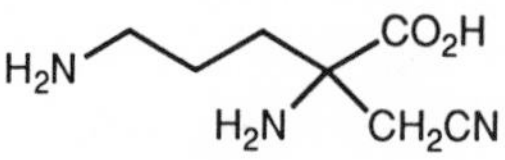

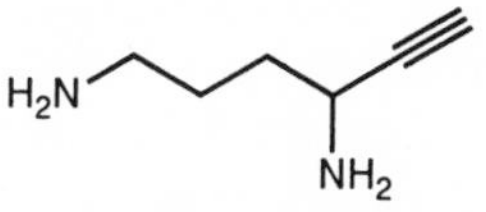

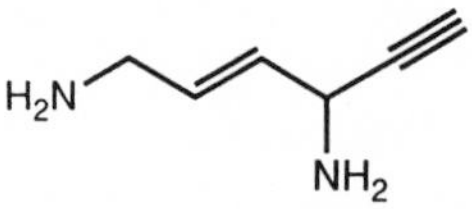

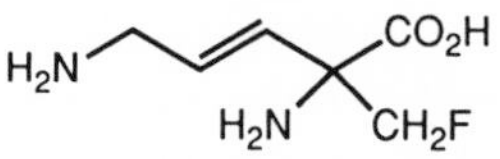

and DFMO has subsequently proven to be a valuable tool for the study of the pharmacological consequences of ODC inhibition. The potential clinical relevance of DFMO has also been investigated extensively. Although results using animal tumor models were very promising, initial human clinical studies failed to demonstrate clear antitumor activity for DFMO. In contrast, DFMO has shown considerable promise in the treatment of protozoal infections, in particular, African trypanosomiasis (sleeping sickness). A cytostatic action on rapidly replicating cells appears to be a unifying consequence of DFMO treatment; this, in turn, can be related to the absolute dependence of these cells on ODC for putrescine biosynthesis. DFMO exhibits little host toxicity. The chemotherapeutic implications of polyamine depletion caused by DFMO inhibition of ODC have been reviewed by Sjoerdsma and Schechter (1984).

The success of DFMO as a pharmacological tool and its potential effectiveness as a chemotherapeutic agent have prompted the search for even more effective mechanism-based inhibitors of ODC. In both substrate and product analogues, introduction of a double bond increases the effectiveness of inhibition [reviewed by Sjoerdsma and Schechter (1984)]. (E)-2-(Fluoromethyl)dehydroornithine (18) (Fig. 7-38) is a particularly effective ODC inhibitor in vitro, binding 14-fold more strongly to the enzyme than does DFMO. Several esters of 18 have been shown to serve as prodrugs, producing effective and prolonged inhibition of mammalian ODC *in vivo* (Mamont *et al.*, 1986, and references therein).

In bacteria and higher plants, arginine decarboxylase (ADC) catalyzes the decarboxylation of arginine to give agmatine, which is then enzymatically processed to putrescine (Fig. 7-39). The ADC pathway thus provides these cells with an alternative polyamine biosynthetic pathway. D,L-α-(Difluoromethyl)arginine (DFMA) (19) (Fig. 7-39) was synthesized by the Merrel Dow research group and found to be a potent mechanism-based inhibitor of bacterial arginine decarboxylases (Kallio *et al.*, 1981). This analogue has played an important role in the elucidation of the role of ADC in normal bacterial and plant cell physiology. For example, inhibition either of ADC with DFMA or of ODC with DFMO in *E. coli* appears to induce a compensatory increase in the activity of the other decarboxylase. Use of both inhibitors in combination with cyclohexylamine (an inhibitor of spermidine synthase) causes a significant decrease in polyamine concentration (Bitonti *et al.*, 1987, and references therein). Several additional ADC substrate and product analogues synthesized subsequently are more potent inhibitors than DFMA (Bitonti *et al.*, 1987).

In related research, D,L-α-(difluoromethyl)lysine (DFML) (20) was found to be a selective irreversible inhibitor of lysine decarboxylase (LDC),

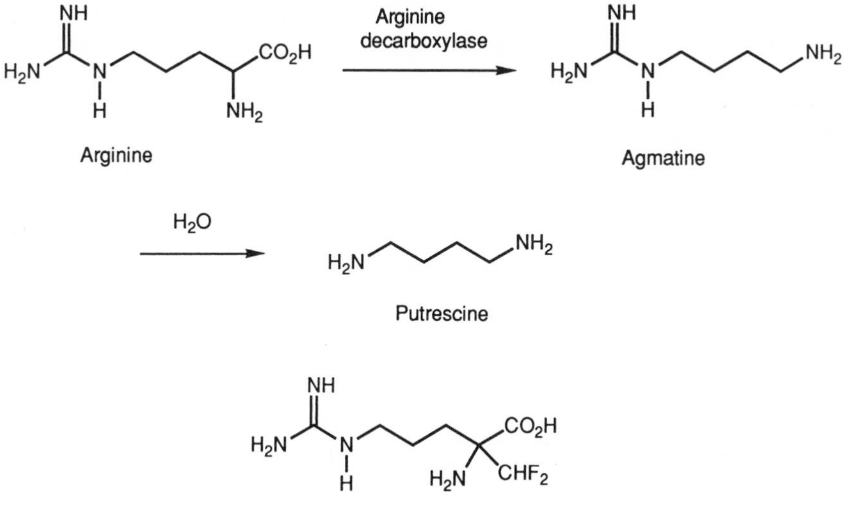

Figure 7-39. Biosynthesis of putrescine from arginine (the ADC pathway) and structure of DFMA (19), a potent mechanism-based inhibitor of bacterial arginine decarboxylases (Kallio *et al.*, 1981).

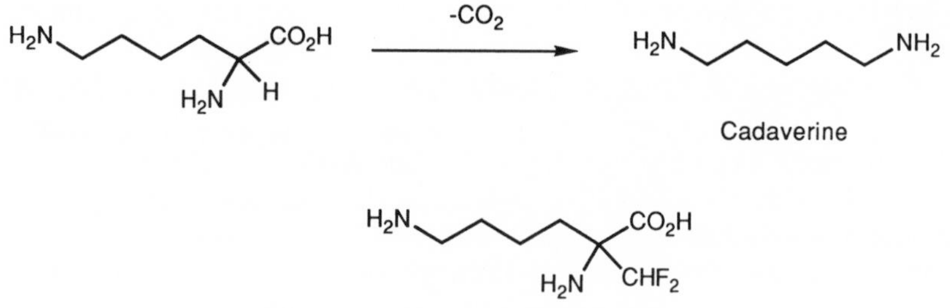

DFML
20

Figure 7-40. Biosynthesis of cadavarine and structure of DFML (**20**), a selective irreversible inhibitor of lysine decarboxylase (Pösö *et al.*, 1984).

the enzyme responsible for the synthesis of cadaverine (Fig. 7-40). This analogue has been examined as a potential drug for the treatment of certain mycoplasmic infections (Pösö *et al.*, 1984).

7.5.5 Inhibitors of γ-Aminobutyrate Aminotransferase (GABA-T)

The neurotransmitter γ-aminobutyrate (GABA) increases the permeability of postsynaptic membranes to K^+ and thus functions as an inhibitory neurotransmitter. GABA is formed by glutamate decarboxylase-catalyzed decarboxylation of glutamate and is inactivated by GABA-T-catalyzed transamination to succinate semialdehyde. Both enzymes are PLP dependent. Drugs that cross the blood–brain barrier and inhibit the activity of GABA-T are effective anticonvulsants, since such drugs increase the brain levels of GABA. An early example of a PLP-targeted mechanism-based inhibitor of GABA-T is 4-amino-5-hexenoic acid (**21**) (Fig. 7-41), effective in the treatment of epilepsy.

Similar to the research described in the previous sections, introduction of α-halomethyl substitution as a site for latent reactivity has been explored extensively to develop further inhibitors of GABA-T. Thus, (*S*)-4-amino-5-fluoropentanoic acid (**22**; Fig. 7-41) was shown to be an effective and selective irreversible inhibitor of GABA-T *in vivo* (Silverman and Levy, 1981). In recognition of the fact that unsaturation in the backbone of enzyme substrates and inhibitors frequently increases binding to the enzyme, Silverman and co-workers prepared a series of fluorinated analogues of 4-aminobutyric acid, both saturated and unsaturated, in a strategy to develop additional inhibitors of GABA-T. In this work, interesting struc-

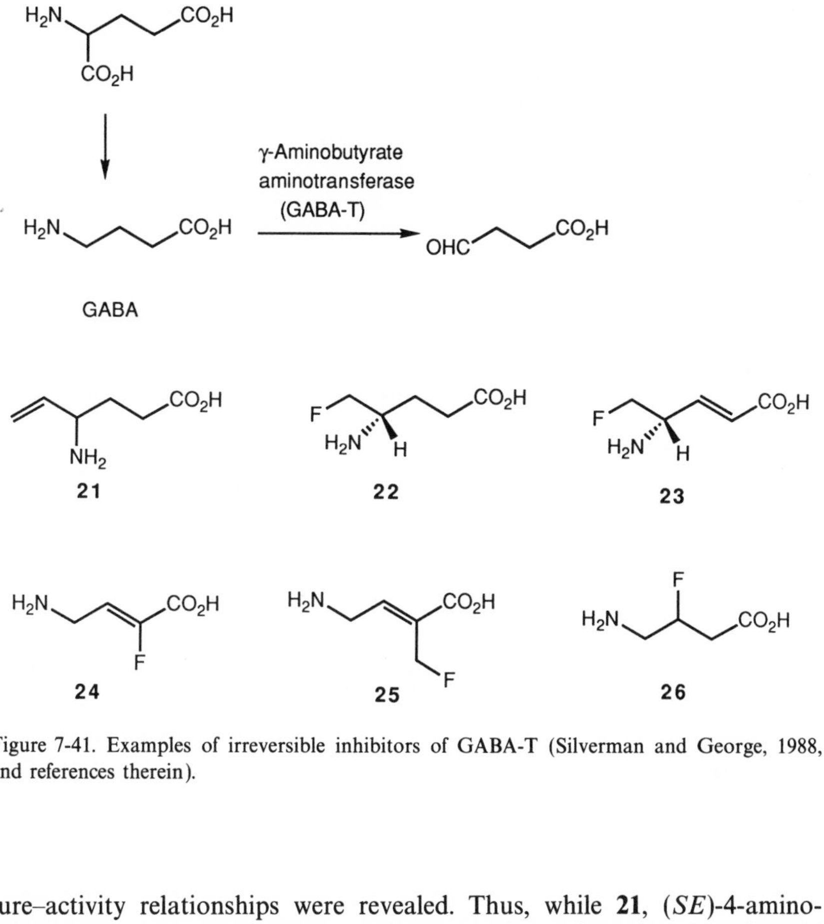

Figure 7-41. Examples of irreversible inhibitors of GABA-T (Silverman and George, 1988, and references therein).

ture–activity relationships were revealed. Thus, while **21**, (*SE*)-4-amino-5-fluoropenten-2-oic acid (**23**), and (*Z*)-4-amino-2-fluorobut-2-enoic acid (**24**) irreversibly inhibit GABA-T, the related analogues 4-amino-2-(fluoromethyl)but-2-enoic acid (**25**) and 4-amino-3-fluorobutanoic acid (**26**) are reversible inhibitors and substrates (Fig. 7-41). GABA-T catalyzes elimination of HF from each of the analogues. Inactivation occurs for *every* turnover of **22** and once every five turnovers for **23** (Silverman and George, 1988, and references therein). Inactivation of GABA-T by **22** and **23** apparently occurs through the enamine mechanism proposed by Metzler and co-workers while **24** operates through an isomerization–Michael addition mechanism operating on the PLP-bound substrate prior to fluoride release (Silverman and George, 1988). These studies have revealed subtle and sometimes unexpected factors which can complicate the rational design of irreversible enzyme inhibitors.

7.6 Transport and Storage of Halogenated Amino Acids

Marquis (1970) has reviewed earlier work on the transport of fluorinated amino acids in microorganisms. From this work, the conclusion was drawn that these analogues can be transported by a normal transport mechanism, but that they have relatively low affinities for transport. A subsequent analysis of transport of Phe analogues by several cell systems, including microorganisms, is given in the review by Wheatley (1978).

The relationship between the immediate fate of transported amino acids and protein synthesis has been a subject of considerable controversy. Wheatley (1978) has discussed in detail the relationship between amino acid transport and protein synthesis, including the several issues concerning the availability and importance of intracellular amino acid pools to protein synthesis in various cell types. A point of debate has been the source of precursor molecules for protein synthesis. Evidence had accumulated which suggested that amino acids are taken up from the extracellular medium, directly activated, and loaded onto tRNA, thus bypassing an existing trichloroacetic-acid (TCA) soluble intracellular pool. Other data supported the claim that this intracellular acid-soluble pool provides the precursors. For example, the ability of 4-FPhe to become incorporated into protein of HeLa S-3 cells despite preloading of cells with Phe was cited by Robertson and Wheatley (1979) as compelling evidence that the acid-soluble pool could be bypassed and that this pool had little relationship to incorporation, at least in mammalian cells. The existence of an amino acid pool in free and rapid equilibrium with the extracellular medium was proposed as the source of substrates for protein biosynthesis. This theory has been extended to question the traditional concept of active transport of amino acids across the cell membrane (Wheatley *et al.*, 1986, and references therein). For recent reviews of amino acid transport mechanisms, see Antonucci and Oxender (1986) (bacteria) and Shotwell *et al.* (1983) (mammalian).

7.7 Incorporation of Halogenated Amino Acids into Protein

7.7.1 Editing Mechanisms in Protein Biosynthesis

Survival of an organism requires the transfer of genetic information to produce proteins with a minimal possibility of an incorrect sequence of peptide bond formation. Thus, in order to be incorporated into protein, an amino acid analogue must resemble its natural relative to the extent that it can pass quite stringent proofreading mechanisms.

7.7.1.1 Selection Is Mediated by Two Recognition Steps

Fidelity in protein biosynthesis is accomplished through two sequential intermolecular recognition phenomena. The second of these is the base pairing involved in codon–anticodon interactions of aminoacyl-charged tRNA with mRNA according to the genetic code. A topic of much current research concerns the fidelity with which individual amino acids are matched with the correct tRNAs in the first recognition step. Thus, it has been demonstrated that the codon–anticodon interaction is independent of the nature of the aminoacyl group carried by the tRNA, so that proper charging of the tRNA is vital to accurate protein synthesis. This matching, mediated by aminoacyl-tRNA synthases, requires, first, a normal substrate–enzyme recognition of the amino acid by the synthase—an area in which amino acid analogue studies have been important in determining specificity—and, second, recognition of the tRNA by the synthase. With respect to the tRNA recognition step, there recently has been proposed a second genetic code, still largely undeciphered, mediated by specific interactions between the synthase enzyme and a specific base pair on the tRNA (a "bilingual" interaction) (Hou and Schimmel, 1988). The characterization of this interaction as a "second genetic code" has been criticized on the basis of the implied existence of a common set of rules for tRNA recognition of aminoacyl-tRNA synthases, a commonality which has been questioned (Schulman and Abelson, 1988).

7.7.1.2 Fidelity of Translation Is Increased by Amino Acyl-tRNA Synthase-Mediated Proofreading

Aminoacylation of tRNA is initiated by condensation of an amino acid with ATP to form an enzyme-bound aminoacyl adenylate. This species reacts with the 2'- or 3'-hydroxyl of tRNA to give aminoacyl-tRNA. Further fidelity in peptide bond formation is achieved by an editing process involving synthase-catalyzed hydrolysis of incorrect aminoester bonds that may be formed at this stage. Thus, acceptance of fraudulent amino acids by aminoacyl-tRNA synthases can catalyze ATP–PP$_i$ exchange without necessarily leading to peptide bond formation (see Fig. 7-42 for a summary of these steps; for a recent review, see Schimmel, 1987).

The importance for species survival of fidelity in translation of genetic information to protein synthesis has prompted several studies into the evolutionary aspects of the accuracy of amino acid tRNA synthases. With respect to Phe tRNA synthase, many characteristics appear to have been conserved through evolution. Indeed, as Santi and Webster (1976) have pointed out, it is reasonable to expect that binding regions important for

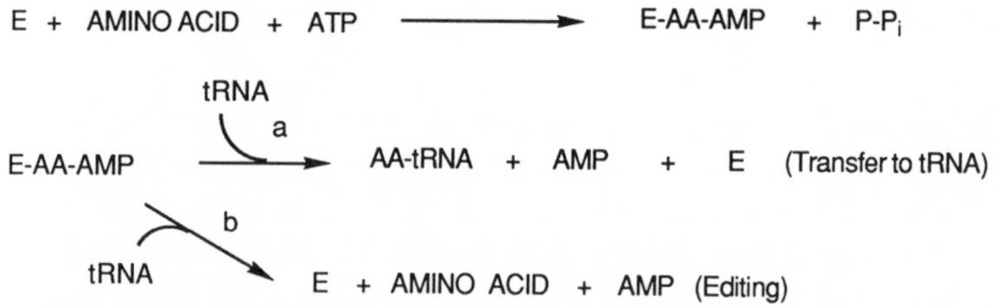

E + AMINO ACID + ATP   ⟶   E-AA-AMP + P-P$_i$

tRNA

$\searrow$ a

E-AA-AMP ⟶ AA-tRNA + AMP + E (Transfer to tRNA)

$\nearrow$ b

tRNA $\searrow$ E + AMINO ACID + AMP (Editing)

Figure 7-42. As an alternative to transfer to tRNA (path a), acceptance of a fraudulent amino acid by tRNA can initiate an editing process involving synthase (E)-catalyzed hydrolysis of the incorrect aminoester bond (path b).

specificity and catalysis would be quite similar, since identical substrates and reactions are involved. Interspecies differences, however, become evident in altered behavior of the different Phe tRNA synthases toward substrate analogues, behavior which is influenced more readily by structural dissimilarities in the enzymes outside the active site. For example, rat liver Phe tRNA synthase is significantly more tolerant than *E. coli* enzyme for phenyl ring substituents, although it has the same qualitative preference for an unsubstituted phenyl ring (Santi and Webster, 1976). Detailed studies of proofreading mechanisms of Phe tRNA synthases from *E. coli*, *Saccharomyces cerevisiae* (yeast), *Neurospora crassa*, turkey and hen liver, and *Methanosarcina barkeri* (an archaebacteria representing a third aboriginal line of cellular descent) have revealed major interspecies differences in the relative importance of each proofreading step (Rauhut *et al.*, 1985; Gabius *et al.*, 1983). Thus, observed differences in the degree of incorporation of amino acid analogues in different systems can be a result of differences in initial discrimination by the amino acid tRNA synthase or different rates of synthase-catalyzed hydrolysis of the analogue either prior or subsequent to transfer of the activated amino acid to tRNA.

7.7.2 Incorporation of Halogenated Amino Acids into Microbial Proteins

From a consideration of steric factors, it is apparent that, among halogenated amino acids, a fluorine-substituted amino acid will be most likely to be accepted by the synthase and to survive the proofreading steps. Accordingly, the vast majority of research in this area has involved fluorinated analogues.

Since the early demonstrations of incorporation of fluorinated amino acids into protein by Munier and Cohen (1959) and by Westhead and Boyer (1961), the protein biosynthetic machinery of many cell types has been exploited very effectively to insert fluorinated analogues into protein sequences for a variety of research purposes. These include, for example, the study of the incorporation process itself, as exemplified in analogue studies of amino acid tRNA synthases, the study of protein biosynthetic mechanisms as determined by analogue incorporation, and, of major impact, the use of incorporated analogues as biological tracers for ^{19}F-NMR studies. In the course of these studies, several strategies have been developed to enhance the degree of incorporation of the analogue [reviewed by Sykes and Weiner (1980)] and, more recently, to identify precisely the position of incorporation.

7.7.2.1 Use of Auxotrophs to Enhance Analogue Incorporation into Bacterial Protein

To increase the level of analogue incorporation, auxotrophic mutants of bacteria (mutants unable to synthesize particular amino acids) have been used extensively to circumvent problems associated with analogue toxicity and preferential incorporation of the natural amino acid. In the simplest approach, an auxotrophic bacterium can be starved for a particular amino acid. Addition of the analogue of that amino acid can result in high levels of incorporation, but the yield of protein is usually low because of decreased cell viability. In practice, a balance of the concentrations of analogue and natural amino acid is sought which allows a reasonable degree of bacterial growth and cell viability but which also leads to a significant degree of incorporation (Sykes and Weiner, 1980, and references therein).

Other strategies have been combined with the use of auxotrophs to increase efficiency of incorporation. A particularly effective tactic has been to induce the synthesis of a specific protein at the same time as the analogue is added to the culture medium. An example of this is seen in the labeling of bacterial alkaline phosphatase with 3-FTyr. Conditions were developed for growing a tyrosine auxotroph of E. coli such that the culture simultaneously became depleted of phosphate and tyrosine. Addition of both phosphate and 3-FTyr resulted in the induction of alkaline phosphatase synthesis with a 10% incorporation of 3-FTyr selectively into the enzyme. The ^{19}F-NMR spectrum of the purified enzyme contained resolved fluorine resonances corresponding to 11 tyrosines in chemically different environments (Sykes et al., 1974).

A similar strategy has been used to label aspartate transcarbamoylase, the enzyme which catalyzes the first committed step in the pyrimidine biosynthetic pathway in E. coli. A Tyr auxotroph that overproduces this enzyme upon uracil starvation was used. Addition of 3-FTyr to the uracil-depleted cells resulted in an 85% replacement of Tyr (Wacks and Schachman, 1985a). Despite this high level of incorporation, high enzyme activity was retained, and the labeled protein was used to study "communication" between the catalytic and regulatory subunits by ^{19}F-NMR spectroscopy (Wacks and Schachman, 1985b).

Bacteriophage-coded proteins labeled with fluorinated amino acids have been prepared in a similar fashion by simultaneous initiation of bacteriophage infection and analogue addition. In a recent study, a Tyr auxotroph was grown in a medium containing limiting Tyr along with optimal concentrations of the 19 other natural amino acids until early log phase was reached. 3-FTyr was added, followed 10 minutes later by wild-type bacteriophage fd. Subsequent investigation by ^{19}F-NMR spectroscopy of isolated gene 5 protein—a protein that forms a stoichiometric complex with single-stranded DNA from daughter virions, thus preventing their use as templates—revealed the presence of three surface and two buried 3-FTyr residues (Anderson et al., 1975). This 3-FTyr-labeled protein was used in subsequent NMR studies of gene 5 protein–oligodeoxynucleotide complexes (Coleman et al., 1976).

The effective use of auxotrophs under most circumstances is dependent on the establishment of an artificial bias to favor incorporation of the analogue over that of the natural amino acid, a reflection of the discrimination which accompanies membership in the ensemble of amino acids recognized by the genetic code. To test the mutability of the genetic code, Wong (1983) carried out serial mutation of a Trp auxotroph of Bacillus subtilis and obtained a new strain which grew much better on 4-FTrp than on Trp. This change in fitness to the genetic code was attributed to the ability of the new mutant strain to function better with protein containing 4-FTrp than with the unsubstituted protein. Wong suggests that this strain may be the first free-living organism in the past two billion years to have learned to thrive on an altered genetic code. Since this auxotroph can be supported on 4-FTrp, it provides a convenient source of 4-FTrp-labeled proteins endogeneous to or clonable into B. subtilis, including proteins that are vital to cell growth. This was demonstrated in a study of B. subtilis ribosomal proteins, a study which was facilitated by suppressed Trp fluorescence as a consequence of 4-FTrp incorporation (Bronskill and Wong, 1988).

7.7.2.2 Genetics Assignment of ^{19}F-NMR Resonances in Analogue-Labeled Protein

Lu et al. (1976) have developed a method to enhance incorporation of fluorinated amino acids which obviates the use of auxotrophs. In this work, the target protein was the *lac* repressor, a macromolecular component of the well-studied lactose operon. Their choice was dictated by the fact that mutations that occur at each of the eight Tyr residues had been isolated, thus providing an opportunity for assignment of ^{19}F-NMR resonances of incorporated 3-FTyr (see below). In addition, the induction and replication of a prophage in the cell line increased the gene copies of the *lac* repressor and greatly augmented the quantity of labeled protein isolated. In order to further enhance utilization of 3-FTyr in protein synthesis, advantage was taken of the ability of Trp, Phe, and Tyr to feedback inhibit and to repress the synthesis of DAHP (see Section 7.2.2.1). Thus, addition of 1 mM concentrations of 3-FTyr, Trp, and Phe to the culture in the middle of balanced growth led to a 96% drop in the utilization of endogenous Tyr and incorporation of 3-FTyr under conditions of optimum growth. The ^{19}F-NMR spectra of the isolated *lac* repressor revealed that each identical subunit had four or five surface tyrosines, two buried tyrosines, and one tyrosine that had an ionized phenolic group or was involved in a hydrogen bond. Lu *et al.* noted that their procedure should be amenable to a large number of inducible proteins.

A systematic genetic approach was used to assign the NMR resonances of the incorporated 3-FTyr in the *lac* repressor, and this has increased greatly the amount of information derived from the above experiments. Included in the many mutations characterized for the gene encoding the *lac* repressor (the *lacI* gene) are mutations having chain-terminating (nonsense) codons (UAG, UAA, UGA) at positions corresponding to all eight tyrosines and both tryptophans. Chain-terminating mutations in bacteria are correctable by nonsense suppressor tRNA molecules which have altered reading capacities caused by mutations in their anticodons. These nonsense suppressors can insert different amino acids in response to the chain-terminating nonsense codons. In assigning the NMR signals of 3-FTyr-containing *lac* repressor, a nonsense suppressor was used to insert a Leu in response to a UAG triplet. Thus, the systematic replacement of 3-FTyr by Leu allowed precise assignment of all ^{19}F-NMR resonances and permitted functional analyses of two independent domains (Jarema *et al.*, 1981). These studies were extended to the incorporation and assignment of 5-FTrp at both Trp positions of this protein. Using 3-FTyr- and 5-FTrp-containing protein, the effects of binding of both inducers and anti-inducers of the *lac* operon were studied by ^{19}F-NMR spectroscopy (Boschelli *et al.*, 1981). This same strategy has been used to insert fluorinated analogues at specific positions in proteins (Jarema *et al.*, 1981).

The synthesis of β-galactosidase by *E. coli* is controlled by the *lac* operon. An 81.4% replacement of Tyr by 3-FTyr in this enzyme was achieved by growing *E. coli* in Tyr-limited medium, followed by simultaneous addition of 3-FTyr and induction of the *lac* operon. The pH profile of 3-FTyr-containing β-galactosidase so obtained was lowered by about 1.5 pH units, a shift which corresponds to the difference in phenolic pK_a between Tyr and 3-FTyr. These and other data were interpreted as showing that a Tyr residue is at the active site of β-galactosidase, where it can function in general acid/general base catalytic processes (Ring *et al.*, 1985).

Fluorine-19 NMR spectroscopy has been used effectively in several studies of the structure and dynamics of membrane lipids and proteins [reviewed by Ho *et al.* (1985)]. The membrane-bound *E. coli* enzyme D-lactate dehydrogenase (D-LDH) labeled with 4-, 5-, and 6-FTrp has been the subject of one such recent investigation (Rule *et al.*, 1987). In this work, the yield of labeled D-LDH was augmented by the use of a recombinant DNA plasmid which, on temperature induction, produced 300-fold more D-LDH than did wild-type cells. The D-LDH so labeled had five resolved ^{19}F-NMR resonances consistent with the five tryptophans expected from examination of the DNA sequence encoding the enzyme. Using techniques analogues to those used in the work of Jarema *et al.* (1981) cited above, resonances were assigned by generating mutant enzymes having one labeled amino acid replaced by a different residue. Included in the results from NMR studies was the finding that the conformation of D-LDH is similar in aqueous solution and in the presence of lipids and detergents.

7.7.2.3 Additional Studies and Summary

2-Fluoro-L-histidine (2-FHis) is bacteriostatic at 10^{-5} M to wild-type *E. coli*, is incorporated into the bacterial protein, and is a potent feedback inhibitor of ATP:phosphoribosyltransferase, the first committed enzyme in histidine biosynthesis (Kirk and Cohen, 1976; S. Nagshineh, K. L. Kirk, and L. A. Cohen, unpublished results). Despite close structural and electronic similarities to 2-FHis, 4-FHis has no effect on *E. coli*, even at 10^{-3} M. Other halogenated analogues of His were also without activity.

An impressive example of aliphatic amino acid analogue incorporation is seen in the complete adaptation of an *E. coli* leucine auxotroph to growth on 5′,5′,5′-trifluoroleucine (TFLeu) (Fenster and Anker, 1969; Rennert and Anker, 1963).

The above examples have been chosen from many reports as representative of techniques used to incorporate fluorinated amino acids into bacterial protein and as illustrations of the use of analogue-containing proteins in biochemical studies. The review by Gerig (1978) includes a discussion of theory and practice of ^{19}F-NMR spectroscopy as applied to biological macromolecules, as well as many additional examples of fluoroamino acid-labeled proteins. The review by Sykes and Weiner (1980) also cites many additional studies and contains a discussion of NMR spectra, while the review by Wheatley (1978) tabulates several examples of incorporation of 4-FPhe.

7.7.3 Incorporation of Halogenated Amino Acids into Protein of Mammalian Cells

7.7.3.1 In Vitro Incorporation

As in the research discussed in the previous section, many of the studies with mammalian cells have been directed toward the isolation of fluoroamino acid-labeled proteins for NMR studies, although many other goals have also been achieved. Particular attention has been paid to the effect of such analogues on completion of protein chains, to the behavior of the fluoro analogue-containing proteins as substrates for processing, and to the ability of such proteins to carry out specific biological functions, whether as enzymes or as other functional or structural entities.

An example of the latter is seen in the investigations of the effects of 4-FPhe on the biosynthesis of proteins required for cell division. In eucaryotic cells, completion of DNA synthesis (during the S phase) is followed by a gap period (G_2 phase) during which no DNA is synthesized, but during which RNA, proteins, and membranes are synthesized. It has been postulated that, during this phase, RNA messages are synthesized which code for certain proteins essential for mitosis. Wheatley and Henderson (1974) found that entry of HeLa cells into mitosis could be rapidly and almost completely inhibited by the presence of $0.2 \, \text{m}M$ 4-F-D,L-Phe, suggesting that division-related proteins containing 4-FPhe in their sequences may malfunction. This also provided a tool for the study of cells in the G_2 phase of the cell cycle, since a presynchronized cell culture could be held in this phase for several hours by treatment with 4-FPhe (Englis and Wheatley, 1979). Sunkara et al. (1981) also investigated this phenomenon and found indication that incorporation of 4-FPhe into mitotic proteins results in an inhibition of chromosome condensation.

Fluorinated amino acids can alter the normal processing of proteins. For example, Jacobson et al. (1970, and references therein) found that if poliovirus proteins are synthesized in the presence of 4-FPhe, proteolytic cleavage of large precursor proteins, from which many poliovirus polypeptide precursors normally are derived, is blocked. This was attributed to an alteration in tertiary structures of these precursors caused by the incorporation of the analogue during protein biosynthesis.

threo-β-Fluoroasparagine (*threo*-FAsn) is another example of an analogue which alters processing of protein. This analogue is highly toxic to certain mammalian cells in culture, an observation that has therapeutic significance in light of the fact that some lymphoma and leukemia cells require asparagine due to a deficiency in asparagine synthase (Stern *et al.*, 1982; also see Section 7.4.3). As discussed in Chapter 6, asparagine-linked oligosaccharides are an important structural unit of glycoproteins, and interference with protein glycosylation could contribute to toxicity. In a cell-free translation system prepared from Krebs II ascites tumor cells, *threo*-FAsn (1 mM) was found to be an effective inhibitor of protein glycosylation (Hortin *et al.*, 1983). This inhibition was blocked by Asn and thus was attributed to replacement of Asn by *threo*-FAsn. Hydrogen bonding between the Asn amide group and a proximate serine or threonine residue in the protein has been proposed as an important feature in the glycosylation reaction. Two models for this hydrogen bonding have been suggested (A and B, Fig. 7-43). Since the electronic effects of fluorine would favor deprotonation of the amide (a process implied by hydrogen bonding to the carbonyl oxygen as in A), the inability of a *threo*-FAsn residue to function as a site for glycosylation was interpreted as favoring a mechanism for N-glycosylation which involves a neutral, rather than a deprotonated, amide nitrogen (B). *erythro*-FAsn exhibited no toxicity and also had no effect on N-glycosylation in the cell-free system. At 30–50 μM concentrations, *threo*-FAsn inhibited the growth of three human leukemia cell lines in culture and, at a dose of 250 mg/kg, increased the lifetime by 60% of mice bearing L1210 tumor cells. Neither *erythro*-FAsn nor *threo*- or *erythro*-FAsp had any effect even at higher concentrations. It is noteworthy that L1210 cells have no nutritional requirement for Asn, nor do these cells respond to asparaginase therapy. While the mechanism for the selective toxicity of *threo*-FAsn has not been established, a higher turnover of

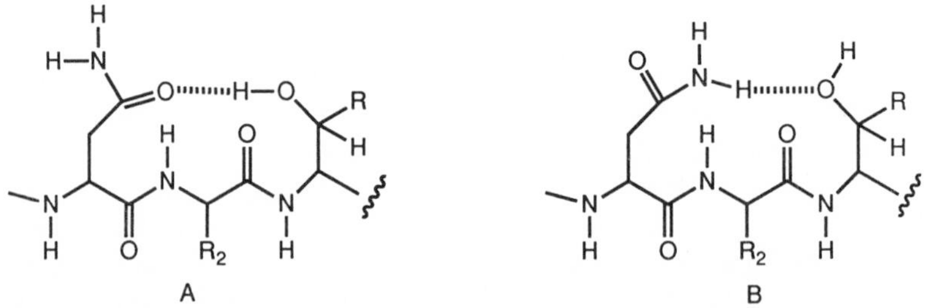

Figure 7-43. Two models for hydrogen bonding between an asparagine amide group and a proximate serine or threonine residue that have been proposed as an important interaction required for protein glycosylation (Hortin *et al.*, 1983).

protein synthesis in the tumor cells was proposed as one explanation (Stern et al., 1984).

A His residue is often found in the active site of enzymes, where it functions as a catalyst in acid–base and nucleophilic processes. Substitution of fluorine on His reduces the pK_a of the imidazole ring by approximately 5 pH units, and this dramatic drop in imidazole ring basicity is reflected in altered biological properties of FHis-containing proteins (Kirk and Cohen, 1976; Klein and Kirk, 1976, and references therein). In parallel with the behavior noted above in bacterial systems, 2-F-, but not 4-FHis, is incorporated readily into mammalian protein in several systems. Thus, the presence of 2-FHis inhibits the induction of several enzymes, for example, the stimulation of pineal gland N-acetyltransferase activity, in cell culture and in vivo. This inhibition of enzyme stimulation was accompanied by a His- and cycloheximide-sensitive incorporation of the analogue into protein, as demonstrated by the recovery of [^{3}H]-2-FHis following pronase digestion of the isolated protein (Klein et al., 1977; Klein and Kirk, 1976, and references therein). While it has not been established unequivocally, a possible explanation for the inhibition of N-acetyltransferase is the synthesis of defective or inactive enzyme containing 2-FHis as a replacement for His.

2-FHis also inhibits the cytopathogenic effects of a number of RNA and DNA viruses on cultured primary rabbit kidney (PRK), HeLa, or Vero cells and, in skin fibroblasts, inhibits the poly(I)poly(C) induction of interferon (De Clercq et al., 1978). 4-FHis, again, is without activity. Inhibition of cell protein synthesis, as measured by the decrease of [^{3}H]leucine incorporation, may be the cause of these antimetabolic effects. At nontoxic doses, 2-FHis in vivo had no effect on the survival of mice infected with HSV-1 or VSV virus.

The effects of 2-FHis on protein synthesis were investigated further in cell-free systems and in mouse L cells (Torrence et al., 1979). In cell-free systems, 2-FHis had no apparent effect on total protein synthesis, as determined by [^{3}H]Leu or [^{3}H]Phe incorporation, although incorporation of [^{3}H]His was inhibited. A direct comparison of the levels of incorporation of His and 2-FHis in mouse L cells revealed that the latter is incorporated at 17% of the efficiency of incorporation of the former. In the cell-free system, the polypeptides synthesized in the presence of 2-FHis were similar in number and size to those synthesized in the presence of His. Thus, in terms of serving as a substrate for protein synthesis, 2-FHis appears to be a fairly good mimic of His. Nonetheless, the functional integrity of the final product clearly is compromised in many instances. The apparent lack of inhibition of protein synthesis observed in cell-free systems stands in contrast to the behavior seen in cell and organ culture. A direct, or indirect, effect of 2-FHis on amino acid transport across the cell membrane may be indicated by this result.

Plasmodium falciparum, an extremely pathogenic strain of malaria parasite, synthesizes several proteins which contain an unusually high amount (7.5–33%) of His. These His-rich proteins may be involved in the process of knob formation, a process which is critical to the sequestration, and protection, of the parasites in the capillary endothelium. In the hope that analogues of His might selectively disrupt synthesis or function of these His-rich proteins, the effects of several His analogues on the growth of *P. falciparum* in cultured erythrocytes were investigated. The observation was made that low concentrations of 2-FHis not only inhibited cyto-adherence, but also prevented knob formation and parasite maturation. A general inhibition of protein synthesis was seen, but only low levels of [^{3}H]-2-FHis were incorporated into protein fractions. Further, 2-FHis was much more effective in the inhibition of the incorporation of Leu and Ile than in the inhibition of the incorporation of His, suggesting that the effect of 2-FHis may be mediated partially through altered amino acid transport mechanisms. 2-IHis was equipotent to 2-FHis with respect to inhibition of His incorporation but was less effective in blocking Leu and Ile incorporation, and this analogue had no effect on knob formation. All other halogenated His analogues tested (including 4-F, 2-Cl, 4-Cl, 2-Br, 4-Br, 4-I, 2,4-diCl, 2,4-diBr, and 2,4-diI) were without effect. While 2-FHis is incorporated to a small extent into the parasite protein, the likelihood that the synthesis of defective His-rich protein is the primary mechanism of antiparasitic action is diminished by the observation that 2-FHis shows comparable activity against simian malaria parasites. This strain appears not to synthesize His-rich proteins (Howard *et al.*, 1986; Panton *et al.*, 1988).

A more common protein containing an unusual amino acid composition is collagen, which has a very high incidence of Gly, Pro, and 4-hydroxy-Pro (synthesized by hydroxylation of Pro in large polypeptide precursors of collagen). Several Pro analogues have been incorporated during the biosynthesis of collagen *in vitro*, including *cis*-4-FPro and *cis*-BrPro. The presence of the analogues in collagen appears to prevent the polypeptide chains from forming the normal triple-helical conformation of collagen, leading to more rapid degradation (Uitto and Prockop, 1974, and references therein).

7.7.3.2 In Vivo Incorporation

The first report of *in vivo* incorporation of a fluorinated amino acid analogue appears to be that of Westhead and Boyer (1961), who showed that 4-FPhe could be incorporated to a significant degree into rabbit protein by maintaining the animal on a diet containing the analogue. Gerig *et al.* (1983) subsequently isolated and purified carbonmonoxylhemoglobin from rabbits maintained on a diet containing 0.3% by weight 4-FPhe and found that about one replacement for each $\alpha_2\beta_2$ tetramer had occurred. The ^{19}F-NMR spectrum of the cyanomet form of the protein contained 16 resonances, one for each Phe residue in the α- and β-globins. In order to assign resonances to each chain, hybrid hemoglobins having only one chain labeled with fluorine were prepared. Based on these and other data, including the effects of complexation with oxygen and of modification of the β-93 sulfhydryl group with a nitroxide spin label, tentative assignments of most FPhe resonances were made. These fluorinated analogues should be useful tools for ^{19}F-NMR studies of the solution behavior of hemoglobin as a function of its role as a regulated agent for oxygen transport (Gerig *et al.*, 1983; Gamcsik *et al.*, 1987). Similar research is being pursued with primate (chimpanzee) hemoglobin (Gamcsik *et al.*, 1986) and with rabbit carbonic anhydrase (Gamcsik and Gerig, 1986).

The toxicity of certain fluorinated amino acids may be associated with protein biosynthesis, although, in some cases, other mechanisms are probably more important (see below). After subcutaneous administration to mice, 2-FHis, but not 4-FHis, becomes incorporated into the protein of all tissue examined. The most notable physiological response to the treatment of mice with 2-FHis is a dramatic reduction in leukocytopoiesis, as evidenced by lowered blood leukocyte levels and atrophy of the spleen (Creveling *et al.*, 1977). While protein synthesis and leukocyte reduction may be linked, particularly since His is often critical to enzyme active-site catalytic activity, this has not been proven. The action of 2-FHis on specific enzymes may be important. In this regard, the demonstration that 2-FHis is a potent inhibitor of bacterial PRPP-ATP synthase may be significant, since this enzyme links histidine and purine biosynthesis through the intermediate 4-aminoimidazole carboxamide riboside.

7.7.4 Total Synthesis of Peptides Containing Halogenated Amino Acids

Early ^{19}F-NMR studies of proteins were carried out using semi-synthetic material having fluorine-labeled probes attached to the polypeptide. For example, Huestis and Raftery (1971) modified ribonuclease by *N*-trifluoroacetylation of the Lys residues 1 and 7 and used ^{19}F-NMR spectroscopy to study conformational changes brought about by the presence of inhibitors. The review by Gerig (1978) contains several other examples of this strategy. More recently, with the rapid advancements made in protein synthesis methodology, total synthesis has become an increasingly viable alternative to biosynthetic incorporation of amino acid analogues into polypeptides.

Several oligopeptides have been synthesized containing fluorinated analogues—for example, oxytocin (C-terminal Tyr replaced with 4-FPhe) and angiotensin II (Phe or Tyr replaced with 4-FPhe). These and several other examples of fluoro-analogue-containing oligopeptides synthesized for ^{19}F-NMR studies have been reviewed by Gerig (1978). In spite of apparent structural similarities between 4-FPhe and Phe, incorporation of 4-FPhe into oligopeptides can have significant effects on biological activity, as shown by Fisher et al. (1977) with 4-FPhe-containing bradykinin.

$\gamma,\gamma,\gamma,\gamma',\gamma',\gamma'$-Hexafluorovaline (HfVal) was incorporated into angiotensin II (Asp-Arg-Val-Tyr-Ile-His-Pro-Phe) (AII) in order to develop a ^{19}F-NMR probe for receptor–hormone interactions (Vine et al., 1981). The analogue retained activity (133%) and, further, was resistant to proteolytic digestion. In a subsequent study, HfVal was incorporated into sarcosine^1-containing analogues of AII to produce long-acting inhibitors of AII (Hsieh et al., 1987).

The incorporation of the analogue can have particularly profound effects on the biological behavior of the protein if the substitution occurs at residues important for catalytic activity or compromises structural integrity. For example, the amino-terminal pentadecapeptide of ribonuclease has been synthesized containing 4-FHis as a replacement for His at position 12. This synthetic peptide served as an active-site analogue of the natural ribonuclease-S-(1–20), as demonstrated by facile formation of a stable, noncovalent complex with the native ribonuclease-S-(21–24) (Dunn et al., 1974). However, the complex is devoid of activity, as a consequence of the lowered imidazole pK_a of the 4-FHis residue now present at the active site (Taylor and Chaiken, 1977).

Thyrotropin-releasing factor (TRH) is a simple tripeptide (L-pyroglutamyl-L-histidyl-L-proline amide) that has been shown to regulate a wide variety of biological functions. The principal biological role of TRH, one of several hormonal releasing factors elaborated by the hypothalamus, initially was believed to be confined to stimulation of release of thyrotropin (TSH) and prolactin (PRL) from the pituitary gland of several mammalian species. Subsequently, TRH was found also to produce effects unrelated to functions related to pituitary activity, effects which include activation of central noradrenergic neurons and related cardiovascular stimulation. The range of responses elicited by TRH has indicated that more than one TRH receptor may be involved. To assess structural requirements for biological activity, several synthetic analogues of TRH have been made by the replacement of the His residue with variously substituted His residues. In contrast to TRH, the 4-FHis- and 2-CF$_3$-His-containing analogues did not bind to pituitary GH$_4$ cells *in vitro*, nor did they stimulate prolactin release from them. However, both analogues were able to elicit cardiovascular responses *in vivo* comparable to those elicited by TRH, and prolactin release *in vivo* was two to three times the level observed with TRH. Receptors outside the pituitary and/or indirect action of the analogues may be responsible for these unexpected observations. Results from research with these and other analogues are revealing that the central effects of TRH are mediated by complex, and yet to be clarified, mechanisms (Labroo *et al.*, 1987; Feuerstein *et al.*, 1984, and references therein).

7.8 Summary

Halogenated amino acid analogues are valuable tools for the study of a host of biochemical processes, and an enormous amount of research has been carried out to exploit these tools. It would be futile to attempt a comprehensive review of this work in this volume. Instead, the attempt has been made in this chapter to give an overview of recent progress in several areas, with enough specific examples to illustrate various strategies and methodologies.

References

Alexander, N. M., 1984. Iodine, in *Biochemistry of the Essential Ultratrace Elements* (E. Frieden, ed.), Plenum, New York, pp. 33–53.

Alston, T. A., and Bright, H. J., 1983. Conversion of trifluoromethionine to a cross-linking agent by γ-cystathionase, *Biochem. Pharmacol.* 32:947–950.

Anderson, R. A., Nakashima, Y., and Coleman, J. E., 1975. Chemical modifications of functional residues of fd gene 5 DNA-binding protein, *Biochemistry* 14:907–917.

Antonucci, T. K., and Oxender, D. L., 1986. The molecular biology of amino-acid transport in bacteria, *Adv. Microb. Phys.* 28:145–180.

Arfin, S. M., and Koziell, D. A., 1971. Inhibition of growth of *Salmonella typhimurium* and of threonine deaminase and transaminase B by β-chloroalanine, *J. Bacteriol.* 105:519–522.

Badet, B., Roise, D., and Walsh, C. T., 1984. Inactivation of the *dadB Salmonella typhimurium* alanine racemase by D and L isomers of β-substituted alanines: Kinetics, stoichiometry, active site peptide sequencing, and reaction mechanism, *Biochemistry* 23:5188–5194.

Bitonti, A. J., Casara, P. J., McCann, P. P., and Bey, P., 1987. Catalytic irreversible inhibition of bacterial and plant arginine decarboxylase activities by novel substrate and product analogues, *Biochem. J.* 242:69–74.

Bode, R., Melo, C., and Birnbaum, D., 1985. Regulation of chorismate mutase, prephenate dehydrogenase, and prephenate dehydratase of *Candida maltose*, *J. Basic Microbiol.* 25:291–298.

Borg-Olivier, S. A., Tarlinton, D., and Brown, K. D., 1987. Defective regulation of the phenylalanine biosynthetic operon in mutants of the phenylalanyl-tRNA synthetase operon, *J. Bacteriol.* 169:1949–1953.

Boschelli, F., Jarema, M. A. C., and Lu, P., 1981. Inducer and anti-inducer interactions with the *lac* repressor as seen by nuclear magnetic resonance changes at tyrosines and tryptophans, *J. Biol. Chem.* 256:11595–11599.

Bronskill, P. M., and Wong, J. T.-F., 1988. Suppression of fluorescence of tryptophan residues in proteins by replacement with 4-fluorotryptophan, *Biochem. J.* 249:305–308.

Brown, K. D., 1970. Formation of aromatic amino acid pools in *Escherichia coli* K-12, *J. Bacteriol.* 104:177–188.

Bussey, H., and Umbarger, H. E., 1970. Biosynthesis of the branched-chain amino acids in yeast: A trifluoroleucine-resistant mutant with altered regulation of leucine uptake, *J. Bacteriol.* 103:286–294.

Calvo, J. M., Freundlich, M., and Umbarger, H. E., 1969. Regulation of branched-chain amino acid biosynthesis in *Salmonella typhimurium*: Isolation of regulatory mutants, *J. Bacteriol.* 97:1272–1282.

Cheung, K. S., Wasserman, S. A., Dudek, E., Lerner, S. A., and Johnston, M., 1983. Chloroalanyl and propargylglycyl dipeptides. Suicide substrate containing antibacterials, *J. Med. Chem.* 26:1733–1741.

Cheung, K. S., Boisvert, W., Lerner, S. A., and Johnston, M., 1986. Chloroalanyl antibiotic peptides: Antagonism of the antimicrobial effects by L-alanine and L-alanyl peptides in gram-negative bacteria, *J. Med. Chem.* 29:2060–2068.

Chiueh, C. C., Burns, R. S., Kopin, I. J., Kirk, K. L., Firnau, G., Nahmias, C., Chirakal, R., and Garnett, E. S., 1986. 6-^{18}F-DOPA positron emission tomography visualized degree of damage to brain dopamine in basal ganglia of monkeys with MPTP-induced parkinsonism, in *MPTP: A Neurotoxin Producing a Parkinsonian Syndrome* (S. Markey, ed.), Academic Press, Orlando, Florida, pp. 327–338.

Chou, T.-C., and Talalay, P., 1966. The mechanism of S-adenosyl-L-methionine synthesis by purified preparations of Bakers' yeast, *Biochemistry* 11:1065–1073.

Christopherson, R. I., 1985. Chorismate mutase-prephenate dehydrogenase from *Escherichia coli*: Cooperative effects and inhibition by L-tyrosine, *Arch. Biochem. Biophys.* 240:646–654.

Coleman, J. E., Anderson, R. A., Ratcliffe, R. G., and Armitage, I. M., 1976. Structure of gene 5 protein–oligodeoxynucleotide complexes as determined by ^{1}H, ^{19}F, and ^{31}P nuclear magnetic resonance, *Biochemistry* 15:5419–5430.

Colombani, F., Cherest, H., and De Robichon-Szulmajster, H., 1975. Biochemical and regulatory effects of methionine analogues in *Saccharomyces cerevisial*, *J. Bacteriol.* 122:375–384.

Copié, V., Faraci, W. S., Walsh, C. T., and Griffin, R. G., 1988. Inhibition of alanine racemase by alanine phosphonate: Detection of an imine linkage to pyridoxal 5′-phosphate in the enzyme–inhibitor complex by solid-state ^{15}N nuclear magnetic resonance, *Biochemistry* 27:4966–4970.

Creveling, C. R., and Kirk, K. L., 1985. The effect of ring-fluorination on the rate of O-methylation of dihydroxyphenylalanine (DOPA) by catechol-O-methyltransferase: Significance in the development of ^{18}F-PETT scanning agents, *Biochem. Biophys. Res. Commun.* 130:1123–1131.

Creveling, C. R., Kirk, K. L., and Highman, B., 1977. Effect of 2-fluorohistidine upon leukocytopoiesis in mice, *Res. Commun. Chem. Pathol. Pharmacol.* 16:507–521.

Cumming, P., Boyes, B. E., Martin, W. R. W., Adam, M., Grierson, J., Ruth, T., and McGeer, E. G., 1987. The metabolism of [^{18}F]6-fluoro-L-3,4-dihydroxyphenylalanine in the hooded rat, *J. Neurochem.* 48:601–608.

De Clercq, E., Billiau, A., Edy, V. G., Kirk, K. L., and Cohen, L. A., 1978. Antimetabolic activities of 2-fluoro-L-histidine, *Biochem. Biophys. Res. Commun.* 82:840–846.

Dunn, B. M., DiBello, C., Kirk, K. L., Cohen, L. A., and Chaiken, I. M., 1974. Synthesis, purification, and properties of a semisynthetic ribonuclease S incorporating 4-fluoro-L-histidine at position 12, *J. Biol. Chem.* 249:6295–6301.

Englis, M. S., and Wheatley, D. N., 1979. Reversible retention of synchronized HeLa cells in G_2 of the mammalian cell cycle, *Cell Biol. Int. Rep.* 3:739–746.

Fangman, W. L., and Neidhardt, F. C., 1964. Demonstration of an altered aminoacyl ribonucleic acid synthetase in a mutant of *Escherichia coli*, *J. Biol. Chem.* 239:1839–1847.

Fenster, E. D., and Anker, H. S., 1969. Incorporation into polypeptide and charging on transfer ribonucleic acid of the amino acid analogue 5',5',5'-trifluoroleucine by leucine auxotrophs of *Escherichia coli*, *Biochemistry* 8:269–274.

Feuerstein, G., Lozovsky, D., Cohen, L. A., Labroo, V. M., Kirk, K. L., Kopin, I. J., and Faden, A. I., 1984. Differential effect of fluorinated analogues of TRH on the cardiovascular system and prolactin release, *Neuropeptides* 4:303–310.

Firnau, G., Sood, S., Pantel, R., and Garnett, S., 1981. Phenol ionization in DOPA determines the site of methylation by catechol-*O*-methyltransferase, *Mol. Pharmacol.* 19:130–133.

Firsova, N. A., Selivanova, K. M., Alekseeva, L. V., and Evstigneeva, Z. G., 1986a. Inhibition of the glutamine synthetase activity by biologically active derivatives of glutamic acid, *Biokhimiya* 51:850–855.

Firsova, N. A., Alekseeva, L. V., Selilvanova, K. M., and Evstigneeva, Z. G., 1986b. Substrate specificity of chlorella with respect to biologically active derivatives of glutamic acid, *Biokhimiya* 51:980–984.

Fisher, G. H., Ryan, J. W., and Berryer, P., 1977. Biologically active bradykinin analogues containing *p*-fluorophenylalanine, *Cardiovasc. Med.* 2:1179–1181.

Fiske, M. J., Whitaker, R. J., and Jensen, R. A., 1983. Hidden overflow pathway to L-phenylalanine in *Pseudomonas aeruginosa*, *J. Bacteriol.* 154:623–631.

Fowden, L., 1972. Fluoroamino acids and protein synthesis, in *Ciba Foundation Symposium*: *Compounds: Chemistry, Biochemistry, and Biological Activities*, Associated Scientific Publishers, New York, pp. 141–159.

Freundlich, M., and Trela, J. M., 1969. Control of isoleucine, valine, and leucine biosynthesis, VI. Effect of 5',5',5'-trifluoroleucine on repression in *Salmonella typhimurium*, *J. Bacteriol.* 99:101–106.

Gabius, H.-J., von der Haar, F., and Cramer, F., 1983. Evolutionary aspects of accuracy of phenylalanyl-tRNA synthetase. A comparative study with enzymes from *Escherichia coli*, *Saccharomyces cerevisiae*, *Neurospora crassa*, and turkey liver using phenylalanine analogues, *Biochemistry* 22:2331–2339.

Gál, E. M., 1972. Molecular basis of inhibition of monooxygenases by *p*-halophenylalanines, *Adv. Biochem. Psychopharmacol.* 6:149–163.

Gál, E. M., and Whitacre, D. H., 1982. Mechanism of the irreversible inactivation of phenylalanine-4- and tryptophan-5-hydroxylases by [4-^{36}Cl,2-^{14}C]*p*-chlorophenylalanine: A revision, *Neurochem. Res.* 7:13–26.

Galivan, J., Coward, J. K., and McGuire, J. J., 1985. Effects of D,L-4-fluoroglutamic acid on glutamylation of methotrexate by hepatic cells *in vitro*, *Biochem. Pharmacol.* 34:2995–2997.

Gamcsik, M. P., and Gerig, J. T., 1986. NMR studies of fluorophenylalanine-containing carbonic anhydrase, *FEBS Lett.* 196:71–74.

Gamcsik, M. P., Gerig, J. T., and Swenson, R. B., 1986. Fluorine-NMR studies of chimpanzee hemoglobin, *Biochem. Biophys. Acta* 874:372–374.

Gamcsik, M. P., Gerig, J. T., and Gregory, D. H., 1987. Fluorine nuclear magnetic resonance spectra of rabbit carbonmonoxyhemoglobin, *Biochim. Biophys. Acta* 912:303–316.

Garnett, E. S., Firnau, G., Chan, P. K. H., Sood, S., and Belbeck, L. W., 1978. [^{18}F]Fluoro-DOPA, an analogue of DOPA, and its use in direct external measurements of storage, degradation, and turnover of intracerebral dopamine, *Proc. Natl. Acad. Sci. USA* 75:464–467.

Garnett, E. S., Firnau, G., and Nahmias, C., 1983. Dopamine visualized in the basal ganglia of living man, *Nature* 305:137–138.

Gerig, J. T., 1978. Fluorine magnetic resonance in biochemistry, in *Biological Magnetic Resonance*, Vol. 1 (L. S. Berliner and J. Reuben, eds.), Plenum, New York, pp. 139–203.

Gerig, J. T., Klinkenborg, J. C., and Nieman, R. A., 1983. Assignment of fluorine nuclear magnetic resonance signals from rabbit cyanomethemoglobin, *Biochemistry* 22:2076–2087.

Gollub, E. J., Liu, K. P., and Sprinson, D. B., 1973. A regulatory gene of phenylalanine biosynthesis (*pheR*) in *Salmonella typhimurium*, *J. Bacteriol.* 115:121–128.

Guroff, G., Kondo, K., and Daly, J., 1966. The production of *meta*-chlorotyrosine from *para*-chlorophenylalanine by phenylalanine hydroxylase, *Biochem. Biophys. Res. Commun.* 25:622–628.

Guroff, G., Daly, J. W., Jerina, D. M., Renson, J., Witkop, B., and Udenfriend, S., 1967. Hydroxylation-induced migration: The NIH shift, *Science* 157:1524–1530.

Ho, C., Dowd, S. R., and Post, J. F. M., 1985. Fluorine-19 NMR investigations of membranes, *Curr. Top. Bioenerg.* 14:53–95.

Hortin, G., Stern, A. M., Miller, B., Abeles, R. H., and Boime, I., 1983. DL-*threo-β*-Fluoroasparagine inhibits asparagine-linked glycosylation in cell-free lysates, *J. Biol. Chem.* 258:4047–4050.

Hou, Y.-M., and Schimmel, P., 1988. A simple structural feature is a major determinant of the identity of transfer RNA, *Nature* 333:140–145.

Howard, R. J., Andrutis, A. T., Leech, J. H., Ellis, W. Y., Cohen, L. A., and Kirk, K. L., 1986. Histidine analogues inhibit growth and protein synthesis by *Plasmodium falciparum in vitro*, *Biochem. Pharmacol.* 35:1589–1596.

Hsieh, K., Needleman, P., and Marshall, G. R., 1987. Long-acting angiotensin II inhibitors containing hexafluorovaline in position 8, *J. Med. Chem.* 30:1097–1100.

Huestis, W. H., and Raftery, M. A., 1971. Use of fluorine-19 nuclear magnetic resonance to study conformation changes in selectively modified ribonuclease S, *Biochemistry* 10:1181–1186.

Igarashi, Y., Kodama, T., and Minoda, Y., 1982. Excretion of L-tryptophan by analogue-resistant mutants of *Pseudomonas hydrogenothermophila* TH-1 in autotrophic cultures, *Agric. Biol. Chem.* 46:1525–1530.

Im, S. W. K., and Pittard, J., 1971. Phenylalanine biosynthesis in *Escherichia coli* K-12: Mutants derepressed for chorismate mutase P-prephenated dehydratase, *J. Bacteriol.* 106:784–790.

Jacobson, M. F., Asso, J., and Baltimore, D., 1970. Further evidence on the formation of poliovirus proteins, *J. Mol. Biol.* 49:657–669.

Jarema, M. A. C., Lu, P., and Miller, J. H., 1981. Genetic assignment of resonances in the NMR spectrum of a protein: *lac* Repressor, *Proc. Natl. Acad. Sci. USA* 78:2707–2711.

Jung, M. J., Palfreyman, J., Wagner, J., Bey, P., Ribereau-Gayon, G., Zraika, M., and Koch-Weser, J., 1979. Inhibition of monoamine synthesis by irreversible blockade of aromatic amino acid decarboxylase with α-monofluoromethyldopa, *Life Sci.* 24:1037–1042.

Kallio, A., McCann, P. P., and Bey, P., 1981. DL-α-(Difluoromethyl)arginine: A potent enzyme-activated irreversible inhibitor of bacterial arginine decarboxylases, *Biochemistry* 20:3163–3166.

Kaufman, S., 1961. The enzymic conversion of 4-fluorophenylalanine to tyrosine, *Biochim. Biophys. Acta* 51:619–621.

Kaufman, S., and Fisher, D. B., 1974. Pterin-requiring aromatic amino acid hydroxylases, in *Molecular Mechanism of Oxygen Activation* (O. Hayashi, ed.), Academic Press, New York, pp. 285–369.

Kingsbury, W. D., Boehm, J. C., Mehta, R. J., and Grappel, S. F., 1983. Transport of antimicrobial agents using peptide carrier systems: Anticandidal activity of *m*-fluorophenylalanine–peptide conjugates, *J. Med. Chem.* 26:1725–1729.

Kirk, K. L., and Cohen, L. A., 1976. Biochemistry and pharmacology of ring-fluorinated imidazoles, in *Biochemistry Involving Carbon–Fluorine Bonds* (R. Filler, ed.), ACS Symposium Series, No. 28, American Chemical Society, Washington, D.C., pp. 23–36.

Klee, C. B., Kirk, K. L., Cohen, L. A., and McPhie, P., 1975. Histidine ammoniallyase. The use of 4-fluorohistidine in identification of the rate-determining step, *J. Biol. Chem.* 250:5033–5040.

Klee, C. B., La John, L. E., Kirk, K. L., and Cohen, L. A., 1977. 2-Fluorourocanic acid, a potent reversible inhibitor of urocanase, *Biochem. Biophys. Res. Commun.* 75:674–681.

Klein, D. C., and Kirk, K. L., 1976. 2-Fluoro-L-histidine: A histidine analogue which inhibits enzyme induction, in *Biochemistry Involving Carbon–Fluoride Bonds* (R. Filler, ed.), ACS Symposium Series, No. 28, American Chemical Society, Washington, D.C., pp. 37–56.

Klein, D. C., Weller, J. L., Kirk, K. L., and Hartley, R. W., 1977. Incorporation of 2-fluoro-L-histidine into cellular protein, *Mol. Pharmacol.* 13:1105–1110.

Koe, B. K., and Weissman, A., 1966a. *p*-Chlorophenylalanine: A specific depletor of brain serotonin, *J. Pharmacol. Exp. Ther.* 154:499–516.

Koe, B. K., and Weissman, A., 1966b. Convulsions and elevation of tissue citric acid levels induced by 5-fluorotryptophan, *Biochem. Pharmacol.* 15:2134–2136.

Koe, B. K., and Weissman, A., 1967. Dependence of *m*-fluorophenylalanine toxicity on phenylalanine hydroxylase activity, *J. Pharmacol. Exp. Ther.* 157:565–573.

Kollonitsch, J., 1982. Suicide substrate enzyme inactivators of enzymes dependent on pyridoxal-phosphate: β-Fluoro amino acids and β-fluoro amines. Design, synthesis and application: A contribution to drug design, in *Biomedicinal Aspects of Fluorine Chemistry* (R. Filler and Y. Kobayashi, eds.), Kodansha Ltd., Tokyo; Elsevier Biomedical Press, Amsterdam, pp. 93–122.

Kollonitsch, J., Barash, L., Kahan, F. M., and Kropp, H., 1973. New antibacterial agent *via* photofluorination of a bacterial cell wall constituent, *Nature* 243:346–347.

Kollonitsch, J., Patchett, A. A., Marburg, S., Maycock, A. L., Perkins, L. M., Doldouras, G. A., Duggan, D. E., and Aster, S. D., 1978. Selective inhibitors of biosynthesis of aminergic neurotransmitters, *Nature* 274:906–908.

Kollonitsch, J., Marburg, S., and Perkins, L. M., 1979. Fluorodehydroxylation, a novel method for synthesis of fluoroaminoacids and fluoroamines, *J. Org. Chem.* 44:771–777.

Kuo, D., and Rando, R. R., 1981. Irreversible inhibition of glutamate decarboxylase by α-(fluoromethyl)glutamic acid, *Biochemistry* 20:506–511.

Labroo, V. M., Cohen, L. A., Lozovsky, D., Siren, A.-L., and Feuerstein, G., 1987. Dissociation of the cardiovascular and prolactin-releasing activities of norvaline-TRH, *Neuropeptides* 10:29–36.

Likos, J. J., Ueno, H., Feldhaus, R. W., and Metzler, D. E., 1982. A novel reaction of the coenzyme of glutamate decarboxylase with L-serine-*O*-sulfate, *Biochemistry* 21:4377–4386.

Lu, P., Jarema, M., Mosser, K., and Daniel, W. E., Jr., 1976. *lac* Repressor: 3-Fluorotyrosine substitution for nuclear magnetic resonance studies, *Proc. Natl. Acad. Sci. USA* 73:3471–3475.

Mamont, P. S., Danzin, C., Kolb, M., Gerhart, F., Bey, P., and Sjoerdsma, A., 1986. Marked and prolonged inhibition of mammalian ornithine decarboxylase *in vivo* by esters of (*E*)-(fluoromethyl)dehydroornithine, *Biochem. Pharmacol.* 35:159–165.

Manning, J. M., Merrifield, N. E., Jones, W. M., and Gotschlich, 1974. Inhibition of bacterial growth by β-chloro-D-alanine, *Proc. Natl. Acad. Sci. USA* 71:417–421.

Marquis, R. E., 1970. Fluoroamino acids and microorganisms, in *Handbook of Experimental Pharmacology*, Vol. XX/11 (O. Eicher, A. Farah, H. Herken, and A. D. Welch, eds.), Springer-Verlag, New York, pp. 166–192.

Matsui, K., Miwa, K., and Sano, K., 1987. Two single-base pair substitutions causing desensitization to tryptophan feedback inhibition of anthranilate synthase and enhanced expression of tryptophan genes of *Brevibacterium lactofermentum*, *J. Bacteriol.* 169:5330–5332.

Maycock, A. L., Aster, S. D., and Patchett, A. A., 1980. Inactivation of 3-(3,4-dihydroxyphenyl)alanine decarboxylase by 2-(fluoromethyl)-3-(3,4-dihydroxyphenyl)alanine, *Biochemistry* 19:709–718.

McGeer, E. G., Peters, D. A. V., and McGeer, P. L., 1968. Inhibition of rat brain tryptophan hydroxylase by 6-halotryptophans, *Life Sci.* 7:605–615.

McGuire, J. J., and Coward, J. K., 1985. DL-*threo*-4-Fluoroglutamic acid. A chain-terminating inhibitor of folylpolyglutamate synthesis, *J. Biol. Chem.* 260:6747–6754.

Metcalf, B. W., Bey, P., Danzin, C., Jung, M. J., Casara, P., and Vevert, J. P., 1978. Catalytic irreversible inhibition of mammalian ornithine decarboxylase (E.C. 4.1.1.17) by substrate and product analogues, *J. Am. Chem. Soc.* 100:2551–1553.

Miles, W. W., Phillips, R. S., Yeh, H. J. C., and Cohen, L. A., 1986. Isomerization of (3*S*)-2,3-dihydro-5-fluoro-L-tryptophan and of 5-fluoro-L-tryptophan catalyzed by tryptophan synthase: Studies using fluorine-19 nuclear magnetic resonance and difference spectroscopy, *Biochemistry* 25:4240–4249.

Miwa, S., Fujiwara, M., Lee, K., and Fujiwara, M., 1987. Determination of serotonin turnover in the rat brain using 6-fluorotryptophan, *J. Neurochem.* 48:1577–1580.

Morino, Y., Kojima, H., and Tanase, S., 1979. Affinity labeling of alanine aminotransferase by 3-chloro-L-alanine, *J. Biol. Chem.* 254:279–285.

Munier, R., and Cohen, G. N., 1959. Substitution de la phénylalanine par l'*o*- ou la *m*-fluorophénylalanine dans les protéines d'*escherichia coli*, *C. R. Acad. Sci.* 248:1870–1873.

Neuhaus, F. C., and Hammes, W. P., 1981. Inhibition of cell wall biosynthesis by analogues of alanine, *Pharmacol. Ther.* 14:265–319.

Palfreyman, M. G., Bey, P., and Sjoerdsma, A., 1987. Enzyme-activated mechanism-based inhibitors, *Essays Biochem.* 23:28–81.

Panton, L. J., Rossan, R. N., Escajadillo, A., Matsumoto, Y., Lee, A. T., Labroo, V. M., Kirk, K. L., Cohen, L. A., Aikawa, M., and Howard, R. J., 1988. *In vitro* and *in vivo* studies on the effects of halogenated histidine analogues on *Plasmodium falciparum*, *Antimicrob. Agents Chemother.* 32:1655–1659.

Pappius, H. M., Dadoun, R., and McHugh, M., 1988. The effect of *p*-chlorophenylalanine on cerebral metabolism and biogenic amine formation content of traumatized brain, *J. Cereb. Blood Flow Metab.* 8:324–334.

Park, J. T., 1958. Selective inhibition of bacterial cell-wall synthesis: Its possible applications in chemotherapy, *Symp. Gen. Microbiol.* 8:49–61.

Patchett, A. A., Taub, D., Weissberger, B., Valiant, M. E., Gadebusch, H., Thornberry, N. A., and Bull, H. G., 1988. Antibacterial activities of fluorovinyl- and chlorovinylglycine and several derived dipeptides, *Antimicrob. Agents Chemother.* 32:319–323.

Pauley, R. J., Fredricks, W. W., and Smith, O. H., 1978. Effect of tryptophan analogues on derepression of the *Escherichia coli* tryptophan operon by indole-3-propionic acid, *J. Bacteriol.* 136:219–226.

Payne, J. W., 1986. Drug delivery systems: Optimizing the structure of peptide carriers for synthetic antimicrobial drugs, *Drugs Exp. Clin. Res.* 12:585–594.

Peters, D. A. V., 1971. Inhibition of serotonin biosynthesis by 6-halotryptophans *in vivo*, *Biochem. Pharmacol.* 20:1413–1420.

Pösö, H., McCann, P. P., Tanskanen, R., Bey, P., and Sjoerdsma, A., 1984. Inhibition of growth of *Mycoplasma dispar* by DL-α-difluoromethyllysine, a selective irreversible inhibitor of lysine decarboxylase, and reversal by cadaverine (1,5-diaminopentane), *Biochem. Biophys. Res. Commun.* 125:205–210.

Rauhut, R., Gabius, H.-J., Engelhardt, R., and Cramer, F., 1985. Archaebacterial phenylalanyl-tRNA synthetase. Accuracy of the phenylalanyl-tRNA synthetase from the archaebacterium *Methanosarcina barkeri*, Zn(II)-dependent synthesis of diadenosine $5',5'''-P_1P_4$-tetraphosphate, and immunological relationship of phenylalanyl-tRNA synthases from different urkingdons, *J. Biol. Chem.* 260:182–187.

Relyea, N. M., Tate, S. S., and Meister, A., 1974. Affinity labeling of the active center of L-aspartate-β-decarboxylase with β-chloro-L-alanine, *J. Biol. Chem.* 249:1519–1524.

Rennert, O. M., and Anker, H. S., 1963. On the incorporation of $5',5',5'$-trifluoroleucine into protein of *E. coli*, *Biochemistry* 2:471–476.

Richmond, M. H., 1962. The effect of amino acid analogues on growth and protein synthesis in microorganisms, *Bacteriol. Rev.* 26:398–420.

Ring, M., Armitage, I. M., and Huber, R. E., 1985. *m*-Fluorotyrosine substitution in β-galactosidase: Evidence for the existence of a catalytically active tyrosine, *Biochem. Biophys. Res. Commun.* 131:675–680.

Robertson, J. H., and Wheatley, D. N., 1979. Pools and protein synthesis in mammalian cells, *Biochem. J.* 178:699–709.

Roise, D., Soda, K., Yagi, T., and Walsh, C. T., 1984. Inactivation of the *Pseudomonas striata* broad specificity amino acid racemase by D and L isomers of β-substituted alanines: Kinetics, stoichiometry, active site peptide, and mechanistic studies, *Biochemistry* 23:5195–5201.

Rule, G. S., Pratt, E. A., Simplaceanu, V., and Ho, C., 1987. Nuclear magnetic resonance and molecular genetic studies of the membrane-bound D-lactate dehydrogenase from *Escherichia coli*, *Biochemistry* 26:549–556.

Santi, D. V., and Webster, R. W., 1976. Phenylalanine transfer ribonucleic acid synthetase from rat liver. Analysis of phenylalanine and adenosine 5'-triphosphate binding sites and comparison to the enzyme from *Escherichia coli*, *J. Med. Chem.* 19:1276–1279.

Schimmel, P., 1987. Aminoacyl tRNA synthetases: General scheme of structure–function relationships in the polypeptides and recognition of transfer RNAs, *Annu. Rev. Biochem.* 56:125–158.

Schnackerz, K. D., Ehrlich, J. H., Giesemann, W., and Reed, T. A., 1979. Mechanism of action of D-serine dehydratase. Identification of a transient intermediate, *Biochemistry* 18:3557–3563.

Schulman, L. H., and Abelson, J., 1988. Recent excitement in understanding transfer RNA identity, *Science* 240:1591–1592.

Shiio, I., and Sugimoto, S., 1978. Altered regulatory mechanisms for tryptophan synthesis in fluorotryptophan-resistant mutants of *Brevibacterium flavum*, *J. Biochem.* 83:879–886.

Shotwell, M. A., Kilberg, M. S., and Oxender, D. L., 1983. The regulation of neutral amino acid transport in mammalian cells, *Biochim. Biophys. Acta* 737:267–284.

Silverman, R. B., and Abeles, R. H., 1976. Inactivation of pyridoxal phosphate dependent enzymes by mono- and polyhaloalanines, *Biochemistry* 15:4718–4723.

Silverman, R. B., and Abeles, R. H., 1977. Mechanism of inactivation of γ-cystathionase by β,β,β-trifluoroalanine, *Biochemistry* 16:5515–5520.

Silverman, R. B., and George, C., 1988. Inactivation of γ-aminobutyric acid aminotransferase by (Z)-4-amino-2-fluorobut-2-enoic acid, *Biochemistry* 27:3285–3289.

Silverman, R. B., and Levy, M. A., 1981. Mechanism of inactivation of γ-aminobutyric acid-α-ketoglutaric acid aminotransferase by 4-amino-5-halopentanoic acids, *Biochemistry* 20:1197–1203.

Singh, M., and Sinha, U., 1979. Mode of action of p-fluorophenylalanine in *Aspergillus nidulans*: Effect on the synthesis and activity of phosphatase isoenzymes, *J. Gen. Microbiol.* 115:101–110.

Sjoerdsma, A., 1981. Suicide enzyme inhibitors as potential drugs, *Clin. Pharmacol. Ther.* 30:3–22.

Sjoerdsma, A., and Schechter, P. J., 1984. Chemotherapeutic implications of polyamine biosynthesis inhibition, *Clin. Pharm. Ther.* 35:287–300.

Smith, F. A., 1970. Biological properties of selected fluorine-containing organic compounds, in *Handbook of Experimental Pharmacology*, Vol. XX/11 (O. Eicher, A. Farah, H. Herken, and A. D. Welch, eds.), Springer-Verlag, New York, pp. 253–408.

Stern, A. M., Foxman, B. M., Tashjian, A. H., Jr., and Abeles, R. H., 1982. DL-*threo*-β-Fluoroaspartate and DL-*threo*-β-fluoroasparagine: Selective cytotoxic agents for mammalian cells in culture, *J. Med. Chem.* 25:544–550.

Stern, A. M., Abeles, R. H., and Tashjian, A. H., 1984. Antitumor activity of D,L-*threo*-β-fluoroasparagine against human leukemia cells in culture and L1210 cells in DBA mice, *Cancer Res.* 44:5614–5618.

Stryer, L., 1988. *Biochemistry*, 3rd ed., W. H. Freeman and Company, New York.

Sunkara, P. S., Chakroborty, B. M., Wright, D. A., and Rao, P. N., 1981. Chromosome condensing ability of mitotic proteins diminished by the substitution of phenylalanine with parafluorophenylalanine, *Eur. J. Cell Biol.* 23:312–316.

Sykes, B. D., and Weiner, J. H., 1980. Biosynthesis and ^{19}F-NMR characterization of fluoroamino acid containing proteins, *Magn. Reson. Biol.* 1:171–196.

Sykes, B. D., Weingarten, H. I., and Schlesinger, M. J., 1974. Fluorotyrosine alkaline phosphatase from *Escherichia coli*: Preparation, properties, and fluorine-19 nuclear magnetic resonance spectrum, *Proc. Natl. Acad. Sci. USA* 71:469–473.

Tate, S. S., Relyea, N. M., and Meister, A., 1969. Interaction of L-aspartate β-decarboxylase with β-chloro-L-alanine. β-Elimination reaction with active-site labeling, *Biochemistry* 8:5016–5021.

Taylor, H. C., and Chaiken, I. M., 1977. Inhibitor and substrate binding by an inactive semisynthetic ribonuclease-S' analogue. Studies by affinity chromatography, *Fed. Proc.* 36:864.

Thornberry, N. A., Bull, H. G., Taub, D., Greenlee, W. J., Patchett, A. A., and Cordes, E. H., 1987. 3-Halovinylglycines. Efficient irreversible inhibitors of *E. coli* alanine racemase, *J. Am. Chem. Soc.* 109:7543–7544.

Torrence, P. F., Friedman, R. M., Kirk, K. L., Cohen, L. A., and Creveling, C. R., 1979. 2-Fluorohistidine—effects on protein synthesis in cell-free systems and in mouse L-cells, *Biochem. Pharmacol.* 28:1565–1567.

Ueno, H., Likos, J. J., and Metzler, D. E., 1982. Chemistry of the inactivation of cytosolic aspartate aminotransferase by serine O-sulfate, *Biochemistry* 21:4387–4393.

Uitto, J., and Prockop, D. J., 1974. Incorporation of proline analogues into collagen polypeptides. Effects on the production of extracellular procollagen and on the stability of the triple-helical structure of the molecule, *Biochim. Biophys. Acta* 336:234–251.

Vidal-Cros, A., Gaudry, M., and Marquet, A., 1985. Interaction of L-*threo* and L-*erythro* isomers of 3-fluoroglutamate with glutamate decarboxylase from *Escherichia coli*, *Biochem. J.* 229:675–678.

Vine, W. H., Hsieh, K., and Marshall, G. R., 1981. Synthesis of fluorine-containing peptides. Analogues of angiotensin II containing hexafluorovaline, *J. Med. Chem.* 24:1043–1047.

Vo-Quang, Y., Carniato, D., Vo-Quang, L., Lacoste, A.-M., Neuzil, E., and Le Goffic, F., 1986. (β-Chloro-α-aminoethyl)phosphonic acids as inhibitors of alanine racemase and D-alanine:D-alanine ligase, *J. Med. Chem.* 29:148–151.

Vorhees, C. V., Butcher, R. E., and Berry, H. K., 1981. Progress in experimental phenylketonuria: A critical review, *Neurosci. Biobehav. Rev.* 5:177–190.

Wacks, D. B., and Schachman, H. K., 1985a. ^{19}F nuclear magnetic resonance studies of fluorotyrosine-labeled aspartate transcarbamoylase. Properties of the enzyme and its catalytic and regulatory subunits, *J. Biol. Chem.* 260:11651–11658.

Wacks, D. B., and Schachman, H. K., 1985b. ^{19}F nuclear magnetic resonance studies of communication between catalytic and regulatory subunits in aspartate transcarbamoylase, *J. Biol. Chem.* 260:11659–11662.

Walsh, C., 1982. Suicide substrates: Mechanism-based enzyme inactivation, *Tetrahedron* 38:871–909.

Walsh, C., 1983. Fluorinated substrate analogues: Routes of metabolism and selective toxicity, *Adv. Enzymol.* 55:197–289.

Walsh, C. T., 1984. Suicide substrate, mechanism-based enzyme inactivators: Recent developments, *Annu. Rev. Biochem.* 53:493–535.

Wang, E., and Walsh, C., 1978. Suicide substrates for the alanine racemase of *Escherichia coli* B, *Biochemistry* 17:1313–1321.

Wang, E., and Walsh, C., 1981. Characteristics of β,β-difluoroalanine and β,β,β-trifluoroalanine as suicide substrates for *Escherichia coli* B alanine racemase, *Biochemistry* 20:7539–7546.

Wanner, M. J., Hageman, J. J. M., Koomen, G.-J., and Pandit, U. K., 1980. Potential carcinostatics. 4. Synthesis and biological properties of *erythro*- and *threo*-β-fluoroaspartic acid and *erythro*-β-fluoroasparagine, *J. Med. Chem.* 23:85–87.

Weissman, A., and Koe, B. K., 1967. *m*-Fluorotyrosine convulsions and mortality: Relationship to catecholamine and citrate metabolism, *J. Pharmacol. Exp. Ther.* 155:135–144.

Westhead, E. W., and Boyer, P. D., 1961. The incorporation of *p*-fluorophenylalanine into some rabbit enzymes and other proteins, *Biochim. Biophys. Acta* 54:145–156.

Weygand, F., and Oettmeier, W., 1970. Fluorine-containing aminoacids, *Russ. Chem. Rev.* 39:290–300.

Wheatley, D. N., 1978. Phenylalanine analogues, *Int. Rev. Cytol.* 55:109–169.

Wheatley, D. N., and Henderson, J. Y., 1974. *p*-Fluorophenylalanine and "division-related proteins," *Nature* 247:281–283.

Wheatley, D. N., Inglis, M. S., and Malone, P. C., 1986. The concept of the intracellular amino acid pool and its relevance in the regulation of protein metabolism, with particular reference to mammalian cells, *Curr. Top. Cell. Regul.* 28:107–182.

White, A., Handler, P., and Smith, E. L., 1973. *Principles of Biochemistry*, 5th ed., McGraw-Hill, New York, pp. 604–628.

Wong, J. T.-F., 1983. Membership mutation of the genetic code: Loss of fitness by tryptophan, *Proc. Natl. Acad. Sci. USA* 80:6303–6306.

Biochemistry of Halogenated Neuroactive Amines

8.1 Introduction

Included for discussion in this chapter will be halogenated analogues of amine neurotransmitters, several of which analogues function as receptor agonists, and halogenated amines that can affect neuronal function in other, less direct ways. The medicinal and pharmacological importance of both peripheral and central nervous system (CNS)-acting analogues will be reviewed. Of obvious importance are compounds that can modulate CNS functions, either to the benefit or detriment of the subject. Accordingly, the role of halogenated analogues in the development of psychotherapeutic and psychotomimetic drugs will be discussed. Analogues that have peripheral cardiovascular action also will receive special attention.

8.2 Biochemistry of Halogenated Sympathomimetic Amines

8.2.1 Introduction

The importance of the naturally occurring catecholamines—dopamine (DA), norepinephrine (NE), and epinephrine (EPI) (Fig. 8-1)—as neurotransmitters and hormones, both in the periphery and the CNS, has stimulated an enormous amount of research toward development of drugs to modulate their activities. NE is the primary neurotransmitter of the sympathetic nervous system, neuronal release of which is stimulated in response to stress. Stress also stimulates release of EPI and NE from the adrenal medulla, elevating levels of circulating catecholamines. DA is an important CNS neurotransmitter, particularly with respect to control of motor functions, and its CNS actions are mediated by dopaminergic receptors. DA also has peripheral functions, mediated by both dopaminergic

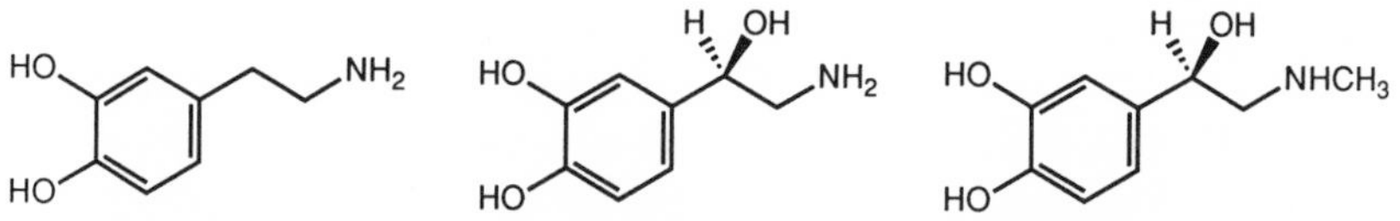

HO — [ring] — NH₂

Dopamine
(DA)

HO — [ring] — with H, OH — NH₂

R-Norepinephrine
(R-NE)

HO — [ring] — with H, OH — NHCH₃

R-Epinephrine
(R-EPI)

Figure 8-1. The naturally occurring catecholamines.

and adrenergic receptors. Examples of the diverse physiological processes affected by stimulation of the sympathetic nervous system are heart rate, force of cardiac contraction, blood pressure, smooth muscle tone, fatty acid and lipid metabolism, and appetite [for a review of catecholamines and the sympathetic nervous system, see Weiner (1985a) and Triggle (1981)].

A schematic diagram of a noradrenergic neuron is given in Fig. 8-2. Modulation of the adrenergic pathway can be effected at any of the steps depicted in this diagram. Thus, drugs have been developed that inhibit the biosynthesis of NE, stimulate release of NE from presynaptic storage vesicles, mimic the direct action of NE at adrenergic receptors (agonists),

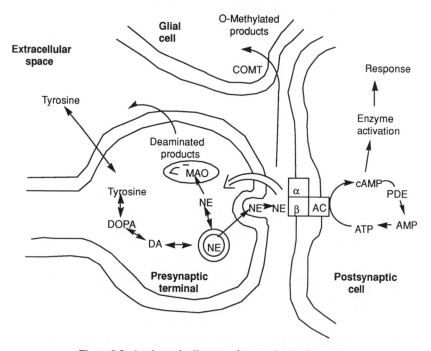

Figure 8-2. A schematic diagram of a noradrenergic neuron.

bind to the receptor but elicit no response (antagonists), inhibit the pre-synaptic reuptake of NE, or inhibit NE catabolic enzymes. While Fig. 8-2 depicts NE as the neurotransmitter, these same principles hold for the remaining noradrenergic and other neurotransmitters.

The physiological responses to sympathomimetic amines are mediated through stimulation of membrane-bound adrenergic receptors. Based on their sensitivities to adrenergic agonists, Ahlquist (1948) classified these receptors as α- or β-adrenergic receptors, while the discovery of more selective agonists and antagonists now has led to the recognition of subclasses of receptors. Thus, β_1-adrenergic receptors are prevalent in cardiac tissue, while β_2-adrenergic receptors are prevalent in smooth muscle tissue, although tissues can possess both types of receptors in different amounts. Similarly, postsynaptic α_1-adrenergic receptors are found in smooth muscle and gland cells, while presynaptic α_2-adrenergic receptors are thought to be responsible for feedback inhibition of NE release. A major field of medicinal chemistry and pharmacology has developed around efforts to find agents which selectively inhibit or stimulate a particular adrenergic receptor subtype. The availability of clinically effective hypotensive agents and of bronchodilators represents two obvious examples of success in this field. Noradrenergic pathways in the CNS also have been targets of drug-mediated modulation. In several instances, halogenated compounds have proven to be valuable pharmacological tools and/or medicinal agents for modulation of noradrenergic function, and progress in this area is reviewed below.

8.2.2 Structure–Activity Relationships of Halogenated Sympathomimetic Amines

8.2.2.1 Overview of Adrenergic Activities of Halogenated Phenethylamines

β-Phenethylamine (Fig. 8-3) can be considered the parent compound of sympathomimetic amines, including the catecholamines. There have been extensive structure–activity relationship (SAR) studies of β-phenethyl-amines variously substituted on the aromatic ring, the side chain, and the terminal amino group. However, many early investigations considered only pressor effects as a measure of activity, and the complexity of receptor systems under study has complicated interpretation of some of these results. Furthermore, whereas direct-acting sympathomimetic amines, such as the catecholamines, elicit their biological response through direct inter-action with receptors, many sympathomimetic drugs, in particular, those

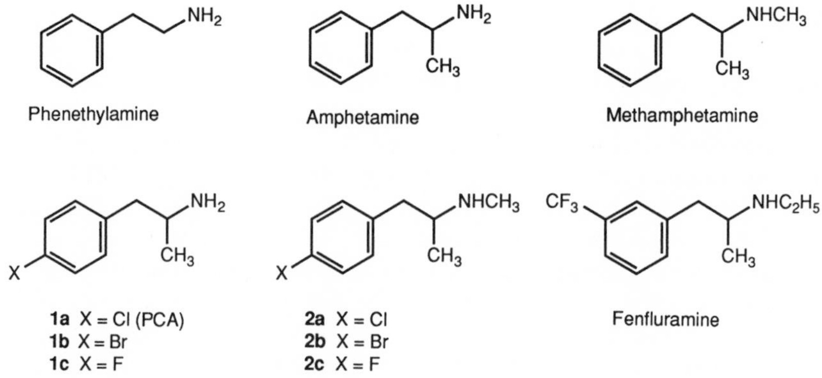

Figure 8-3. β-Phenethylamines and examples of halogenated analogues thereof.

lacking a phenolic hydroxyl group, have an indirect action resulting from the stimulation of release of NE from the noradrenergic neuron.

The presence of hydroxyl groups at the 3 and 4 positions of the aromatic ring of phenethylamine is required for maximal direct adrenergic agonist activity. Thus, compounds having halogen on the aromatic ring in place of phenolic hydroxyl groups have greatly reduced activity, and many function as antagonists at β-adrenergic receptors [for an extensive review of early work, see Smith (1970)]. Ring halogenation can have profound effects on the biological properties of phenethylamines and their simple alkyl derivatives, for example, amphetamine and methamphetamine (Fig. 8-3), amines that exert most of their CNS effects through inhibition of reuptake of NE and EPI into noradrenergic neurons. 4-Chloroamphetamine (PCA) (Fig. 8-3, **1a**) is a particularly potent inhibitor of NE and serotonin (5-hydroxytryptamines, 5HT) uptake by brain tissue. 4-Chloro- and 4-bromo analogues of amphetamine and methamphetamine (**1a, b** and **2a, b**, respectively; Fig. 8-3) have severe and long-lasting neurotoxic effects selective to serotonergic neurons (Section 8.4). 4-Fluoroamphetamine (**1c**) also depletes 5HT levels, but the effects are short-term [reviewed by Fuller (1978)]. The anorectic drug fenfluramine, licensed by the U.S. Food and Drug Administration for the treatment of adult obesity, also can lower 5HT levels (Fuller, 1978). Fuller and Molloy (1976) have studied the biological properties of a series of β-fluoro- and β,β-difluoro-phenethylamines (**3-7**; Fig. 8-4), with particular attention given to the effects of the drastically lowered amine pK_a on biological distribution, interactions with catabolic enzymes, and modulation of endogenous amine levels. For example, unlike amphetamine, β,β-difluoroamphetamine (**6**) is unable to lower NE levels in heart tissue.

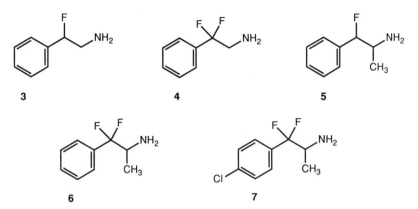

3   **4**   **5**

6   **7**

Figure 8-4. Examples of side-chain-fluorinated phenethylamines (Fuller and Molloy, 1976).

An exception to the general rule that direct action on adrenergic receptors requires the presence of a phenolic hydroxyl is found in the potent α-adrenergic activities of a series of imidazolines. Several halogenated compounds are important members of this series, as exemplified by the selective α-adrenergic agonist clonidine (Fig. 8-5), an extremely potent and long-lasting hypotensive agent. The high lipid solubility provided by the 2,6-dichloro substituent appears to be critical for hypotensive activity [reviewed by Comer *et al.* (1981)].

8.2.2.2 Halogen-Induced Adrenergic Selectivities

Despite extensive research concerning ring-halogenated and side-chain-halogenated neuroactive amines, early reports of ring-halogenated phenolic amines are few. Examples include 3,5-dibromo- and 3,5-dichlorotyramine (**8** and **9**) (Zeynek, 1921), 3-fluorotyramine (**10**) (Schiemann and Winkelmüller, 1932), and 2-(3-fluoro-4-hydroxyphenyl)- and 2-(3-chloro-4-hydroxyphenyl)-2-hydroxy-*N*-methylethylamine (**11a** and **11b**) (Fosdick

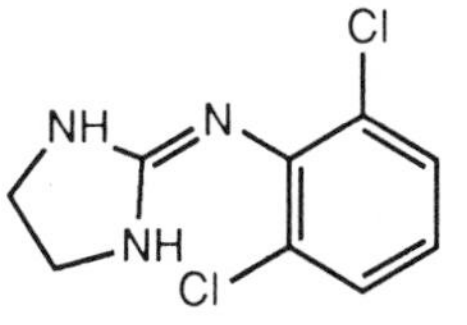

Clonidine

Figure 8-5. Clonidine, a selective α-adrenergic agonist.

et al., 1946) (Fig. 8-6). The latter two examples were viewed as epinephrine analogues (halogen for hydroxyl replacement) and had greatly reduced pressor activity relative to that of EPI. As part of a program to develop selective and potent β-adrenergic agonists, Kaiser *et al.* (1974) prepared 2-, 5-, and 6-chloro analogues (**12b–d**) of the potent and selective β-adrenergic agonist isoproterenol (ISO) (**12a**) (Fig. 8-6) as well as the corresponding analogues of related *N*-alkylethanolamines. The presence of chlorine in the 5 and 6 positions reduced β-adrenergic activity, but 2-chloroisoproterenol was approximately five times more potent than ISO.

Because of the importance of phenolic hydroxyl groups to the agonist properties of catecholamines, several ring-fluorinated catecholamines were synthesized to investigate the importance of such physicochemical factors as phenolic pK_a to receptor binding and agonist activity, the expectation being that steric perturbations would be minimal [reviewed by Kirk and Creveling (1984)]. Examination of the adrenergic agonist properties of 2-, 5-, and 6-fluoro-norepinephrine (2-, 5-, and 6-FNE, **13a–c**; Fig. 8-7) revealed unexpected and dramatic fluorine-induced adrenergic selectivities. 2-FNE had potency comparable to that of NE as a β-adrenergic agonist but was essentially devoid of α-adrenergic activity. 6-FNE mimicked NE as an α-adrenergic agonist but was much less potent as a β-adrenergic agonist. Since the effect of the 6-fluoro substituent on phenolic pK_a is minimal, this result casts doubt on the existence of a direct relationship between phenolic acidity and fluorine-induced agonist selectivity. 5-FNE was comparable to

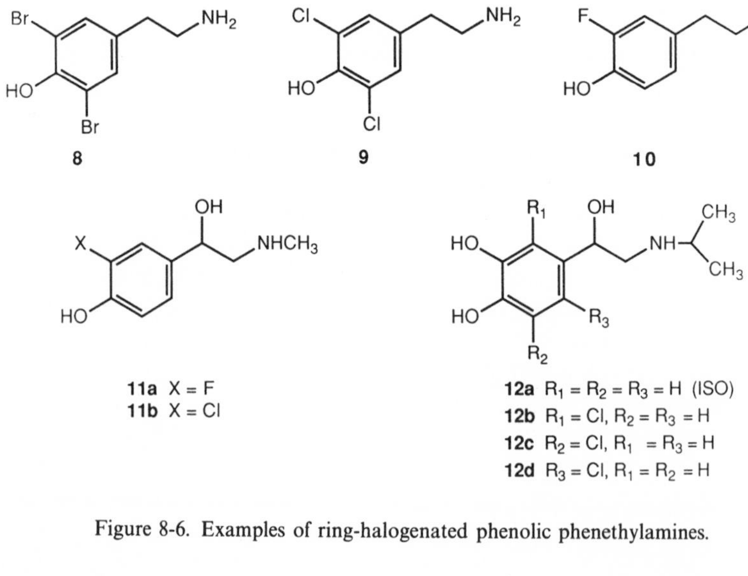

11a X = F
11b X = Cl

12a $R_1 = R_2 = R_3 = H$ (ISO)
12b $R_1 = Cl$, $R_2 = R_3 = H$
12c $R_2 = Cl$, $R_1 = R_3 = H$
12d $R_3 = Cl$, $R_1 = R_2 = H$

Figure 8-6. Examples of ring-halogenated phenolic phenethylamines.

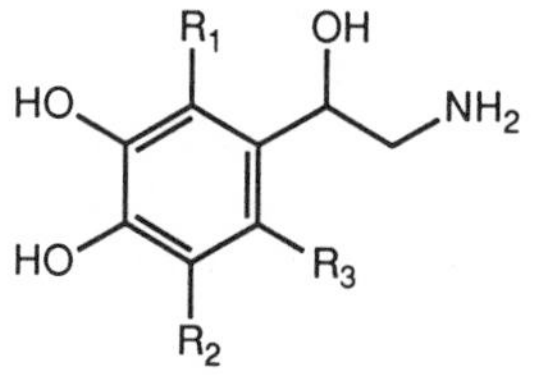

13a $R_1 = F, R_2 = R_3 = H$
13b $R_2 = F, R_1 = R_3 = H$
13c $R_3 = F, R_1 = R_2 = H$

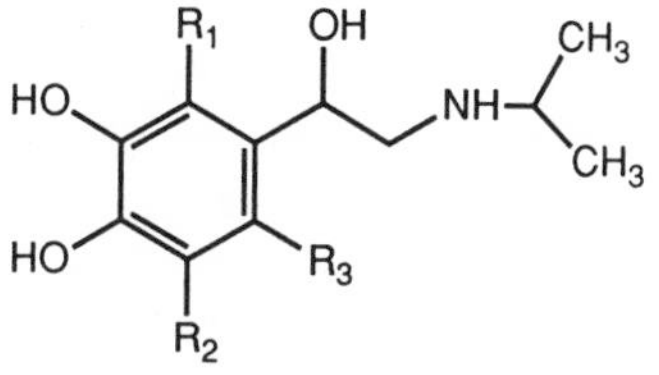

14a $R_1 = F, R_2 = R_3 = H$
14b $R_2 = F, R_1 = R_3 = H$
14c $R_3 = F, R_1 = R_2 = H$

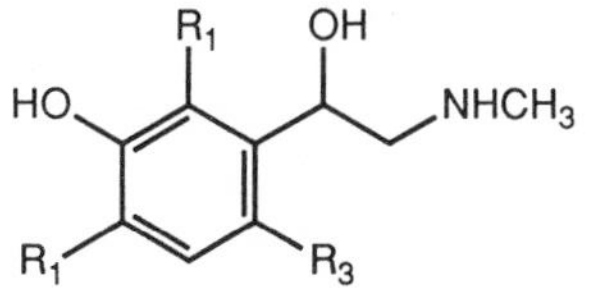

15a $R_1 = R_2 = R_3 = H$ (PE)
15b $R_1 = F, R_2 = R_3 = H$
15c $R_2 = F, R_1 = R_3 = H$
15d $R_3 = F, R_1 = R_2 = H$

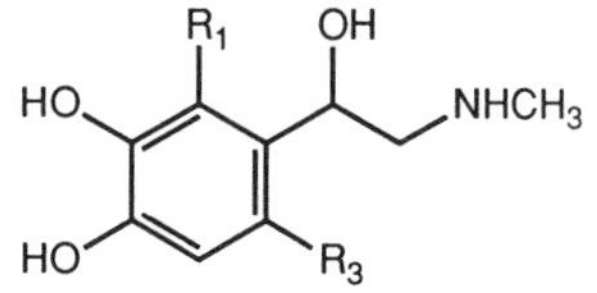

16a $R_1 = F, R_2 = H$
16b $R_2 = F, R_1 = H$

Figure 8-7. Ring-fluorinated analogues of biogenic amines.

NE as an α-adrenergic agonist and, depending on the system, was equal to or somewhat better than NE as a β-adrenergic agonist (Kirk *et al.*, 1979). Ring-fluorinated analogues (**14a–c**; Fig. 8-7) of ISO were examined next. While fluorine at the 6 position greatly reduced β-adrenergic potency, fluorine at the 5 or the 2 position had little effect on activity. None of the analogues showed significant α-adrenergic activity (Kirk *et al.*, 1982). Fluorinated analogues of the α-adrenergic agonist phenylephrine (PE) (**15a**) (2-, 4-, and 6-FPE, **15b–d**; Fig. 8-7) showed comparable fluorine-induced selectivities, but, for the first time, an increase in potency was seen. Thus, 2-FPE had greatly reduced α-adrenergic activity, while 6-FPE had increased α_1-adrenergic activity relative to that of PE (Kirk *et al.*, 1986). The decreased β-adrenergic activity shown by 6-FPE makes this analogue more potent and more selective than PE, which is sold commercially as a nasal decongestant. Similar increases in both α- and β-adrenergic potency accompanied fluorine-induced adrenergic selectivities in ring-fluorinated analogues of epinephrine (2- and 6-FEPI, **16a, b**; Fig. 8-7) (Adejare *et al.*, 1988).

From the above results, it becomes clear that β-adrenergic activity of these phenolic ethanolamines is favored by the absence of fluorine at the 6 position, while α-adrenergic activity is favored by the absence of fluorine at the 2 position (summarized in Table 8-1). From a practical viewpoint, 2- and 6-FNE have proven to be very versatile pharmacological agents and have been used in numerous studies as an aid to the characterization of adrenergic receptor systems (e.g., Daly *et al.*, 1981; Mueller *et al.*, 1983; Stone *et al.*, 1986). In addition to practical considerations, these results have stimulated considerable speculation concerning the biochemical mechanism(s) responsible for these selectivities. Since adrenergic selectivities have been found to correlate with receptor binding (Nimit *et al.*, 1980), fluorine-induced alterations in ligand–receptor interactions appear to be involved. Of agonists examined, only the adrenergic responses of DA appeared insensitive to fluorine substitution at positions 2 and 6 (Goldberg *et al.*, 1980). This observation implicated an interaction of an *ortho*-fluorine with the benzylic hydroxyl group of NE and other ethanolamines. Rotamer stabilization by an F $\cdots$ HO hydrogen bond was initially proposed, wherein one conformation (A, Fig. 8-8) would favor interaction with the α receptor while the other stabilized conformation (B, Fig. 8-8) would favor binding to the β receptor (Cantacuzene *et al.*, 1979). More recently, rotamer

Table 8.1. A Summary of the Effect of Fluorine on the Adrenergic Selectivities of Biogenic Amines

Compound	Adrenergic Receptor Selectivity	
	α	β
Norepinephrine	+	+
2-FNE	−	+
5-FNE	+	+
6-FNE	+	−
Isoproterenol	−	+
2-FISO	−	+
5-FISO	−	+
6-FISO	−	−
Phenylephrine	+	−
2-FPE	−	+ −
4-FPE	+	−
6-FPE	+ +	− −
Epinephrine	+	+
2-FEPI	−	+ +
6-FEPI	+ +	−

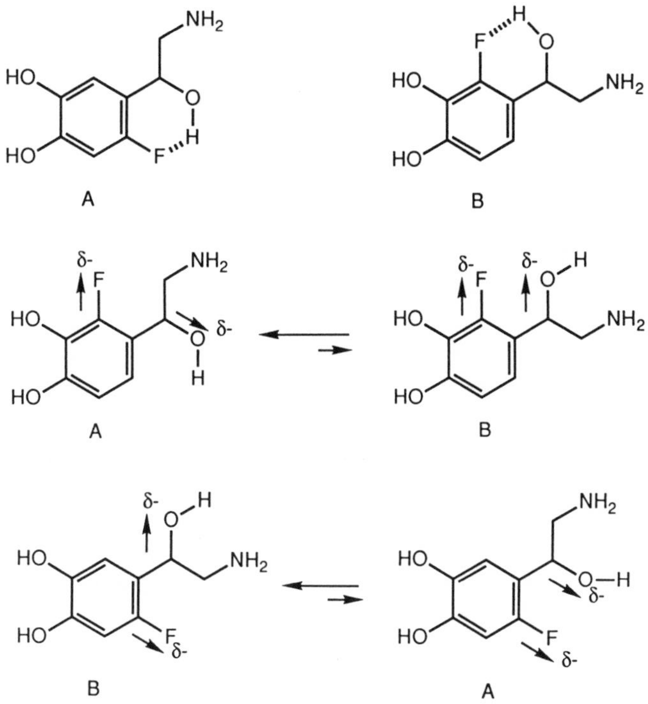

Figure 8-8. Two hypothetical models invoking intramolecular interactions to explain the effect of fluorine substitution on adrenergic selectivities (see text for details).

destabilization caused by dipole–dipole repulsion between F and OH was suggested as a more attractive mechanism involving the benzylic OH, wherein the conformational relationships are reversed relative to those for the initial proposal (Fig. 8-8) (Kirk *et al.*, 1986; DeBernardis *et al.*, 1985). However, explanations which do not directly involve the benzylic hydroxyl are possible, and mechanisms involving fluorine-induced alteration of aromatic electron density have been proposed (for further discussions, see Kirk *et al.*, 1988; Adejare *et al.*, 1988).

Trimetoquinol (TMQ) (**17a**; Fig. 8-9) is a potent β-adrenergic agonist, the S-($-$)-isomer of which is marketed in Japan as a bronchodilator. In view of the fluorine-induced adrenergic agonist selectivities of FNEs, fluorinated analogues of TMQ were prepared as part of a program to develop potent and selective adrenergic stimulants. The 5-fluoro- and 8-fluoro analogues of TMQ (**17b, c**; Fig. 8-9) retained β_1-adrenergic activity but had significantly reduced β_2-adrenergic activity, producing a rank order of β_1/β_2 selectivity of 8-fluoro-TMQ > 5-fluoro-TMQ > TMQ.

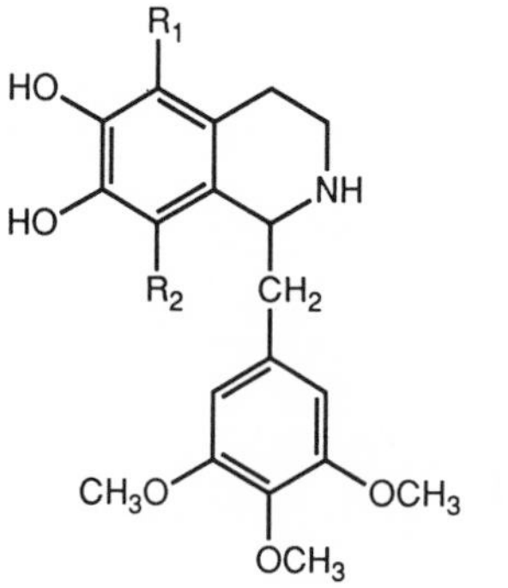

17a $R_1 = R_2 = H$ (TMQ)
17b $R_1 = F, R_2 = H$
17c $R_2 = F, R_1 = H$

Figure 8-9. Trimetoquinol (TMQ) and fluorinated analogues thereof (Clark *et al.*, 1987; Miller *et al.*, 1988).

Fluorine substitution in TMQ also led to selectivity in the antagonism of the muscle contractile and platelet aggregatory actions of thromboxane A_2 (TXA_2), mediated by α (aggregation) and τ-(tone) receptors, respectively (Clark *et al.*, 1987; Miller *et al.*, 1988).

8.2.2.3 β-Adrenergic Blocking Agents

Compounds that can selectively block β-adrenergic receptors are used clinically for the management of such cardiovascular disorders as hypertension. As noted above, replacement of phenolic hydroxyl groups with halogen often produces adrenergic antagonists, and, indeed, dichloro-isoproterenol (DCI; Fig. 8-10) was the first drug shown to have selective antagonism at β-adrenergic receptors (Powell and Slater, 1958). DCI has been very useful in pharmacological studies for characterization of β-adrenergic responses. However, because it is also a partial β-adrenergic agonist, it has not been used clinically. Other ring-halogenated analogues of β-phenyl-β-hydroxyethylisopropylamine also produce β-adrenergic blockade (Levy and Ahlquist, 1961).

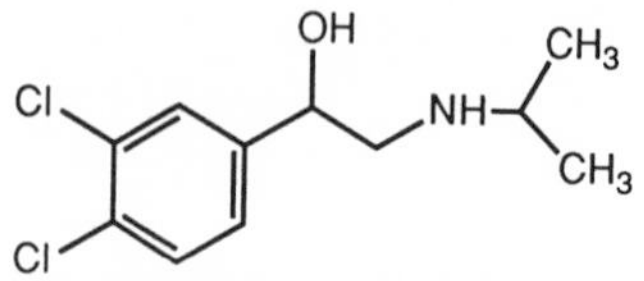

DCI

Figure 8-10. Dichloroisoproterenol, a potent and selective β-adrenergic antagonist.

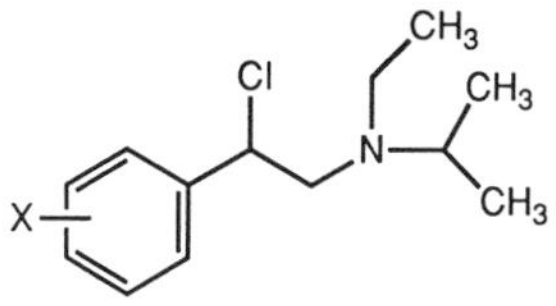

18

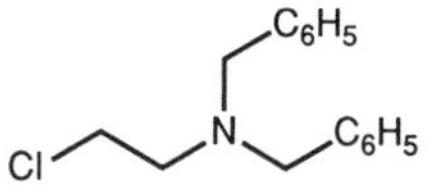

Dibenamine 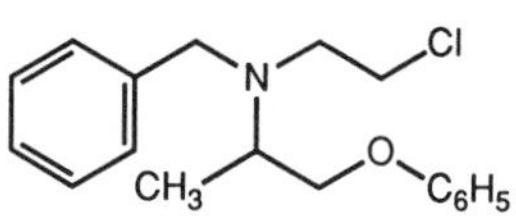   Phenoxybenzamine

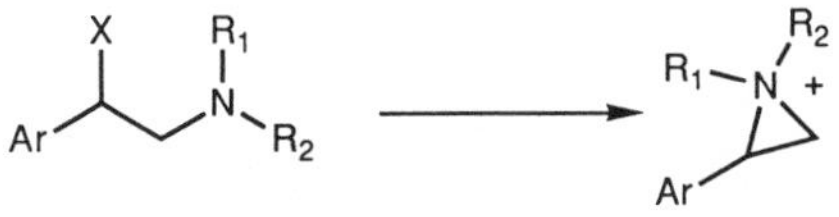

Figure 8-11. α-Adrenergic blocking agents. These are converted *in vivo* to ethyleneiminium ions that react with, and inactivate, the α-adrenergic receptor.

8.2.2.4 α-Adrenergic Blocking Agents

Replacement of the benzylic hydroxyl group of ring-halogenated β-phenyl-β-hydroxyethylisopropylamine with chlorine produces α-adrenergic blocking agents (e.g., **18**; Fig. 8-11) [reviewed by Smith (1970)]. N,N-Dibenzyl-β-chloroethylamine (dibenamine; Fig. 8-11) and phenoxybenzamine (Fig. 8-11) are other examples of a huge number of nitrogen mustard-type analogues which selectively and irreversibly block α-adrenergic receptors. Several lines of evidence support the concept that these compounds are converted *in vivo* to ethyleneiminium ions and that it is this species that reacts with, and inactivates, the α-adrenergic receptor (Fig. 8-11). The importance and diversity of α-adrenergic receptors in cardiovascular reflexes has limited practical therapeutic application of these drugs. Several reversible antagonists, for example, the α$_1$-selective antagonist prazosin (Fig. 8-12), are available and have more selective action [reviewed by Weiner (1985b)].

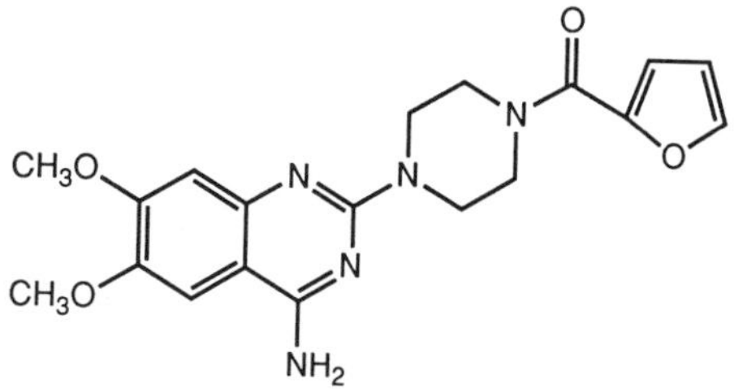

Figure 8-12. The α$_1$-selective antagonist prazosin.

Prazosin

8.3 Dopamine and Dopaminergic Neurons

8.3.1 Halogenated Analogues of Dopamine

In the kidney, low doses of DA cause an increase in glomerular filtration rate, renal blood flow, and sodium excretion, effects mediated by a specific dopaminergic receptor. SAR studies had shown that the agonists activating the DA vascular receptor have much stricter structural requirements than agonists for adrenergic receptors. In view of the effects of fluorine substitution on the adrenergic agonist selectivities of NE, even greater alterations in behavior of fluorinated DAs (FDAs, **19a–c**; Fig. 8-13) (Kirk, 1976a) at the renal vascular receptor might have been expected. In fact, 2-F- and 5-FDA were essentially equipotent with DA and 6-FDA was approximately four times less potent than DA as agonists at this receptor, as determined in the intact dog (Goldberg *et al.*, 1980). This was the first instance in which additional substituents on the aromatic ring of DA did not lead to loss of activity at the DA vascular receptor. Similar behavior was found for CNS dopamine receptors. Thus, 2-, 5-, and 6-FDA and DA all bound with comparable affinity to the D_2 and D_3 DA receptor sites in striatal tissue of rat and calf brain. The fluorinated analogues were somewhat more weakly bound to the D_1 DA receptor (dopamine-sensitive adenylate cyclase). The D_2 receptor, defined as the site blocked by nanomolar concentrations of neuroleptics but occupied by micromolar concentrations of dopamine, is related to various dopaminergic behaviors. The finding that the affinity of the fluorinated analogues of DA for this site is similar to that of DA is particularly relevant to the use of [18F]fluoro-DOPA-derived [18F]fluorodopamine for PET studies of dopamine function (Chapter 7).

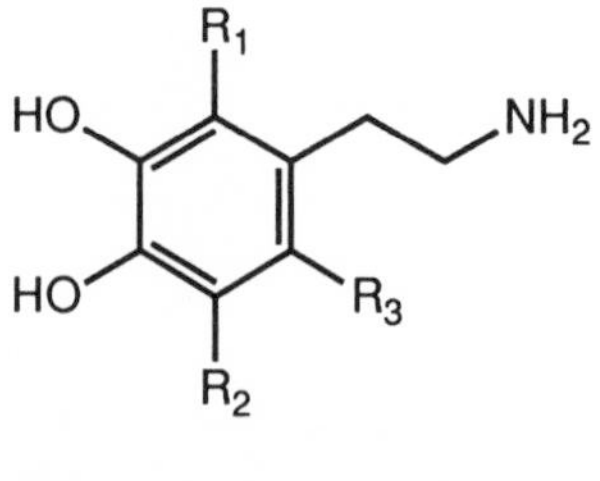

19a $R_1 = F, R_2 = R_3 = H$
19b $R_2 = F, R_1 = R_3 = H$
19c $R_3 = F, R_1 = R_2 = H$

Figure 8-13. Ring-fluorinated dopamines.

8.3.2 Antipsychotic Agents (Neuroleptics)

The "dopamine theory" of schizophrenia is based on, among other things, the fact that the efficacy of certain antipsychotic drugs correlates well with their affinity for dopaminergic receptors (Carlsson, 1978; for a recent discussion of controversial issues related to the biology of schizophrenia, see Barnes, 1987). Of the several classes of antipsychotic drugs currently prescribed, the most widely used are phenothiazines (**20**), thioxanthenes (**21**), and butyrophenones (**22**). The first synthetic antipsychotic drug, chlorpromazine (**20a**; Fig. 8-14), was synthesized in 1950 and is still used clinically. A series of phenothiazines based on this lead have been developed. In the 1960s, Janssen and co-workers developed a series of butyrophenones in an attempt to increase the morphine-like activity of meperidine [1-methyl-4-phenyl-4-carbethoxypiperidine (**23**); Fig. 8-14]. The discovery that a *para*-fluoro substituent on the aryl ring

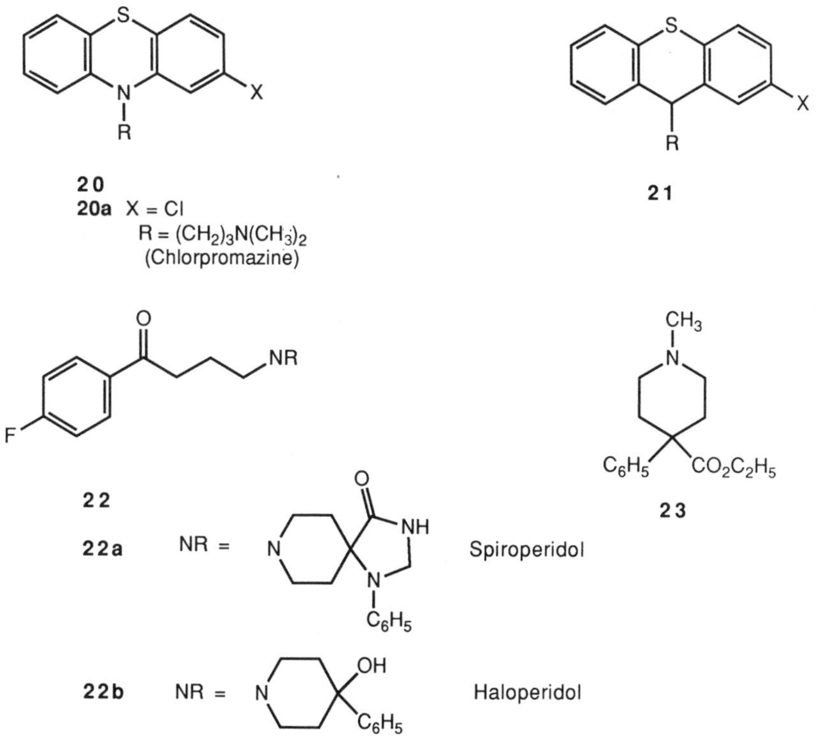

20
20a X = Cl
R = $(CH_2)_3N(CH_3)_2$
(Chlorpromazine)

21

22

22a NR = Spiroperidol

23

22b NR = Haloperidol

Figure 8-14. Examples of antipsychotic drugs: phenothiazines (**20**), thioxanthenes (**21**), and butyrophenones (**22**).

contributed to a diminution of morphine-like activity but accentuated the tranquilizing properties prompted the screening of a series of 4-fluoro-butyrophenones as potential neuroleptic agents [reviewed by Smith (1970)]. Examples of the many compounds examined are spiroperidol (spiperone) and haloperidol (**22a** and **22b**, respectively; Fig. 8-14), extremely potent neuroleptics. A 4-fluorobenzoyl group appears to be an absolute structural requirement for optimum potency. Kaiser and Setler (1981) have provided a particularly extensive discussion of SAR studies in this series, and Elliot (1982) has reviewed the role of fluorine in the development of neuroleptics.

8.4 Halogenated Compounds and Indoleamine Biochemistry

8.4.1 Halogenated Analogues of Serotonin (5-Hydroxytryptamine, 5HT) and of Melatonin (*N*-Acetyl 6-Methoxytryptamine)

Serotonin (5HT) (Fig. 8-15) is found in many cell types, for example, mast cells, platelets, and enterochromaffin cells, and is involved in a broad range of responses related to cardiovascular, respiratory, and gastro-intestinal systems. Although only 1–2% of the total 5HT is found in the brain, this amine nonetheless is linked intimately with neuropsycho-pharmacology. Indeed, the structural similarities between the potent hallucinogen LSD and 5HT was instrumental in focusing early attention on the potential involvement of 5HT in mental illness. Included in the

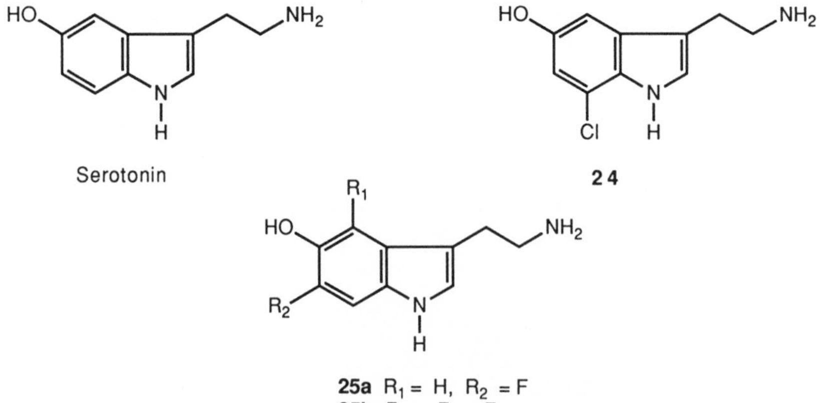

25a R_1 = H, R_2 = F
25b R_1 = R_2 = F

Figure 8-15. Serotonin and ring-halogenated analogues thereof.

many functions associated with central serotonergic neurons are pain perception, sleep, and behavior—both normal and abnormal. 5HT also serves as the biological precursor of melatonin (for reviews, see Glennon, 1987; Cooper et al., 1986; Douglas, 1985).

Central 5HT receptors are located almost exclusively on the raphe nuclei of the brain stem. As with adrenergic functions, multiple receptor types appear to mediate 5HT responses. $5HT_1$ (inhibitory), $5HT_2$ (excitatory), and $5HT_3$ receptors have been proposed, and these receptor types have been further subclassified (Glennon, 1987; Cooper et al., 1986). Agents that can affect endogenous brain 5HT levels have been powerful experimental tools in psychopharmacology and have been targets for the development of drugs for the treatment of mental disorders. Antagonists targeted to the peripheral functions of 5HT also have clinical applications (Douglas, 1985).

8.4.1.1 6-Fluoroserotonin and 4,6-Difluoroserotonin

Similar to the situation with catecholamines, there are relatively few reports of ring-halogenated tryptamine derivatives that retain the critical 5-hydroxyl group. 7-Chloro-5-hydroxytryptamine (24; Fig. 8-15) is an example (Lee et al., 1973). 6-Fluoroserotonin (6F-5HT) (25a) and 4,6-difluoroserotonin (4,6-diF-5HT) (25b) were synthesized to probe the effects of such parameters as phenolic pK_a on the biological behavior of 5HT (Kirk, 1976b). Preliminary results have indicated that there are minimal differences in behavior between the fluorinated analogues and 5HT in interaction with 5HT receptors (E. Hollingsworth, J. W. Daly, and K. L. Kirk, unpublished results).

The effects of ring fluorination on 5HT transport and storage mechanisms in platelets have been studied extensively. Despite having phenolic pK_a values (6F-5HT, 9.1; 4,6-diF-5HT, 8.0) lower than that of 5HT (10.7), at pH 7.4 6F-5HT and 4,6-diF-5HT are transported into the vesicular compartments of human platelets at essentially the same rate as 5HT (Costa et al., 1982). A more recent demonstration that 4,6-diF-5HT and 5HT are also transported at the same rate at pH 9, where the former has a largely dissociated hydroxyl group, indicates that the platelet serotonin transporter uses both the zwitterionic and cationic forms of the amines. In contrast, in bovine chromaffin granules the apparent K_m for the transport of 4,6-diF-5HT increases relative to that for 5HT on going from pH 7 to pH 9, indicating that the chromaffin granule amine transporter uses the protonated (hydroxyl) form of the substrate (Rudnick et al., 1989).

The ready uptake of 4,6-diF-5HT into platelets has provided an opportunity to apply spectroscopic techniques to the investigation of storage mechanisms. Thus, electron energy-loss spectroscopy was used to localize storage vesicles preloaded with 4,6-diF-5HT in one of the first demonstrations of the use of this technique for the detection and localization of light elements (Costa et al., 1978). In addition, several ^{19}F-NMR investigations of human and pig platelets preincubated with 6F-5HT and 4,6-diF-5HT have provided information on the motional states of the vesicularly stored amines (Costa et al., 1981, and references therein).

Research on classification, distribution, and functioning of 5HT receptors has benefited from the availability of agents that bind selectively to certain of the subtypes of receptors. For example, 1-(2,5-dimethoxy-4-bromophenyl)-2-aminopropane (DOB) (Fig. 8-16) is a particularly potent member of a series of phenylisopropylamine derivatives selective for the $5HT_2$ receptor, and [^{3}H]DOB has been studied as a label for these sites. Similarly, several alkylpiperidine neuroleptics such as spiperone (**22a**, Fig. 8-14), in addition to possessing high affinity for DA receptors, also have high affinity for $5HT_2$ and for $5HT_{1A}$ sites (Glennon, 1987).

8.4.1.2 Halogenated Analogues of Melatonin

The pineal hormone melatonin (Fig. 8-17) has several physiological functions in vertebrates, including regulation of reproduction of seasonally breeding animals and control of circadian rhythms in birds and reptiles. One of the most extensively studied effects of melatonin in mammals is its gonadotropic action. While melatonin is a potent inhibitor in vitro of the pituitary response to luteinizing hormone-releasing hormone (LHRH), the antigonadotropic response in vivo is complex and varies from species to species.

A major difficulty in studies done in vivo has been the extremely short biological half-life of melatonin (12–15 min in rats), and there has been considerable interest in producing analogues with greater biological stability. In addition, such analogues would have several potential applica-

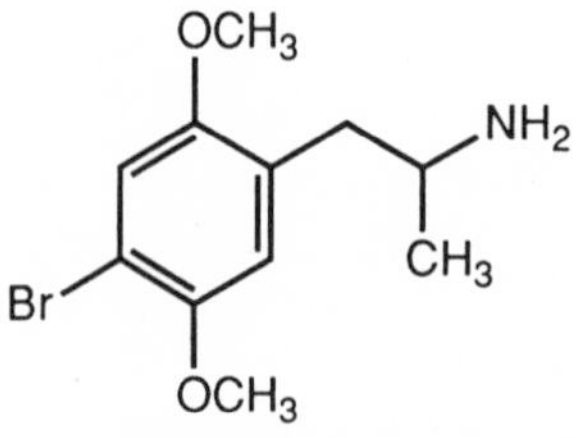

Figure 8-16. DOB, a potent and selective ligand for the $5HT_2$ receptor.

DOB

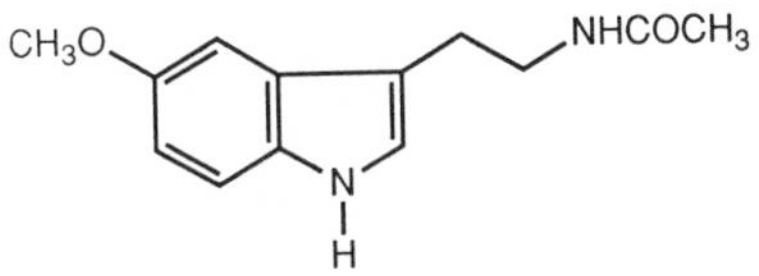

Melatonin

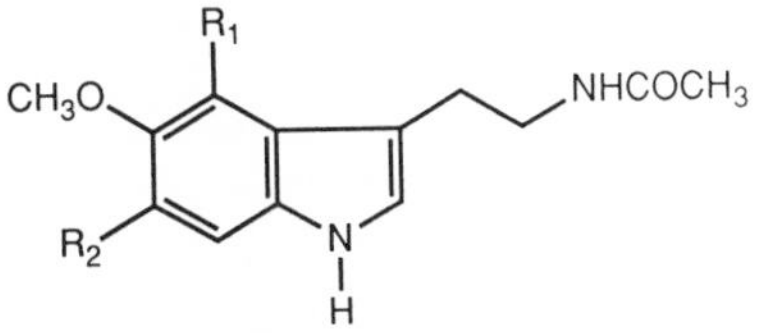

26a R_1 = H, R_2 = F
26b R_1 = R_2 = F
26c R_1 = H, R_2 = Cl
26d R_1 = H, R_2 = CH$_3$

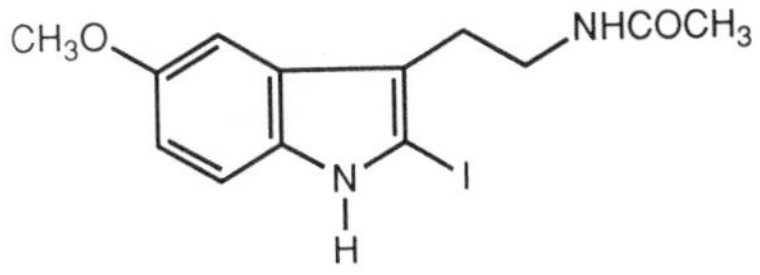

27

Figure 8-17. Melatonin and ring-halogenated analogues thereof.

tions related to reproduction control. In the intact animal, the major route of melatonin metabolism is hydroxylation at position 6, and blockade of this hydroxylation has been studied as a potential method for increasing biological half-life. The effects of 6-fluoro- and 4,6-difluoromelatonin (**26a, b**; Fig. 8-17) on the pituitary LH response to LHRH *in vitro* were studied and found to mimic the effects of melatonin (Martin *et al.*, 1980). Despite the expectation that fluorine at position 6 would block enzymatic hydroxylation, preliminary data indicated that biological response in the intact animal was not prolonged (J. E. Martin, unpublished results). Clemens *et al.* (1980) have studied the inhibition of LH and prolactin release and ovulation-blocking activity of 6-fluoro-, 6-chloro- (**26c**), and 6-methylmelatonin (**26d**) in the rat. While the 6-methyl analogue was inactive, the 6-halo analogues were more effective than melatonin in inhibiting LH release and causing the related blockade of ovulation. Unlike melatonin, they had no effect on prolactin levels. While 6-chloromelatonin had a significantly longer biological half-life (27 min), this still rapid rate of metabolism suggests that dioxygenase cleavage of the indole ring may provide a ready route for inactivation. In hamsters, long-term injections or implants of 6-chloromelatonin showed comparable antigonadotropic and countergonadotropic effects to those produced by similar treatment with melatonin (Vaughan *et al.*, 1986).

Progress in understanding the mechanism of melatonin action has also been hampered by difficulties in detection of high-affinity binding sites in tissues having low densities of receptors. A major advance in melatonin pharmacology has come with the discovery that 2-[^{125}I]iodomelatonin (Vakkuri et al., 1984) is a selective, high-affinity ligand suitable for autoradiographic visualization of melatonin binding sites (Dubocovich and Takahashi, 1987). Of the analogues examined, 2-iodomelatonin (27; Fig. 8-17) has the highest affinity for the melatonin receptor and binds significantly better than does melatonin. The value of this radioligand has been increased by the demonstration that 2-iodomelatonin mimics the action of melatonin in vivo (Weaver et al., 1988). Among recent results obtained using this radioligand is evidence that melatonin receptors are localized in the suprachiasmatic nuclei of the human hypothalamus, the site of a putative human biological clock (Reppert et al., 1988).

8.4.3 p-Chloroamphetamine and Depletion of Serotonin

As noted briefly above, p-chloroamphetamine [PCA (1a); Fig. 8-3] has long-lasting acute toxicity to serotonergic neurons. One injection of PCA can deplete brain serotonin (5HT) and its principal metabolite, 5-hydroxyindoleacetic acid, for as much as four months, a decrease accompanied by decreases in high-affinity 5HT uptake and in tryptophan hydroxylase activity. The depression of all the several parameters of serotonergic function is indicative of serotonin neuron destruction [reviewed by Fuller (1978)]. In contrast, the depletion of 5HT caused by p-chlorophenylalanine (Chapter 7) has been related to irreversible inhibition of tryptophan hydroxylase. Despite the extensive use of PCA as a research tool for the study of various aspects of serotonergic functions, the mechanism of toxicity has remained elusive. Recently, Commins et al. (1987a, b) have related the toxic effects of PCA to a similar toxicity shown by such drugs as methamphetamine, fenfluramine, and 3,4-methylenedioxy-amphetamine. A common mechanism of toxicity has been proposed based on the demonstration of methamphetamine-induced production of the neurotoxin 5,6-dihydroxytryptamine (5,6-DHT). Similarly to 6-hydroxy-dopamine, 5,6-DHT is oxidized in vivo to a highly reactive quinone which irreversibly reacts with and cross-links macromolecular neuronal structures. The presence of 5,6-DHT in the rat hippocampus following PCA administration suggested that this metabolite may also be responsible for the toxicity of PCA. According to this proposal, the ability of PCA and related drugs to function also as monoamine oxidase (MAO) inhibitors is critical to 5,6-DHT formation. Thus, a combination of stimulation of 5HT release by these drugs and the accompanying MAO inhibition leads to

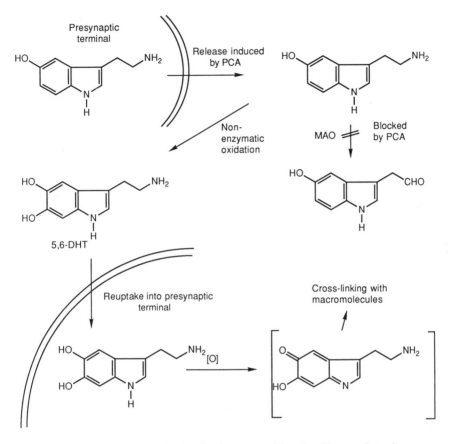

Figure 8-18. Proposed mechanism for the neurotoxicity of *p*-chloroamphetamine.

nonenzymatic oxidation of 5HT to 5,6-DHT at serotonergic synapses. Subsequent uptake of 5,6-DHT into the presynaptic terminal results in destruction of those neurons (summarized in Fig. 8-18) (Commins *et al.*, 1987a, b).

8.5 Halogenated Compounds and Histamine Response

Endogenous histamine (Fig. 8-19) is found in almost all mammalian tissue, where it is synthesized by histidine decarboxylase-catalyzed decarboxylation of histidine. Important biological responses to histamine include contraction of smooth muscle (e.g., gut and bronchi), relaxation of smooth

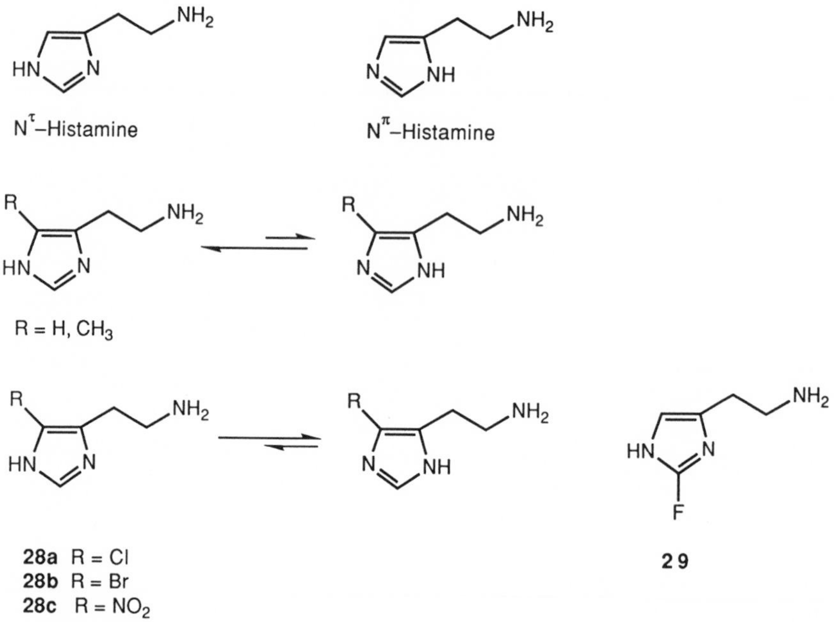

Figure 8-19. The effect of polar substituents on the tautomeric equilibrium of histamine (see text for details). 2-Fluorohistamine (**29**), an analogue having affinity for the H_1 receptor, is shown at the lower right.

muscle (e.g., those in fine blood vessels), and stimulation of gastric acid secretion. Based on studies of compounds capable of selective blockade of histamine H_1 and H_2 receptors, it has been determined that broncho-constriction and contraction of the gut are mediated by the H_1 histamine receptor, whereas gastric acid secretion is mediated by the H_2 hista-mine receptor. As in other systems involving receptor subtypes, the develop-ment of drugs that can antagonize selectively the response at a particular receptor type has had several clinical applications. The presence of hista-mine in the CNS has implicated a neurotransmitter role for this amine.

8.5.1 Halogenated Analogues of Histamine

4-Chlorohistamine (**28a**; Fig. 8-19) was included in a series of histamine analogues having weakly basic rings used to investigate the ionization state of the receptor-active form of the imidazole ring. Despite an extremely low concentration of protonated imidazole at physiological pH, these compounds retained considerable potency, suggesting that the

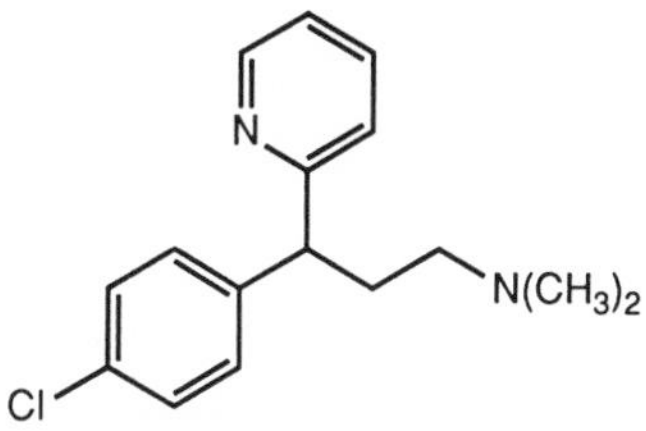

Figure 8-20. Chlorpheniramine (**30**) and chlorcyclizine (**31**).

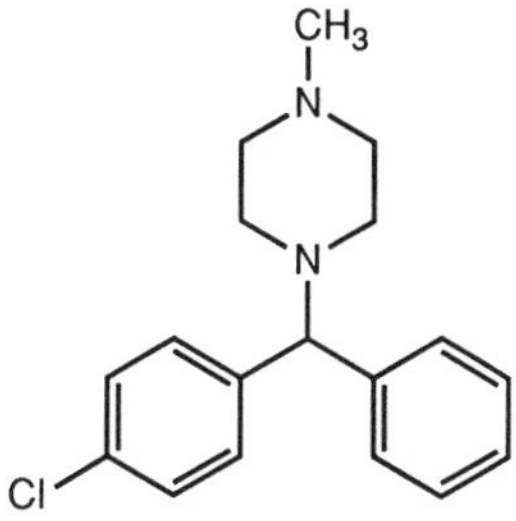

receptor recognizes the unprotonated imidazole ring [reviewed by Ganellin (1982)]. Polar substituents, for example, Cl, Br, and NO_2 (in compounds **28a–c**; Fig. 8-19), affect the tautomeric equilibrium of the imidazole ring, shifting it in favor of the N^π tautomer (Fig. 8-19). The lower potency of 4-chloro- and 4-bromohistamine, and other 4-substituted imidazoles, at H_2 histamine receptors was taken as evidence that the H_2 receptor recognizes the N^τ tautomer, although alterations in steric and lipophilic properties also could be important (Ganellin, 1982). 2-Fluorohistamine (**29**) has affinity for H_1 receptors (guinea pig ileum) comparable to that of histamine and was found to stimulate cyclic AMP production in brain slices, which have both H_1 and H_2 histamine receptors (Kirk and Cohen, 1976).

8.5.2 Halogenated Compounds as Histamine Antagonists

Over the past several decades, a vast number of antagonists selective for the histamine H_1 receptor have been developed. These antihistamines are characterized by the presence of lipophilic aryl groups and a basic amino side chain. Included in structural details shown to affect antihistamine activity is the presence of a *para*-chloro substituent on the aryl group, as evidenced by the 10-fold increase in potency of certain *para*-chloro-substituted compounds, for example, chlorpheniramine (**30**) and chlorcyclizine (**31**) (Fig. 8-20), over their unsubstituted counterparts. Imidazoles sub-

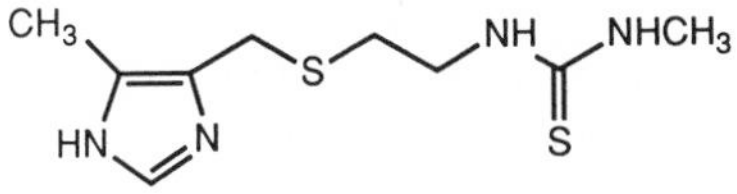

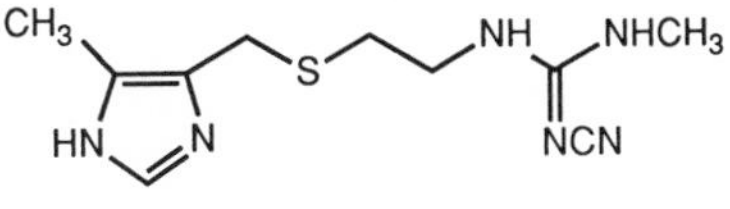

Figure 8-21. Metiamide (**32**) and cimetidine (**33**).

stituted with an extended nonbasic side chain, for example, metiamide and cimetidine (**32** and **33**; Fig. 8-21), constitute a series of H_2-selective antagonists (Ganellin, 1982).

8.6 Effects of Halogenation on Amine Metabolism

8.6.1 Monoamine Oxidase (MAO)

8.6.1.1 Substrates and Reversible Inhibitors: MAO A versus MAO B Selectivity

Reuptake of released amine neurotransmitters is the principal mechanism for the rapid curtailment of receptor stimulation. Two important catabolic routes are also available for the regulation of amine levels. The first of these to be discussed is the oxidation of monoamines to carbonyl groups, catalyzed by flavin-linked mitochondrial monoamine oxidases (MAO). MAO has several important physiological roles. Thus, MAO inactivates serotonergic and catecholaminergic neurotransmitters, intestinal MAO metabolizes pressor amines ingested in food, vascular MAO protects organs from circulating pressor amines, and liver MAO controls blood levels of these amines. MAO will oxidize a wide range of amine substrates, a requirement being the presence of an α hydrogen. α-Alkylamines frequently are effective reversible inhibitors of MAO.

Recognition of the mood-elevating properties of MAO inhibitors led to their extensive use for the treatment of depressive illness between 1959 and 1962. However, blockade of the metabolism of tyramine (present in certain foods) often led to severe blood pressure elevation (the "cheese" effect), and the use of these drugs declined. The discovery in 1968 of two forms of MAO—MAO A and MAO B, characterized by different substrate and inhibitor specificities—now has led to a dramatic resurgence of interest in these compounds. For example, based on organ distribution—for example, intestinal enzyme is primarily MAO A—and substrate selectivities of MAO A and B, the proposal has been advanced that an MAO B-specific enzyme would be free of the "cheese" effect (Knoll, 1979). For this and other reasons, the study of substrates and inhibitors selective for one or the other of the two enzymes is now an important area of medicinal chemistry (for a recent review, see Fowler and Ross, 1984).

As seen frequently with other enzymatic reactions, introduction of halogen can have dramatic effects on the properties of substrates and inhibitors of MAOs. For example, preliminary studies have indicated that

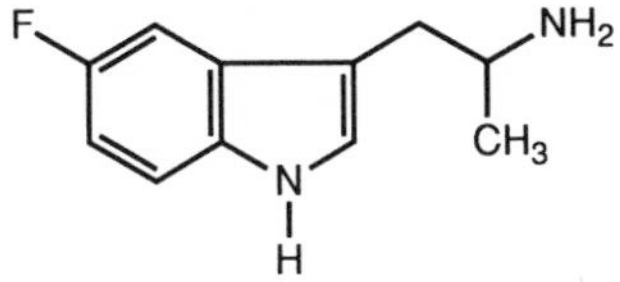

Figure 8-22. 5-Fluoro-α-methyltryptamine (**34**), an MAO A-selective inhibitor, and *p*-chloro-β-methylphenethylamine (**35**), an MAO B-selective inhibitor.

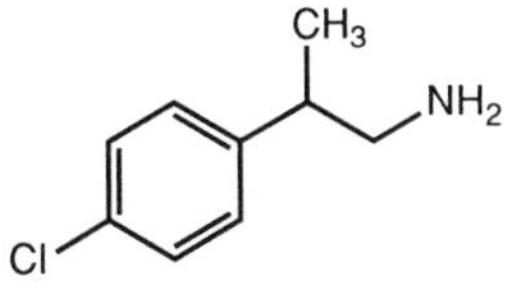

ring fluorination of 5HT causes this predominantly MAO A substrate to be metabolized significantly by platelet MAO B (Costa *et al.*, 1982). *p*-Fluorobenzylamine is a better substrate for MAO than is benzylamine, presumably because the greater acidity of the benzyl group of the former facilitates proton removal—an initial step in the oxidation process (Williams, 1973). Participation of the protonated form of the amine as the MAO substrate caused β,β-difluorophenethylamine to be oxidized by MAO more slowly than phenethylamine (Fuller and Molloy, 1976).

While many reversible competitive inhibitors selective for MAO A have been developed, including a large number of α-methylamines, very few effective MAO B-selective reversible inhibitors have been reported to date. Accumulated experience has indicated that, while many inhibitors, such as α-alkylamines, are inhibited sterically from binding to MAO B—and thus are A selective—no complementary steric inhibition selective for the A enzyme is available [reviewed by Fowler and Ross (1984)]. However, recent results indicate that, while α-methylated substrate analogues, such as

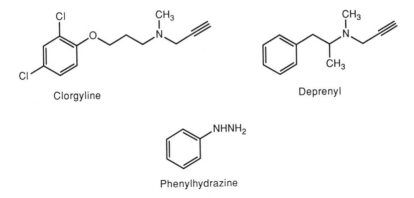

Clorgyline

Deprenyl

Phenylhydrazine

Figure 8-23. Examples of irreversible MAO inhibitors.

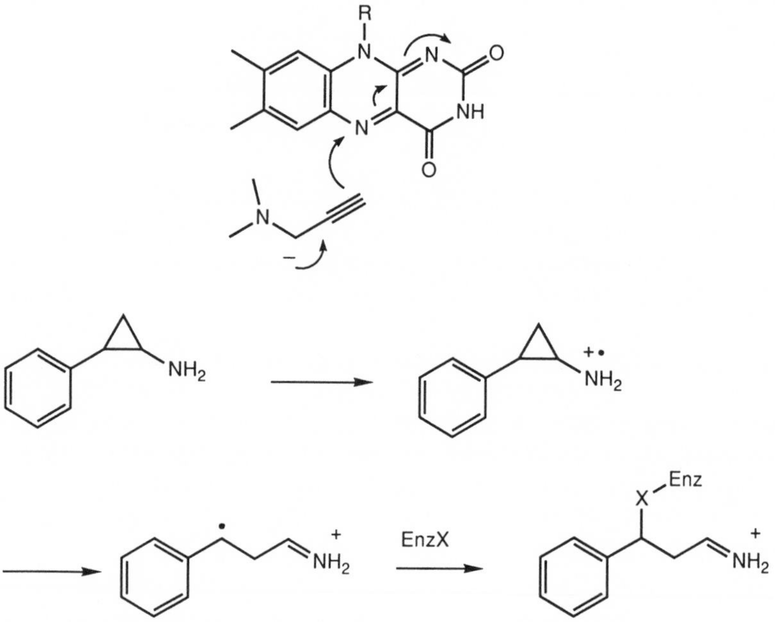

Figure 8-24. Proposed mechanisms for the inactivation of MAO by acetylenic amines (above) and by cyclopropylamines (below).

5-fluoro-α-methyltryptamine (**34**; Fig. 8-22), are A selective, β-substitution favors B selectivity. For example, *p*-chloro-β-methylphenethylamine (**35**; Fig. 8-22) has a 618-fold selectivity for MAO B (Kinemuchi *et al.*, 1987).

8.6.1.2 Halogenated Irreversible Inhibitors of MAO

Indicative of the interest in MAO inhibitors as pharmacological and medicinal agents, several hundred irreversible inhibitors of MAO have been

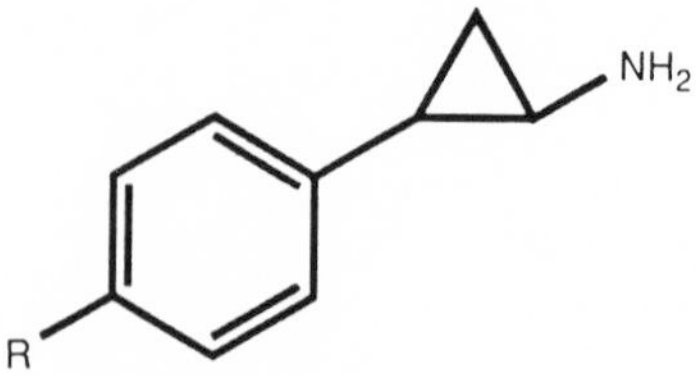

36a R = H
36b R = F

Figure 8-25. Tranylcypromine (TCP) (**36a**) and 4-fluorotranylcypromine (**36b**).

developed. Most are either acetylenic compounds, as exemplified by the MAO A-selective inhibitor clorgyline and the MAO B-selective inhibitor 1-deprenyl (Fig. 8-23), cyclopropylamines, or hydrazines, such as the MAO B-selective phenylhydrazine (Fig. 8-23). Inactivation by acetylenic compounds involves addition of a stabilized enzyme-derived carbanion to N-5 of the flavin cofactor (Fig. 8-24). Initial one-electron oxidation of cyclopropylamines leads to ring opening and covalent attachment to an active-site SH group as shown in Fig. 8-24 (Palfreyman *et al.*, 1987).

Tranylcypromine (TCP) (**36a**; Fig. 8-25) has been shown to be effective in the treatment of certain depressive illnesses. A search for analogues of TCP with improved pharmacodynamics has produced 4-fluorotranylcypromine (**35b**), some 10 times more potent than TCP. This analogue also achieved higher brain levels at lower doses in intact animals. Neither **36a** nor **36b** showed A or B selectivity. Blockade of metabolic *para*-hydroxylation or increased lipophilicity, or both, may contribute to the increased *in vivo* activity (Coutts *et al.*, 1987).

The search for new MAO B-selective inhibitors—agents that do not induce the "cheese" effect—has produced a series of 3-haloallylamines

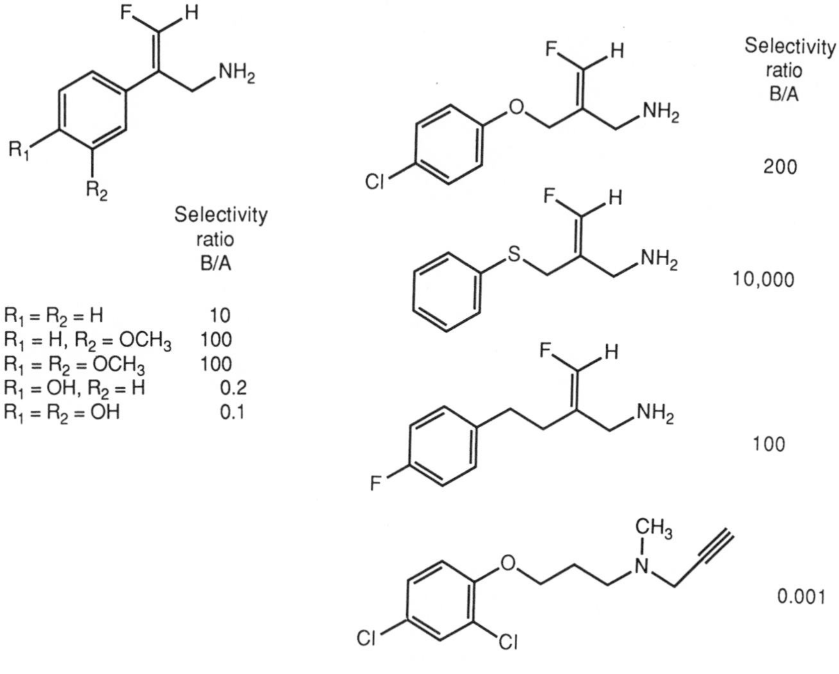
F
H
NH_2
R_1
R_2
Selectivity ratio B/A
$R_1 = R_2 = H$ 10
$R_1 = H, R_2 = OCH_3$ 100
$R_1 = R_2 = OCH_3$ 100
$R_1 = OH, R_2 = H$ 0.2
$R_1 = R_2 = OH$ 0.1
Selectivity ratio B/A
O
Cl
200
S
10,000
100
CH_3
N
0.001

Figure 8-26. 3-Haloallylamines, selective MAO B inhibitors.

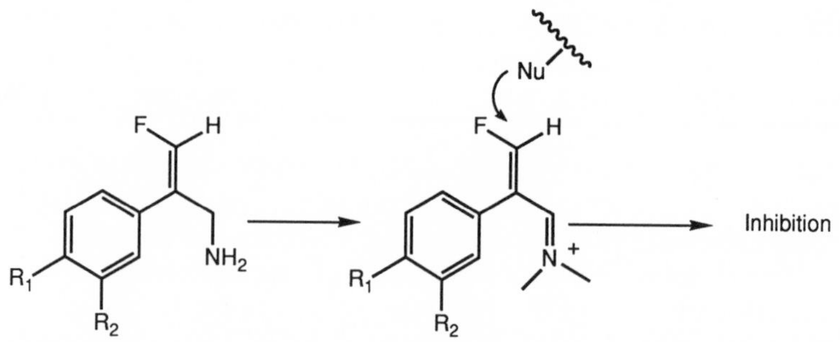

Figure 8-27. Proposed mechanism for the inactivation of MAO by 2-phenyl-3-haloallyl-amines.

(Fig. 8-26) that behave as selective MAO B inhibitors, free from indirect sympathomimetic action or uptake blocking properties. The proposed mechanism of inactivation involves addition either of a nucleophile at the active site of the enzyme or of the flavin cofactor to the double bond activated by the intermediate iminium species (Fig. 8-27). Consistent with this proposal are results from SAR studies which show that a halogen on the double bond, preferentially *cis* to the aromatic ring, is required for efficient inactivation. Ring methoxyl groups favor MAO B selectivity, while hydroxyl groups are detrimental to B selectivity (Palfreyman *et al.*, 1987; McDonald *et al.*, 1985; Bey *et al.*, 1984). 2-Phenyl-3-haloallylamines are also effective irreversible inhibitors of semicarbazide-sensitive amine oxidase (SSAO), enzymes with high activity in blood vessels of many animal tissues. The availability of effective SSAO inhibitors should help in current investigations to define the role of these enzymes (Lyles *et al.*, 1987).

In the treatment of depression, the adverse side effects of MAO inhibitors likely would be reduced if the inhibition could be targeted selectively to brain aminergic neurons. In a novel strategy to achieve this selectivity, (E)-β-(fluoromethylene)-*m*-tyrosine methyl ether (**37**; Fig. 8-28) was

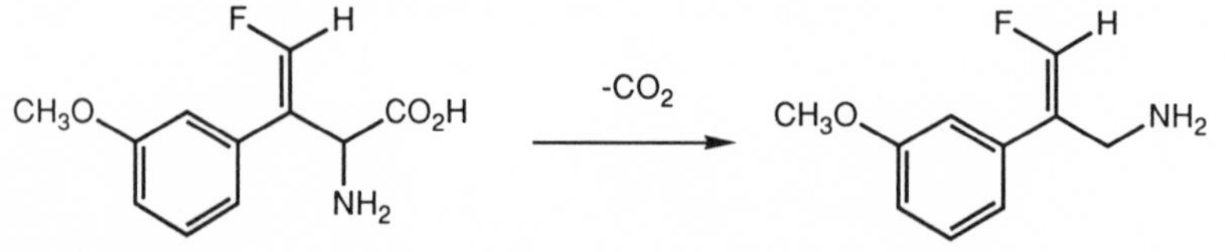

Figure 8-28. (E)-β-(Fluoromethylene)-*m*-tyrosine methyl ether (**37**) and the product of its decarboxylation, (E)-β-fluoromethylene-*m*-tyramine methyl ether (**38**).

prepared (McDonald *et al.*, 1984). This amino acid readily crosses the blood–brain barrier and is decarboxylated in the brain by aromatic amino acid decarboxylase (AADC), with release of the potent MAO inhibitor (*E*)-β-fluoromethylene-*m*-tyramine methyl ether (**38**). When an inhibitor of extracerebral AADC was coadministered, MAO inhibition in the brain was achieved with minimal potentiation of cardiovascular effects of tyramine (Palfreyman *et al.*, 1985).

8.6.2 Catechol *O*-Methyltransferase (COMT)

Catechol *O*-methyltransferase (COMT) transfers a methyl group from *S*-adenosylmethionine to a catechol hydroxyl group to yield a monomethyl catechol. This enzyme has several critical functions including extraneuronal deactivation of catecholamine neurotransmitters, inactivation of such administered drugs as isoproterenol and DOPA, detoxification of catechol derivatives of estrogen, and regulation of catechol transport (Thakker *et al.*, 1988, and references therein).

The effect of ring fluorination on the rate of methylation of DOPA and the implications of this for the development of PET scanning agents were discussed in Chapter 7. The regioselectivity of the methyl transfer also has been a subject of considerable interest. Thus, although the 3- and 4-hydroxyl groups of such natural catecholamines as NE are of comparable acidity, at pH 7.4 NE is methylated predominantly on the 3-hydroxyl (*meta*-methylation), suggesting that side-chain orientation and other factors control regiochemistry. On the other hand, studies with fluoro-DOPAs (Chapter 7) and FNEs have indicated that the more acidic hydroxyl group is methylated preferentially. In addition, a fourfold relative increase of *para*-methylation of NE on going from pH 7 to pH 9 provided evidence that catechol ionization favors *para*-methylation (Thakker *et al.*, 1982).

The increased *para*-methylation of NE appeared to be related kinetically to an ionization process ($pK_a = 8.6$) consistent with phenolic hydroxyl ionization. However, ionization of a functional group of similar pK_a on the active site could not be ruled out. The use of fluorinated analogues is a very effective approach to problems such as this, since fluorine substitution can be used to lower substrate pK_a, often with little perturbation of other factors in an enzymatic process. Indeed, the degree of *para*-methylation of 5-FNE, already favored at pH 7, increased with increasing pH and appeared related to ionization of a functional group with $pK_a = 7.9$, identical to the pK_a of 5-FNE. While *para*-methylation is clearly related to phenol ionization, these and other studies with fluorinated analogues suggest that *meta*-methylation is subject to general base catalysis by the

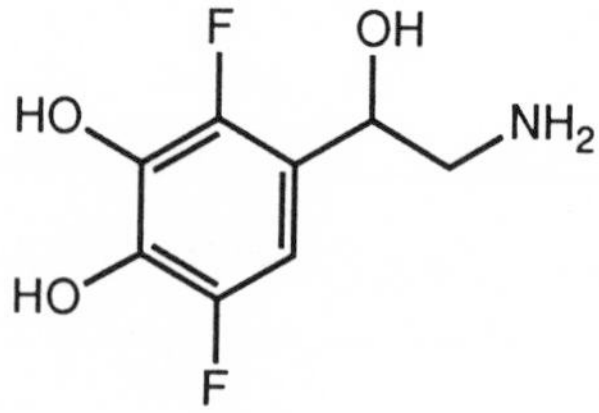

39

Figure 8-29. 2,5-Difluoronorepinephrine.

enzyme (Thakker *et al.*, 1986). Consistent with these results was the finding that the percent of *para*-methylation of 2,5-difluoro-NE (**39**; Fig. 8-29) also increased with ionization of a group with $pK_a = 7.2$ (the pK_a of 2,5-difluoro-NE), but *meta*-methylation was favored at all pH values, a reflection of the increased pK_a of both phenolic hydroxyl groups (Thakker *et al.*, 1988).

8.7 Radiolabeled Amines for Noninvasive Quantitation of Neuronal Function

8.7.1 CNS Studies

The use of 6-[^{18}F]fluoro-DOPA as a bioprecursor of 6-[^{18}F]fluoro-dopamine ([^{18}F]-6-FDA) for imaging of central dopaminergic neurons was discussed in Chapter 7. In this procedure, PET scanning quantitates and maps the regional distribution of [^{18}F]-6-FDA stored in intraneuronal storage vesicles. Over the past few years, several scanning agents have been prepared which, through their high affinity and selectivity for receptors, can directly label specific receptors. Examples include the butyrophenone neuroleptic spiroperidol (**40**; Fig. 8-30) and analogues thereof. 3-[^{18}F]-Acetylcyclofoxy (**41**; Fig. 8-30), a radiofluorinated analogue of naltrexone, has been used for *in vivo* imaging of opiate receptors in primate brain (Pert *et al.*, 1984).

These examples illustrate the use of the relatively new technology of PET for imaging various biochemical processes in the living brain. This is an extremely active area of research, and as more and more PET centers become available to researchers, this already large field undoubtedly will expand at an even faster rate.

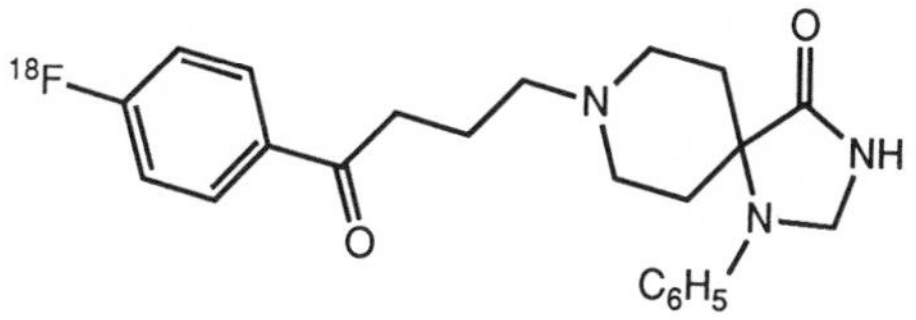

40

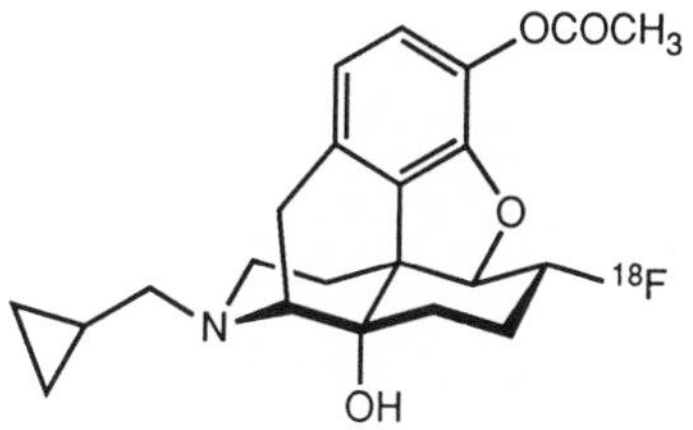

41

Figure 8-30. Examples of ^{18}F-labeled analogues useful as PET scanning agents: ^{18}F-labeled spiroperidol (**40**) and 3-[^{18}F]acetylcyclofoxy (**41**).

8.7.2 Studies in the Peripheral Nervous System

In an example of PET applications in peripheral studies, recent attention has been given to the application of ^{18}F-labeled neurotransmitters for the visualization *in vivo* of noradrenergic innervation. 6-FDA is a substrate for dopamine β-hydroxylase. The resulting 6-FNE displaces NE and is stored in noradrenergic neurons, is released on stimulation of the neuron, and thus functions as a false adrenergic neurotransmitter (Chieuh *et al.*, 1983). Following injection of [^{18}F]-6-FDA, prepared by enzymatic decarboxylation *in vitro* of [^{18}F]-6-FDOPA, PET scans demonstrated organ-specific uptake and retention of the radionuclide and were used to measure *in vivo* biosynthesis and storage of [^{18}F]-6-FNE (Eisenhofer *et al.*, 1988). A similar approach uses 6-[^{18}F]fluorometaraminol as a PET scanning agent for imaging adrenergic nerves in the heart (Mislankar *et al.*, 1988).

8.8 Summary

Research in medicinal chemistry designed to optimize potency and selectivity of drugs often involves the synthesis and biological evaluation of many analogues of a parent compound. Halogenated analogues very often are important members of such series. For this reason, a comprehensive review of the role of halogenated compounds in research related to neuroactive amines—wherein much emphasis has been placed on drug development—would require tabulation of a vast number of compounds with accompanying biological data. To avoid this, the attempt has been made instead to select examples for review wherein halogenated analogues have been critical to biological response or have had other significant roles. While this chapter has been confined to amines related to neurotransmission and hormone function, the ubiquity of the amino group throughout nature would allow expansion of this subject to include a whole host of topics. For example, the use of nitrogen mustards in cancer chemotherapy represents an important medical application of halogenated amines, as does the use of halogenated quinolines such as chloroquine and fluoroquine as antimalarial agents. The effect of fluorine substitution on the carcinogenicity and mutagenicity of N-nitrosamines has received recent attention (Gottfried-Anacker *et al.*, 1985). Reference was made in Chapter 7 to the role of polyamines in cell differentiation, and the use of halogenated substrate and product (amine) analogue irreversible inhibitors of decarboxylases responsible for their synthesis was discussed.

References

Adejare, A., Gusovsky, F., Padgett, W., Creveling, C. R., Daly, J. W., and Kirk, K. L., 1988. Syntheses and adrenergic activities of ring-fluorinated epinephrines, *J. Med. Chem.* 31:1972–1977.

Ahlquist, R. P., 1948. A study of adrenotropic receptors, *Am. J. Physiol.* 153:586–600.

Barnes, D. M., 1987. Biological issues in schizophrenia, *Science* 235:430–433.

Bey, P., Fozard, J., Lacoste, J. M., McDonald, I. A., Zreika, M., and Palfreyman, M. G., 1984. (E)-2-(3,4-Dimethoxyphenyl)-3-fluoroallylamine: A selective, enzyme-activated inhibitor of type B monoamine oxidase, *J. Med. Chem.* 27:9–10.

Cantacuzene, D., Kirk, K. L., McCulloh, D. H., and Creveling, C. R., 1979. Effect of fluorine substitution on the agonist specificity of norepinephrine, *Science* 204:1217–1219.

Carlsson, A., 1978. Antipsychotic drugs, neurotransmission and schizophrenia, *Am. J. Psychiat.* 135:164–173.

Chieuh, C. C., Zukowska-Crojec, Z., Kirk, K. L., and Kopin, I., 1983. 6-Fluorocatecholamines as false adrenergic neurotransmitters, *J. Pharmacol. Exp. Ther.* 225:529–533.

Clark, M. T., Adejare, A., Shams, G., Feller, D. R., and Miller, D. D., 1987. 5-Fluoro- and 8-fluorotrimetoquinol: Selective β_2-adrenoceptor agonists, *J. Med. Chem.* 30:86–90.

Clemens, J. A., Flaugh, M. E., Parli, J., and Sawyer, B. D., 1980. Inhibition of luteinizing hormone release and ovulation by 6-chloro- and 6-fluoromelatonin, *Neuroendocrinology* 30:83–87.

Comer, W. T., Matier, W. L., and Amer, M. S., 1981. Antihypertensive agents, in *Burger's Medicinal Chemistry*, Part III (M. E. Wolff, ed.), John Wiley and Sons, New York, pp. 285–337.

Commins, D. L., Axt, K. J., Vosmer, G., and Seiden, L. S., 1987a. 5,6-Dihydroxytryptamine, a serotonergic neurotoxin, is formed endogenously in the rat brain, *Brain Res.* 403:7–14.

Commins, D. L., Axt, K. J., Vosmer, G., and Seiden, L. S., 1987b. Endogenously produced

5,6-dihydroxytryptamine may mediate the neurotoxic effects of para-chloroamphetamine, *Brain Res.* 419:253–261.

Cooper, J. R., Bloom, F. E., and Roth, R. H., 1986. *The Biochemical Basis of Neuropharmacology*, 5th ed., Oxford University Press, New York, pp. 315–351.

Costa, J. L., Joy, D. C., Maher, D. M., Kirk, K. L., and Hui, S. W., 1978. Fluorinated molecule as a tracer: Difluoroserotonin in human platelets mapped by electron energy-loss spectroscopy, *Science* 200:537–539.

Costa, J. L., Dobson, C. M., Fay, D. D., Kirk, K. L., Poulsen, F. M., Valeri, C. R., and Vecchione, J. J., 1981. Nuclear magnetic resonance studies of amine storage in pig platelets, *FEBS Lett.* 136:325–328.

Costa, J. L., Kirk, K. L., and Stark, H., 1982. Uptake of 6-fluoro-5-hydroxytryptamine and 4,6-difluoro-5-hydroxytryptamine into releasable and nonreleasable compartments of human platelets, *Br. J. Pharmacol.* 75:237–242.

Coutts, R. T., Rao, T. S., Baker, G. B., Micetich, R. G., and Hall, T. W. E., 1987. Neurochemical and neuropharmacological properties of 4-fluorotranylcypromine, *Cell. Mol. Neurobiol.* 7:271–290.

Daly, J. W., Padgett, W., Creveling, C. R., Cantacuzene, D., and Kirk, K. L., 1981. Cyclic AMP-generating systems: Regional differences in activation by adrenergic receptors in rat brain, *J. Neurosci.* 1:49–59.

DeBernardis, J. F., Kerkman, D. J., Winn, M., Bush, E. N., Arendsen, D. L., McClellan, W. J., Kyncl, J. J., and Basha, F. Z., 1985. Conformationally defined adrenergic agents. 1. Design and synthesis of novel α_2 selective adrenergic agents: Electrostatic repulsion based conformational prototypes, *J. Med. Chem.* 28:1398–1404.

Douglas, W. W., 1985. Histamine and 5-hydroxytryptamine (serotonin) and their antagonists, in *Goodman and Gilman's The Pharmacological Basis of Therapeutics*, 7th ed. (A. G. Gilman, L. S. Goodman, T. W. Rall, and F. Murad, eds.), Macmillan, New York, pp. 605–638.

Dubocovich, M. L., and Takahashi, J. S., 1987. Use of 2-[^{125}I]iodomelatonin to characterize melatonin binding sites in chicken retina, *Proc. Natl. Acad. Sci. USA* 84:3916–3920.

Eisenhofer, G., Hovevey-Sion, D., Kopin, I. J., Miletich, R., Kirk, K. L., Finn, R., and Goldstein, D. S., 1988. Neuronal uptake and metabolism of 2- and 6-fluorodopamine: False neurotransmitters for positron emission tomographic imaging of sympathetically innervated tissues, *J. Pharmacol. Exp. Ther.* 248:419–427.

Elliot, A. J., 1982. The role of fluorine in the development of central nervous system agents, in *Biomedicinal Aspects of Fluorine Chemistry* (R. Filler and Y. Kobayashi, eds.), Kodansha Ltd., Tokyo; Elsevier Biomedical Press, Amsterdam, pp. 55–74.

Fosdick, L. S., Fancher, O., and Urbach, K. F., 1946. Pressor amines containing nuclear chlorine and fluorine, *J. Am. Chem. Soc.* 68:840–843.

Fowler, C. J., and Ross, S. B., 1984. Selective inhibitors of monoamine oxidase A and B: Biochemical, pharmacological, and clinical properties, *Med. Res. Rev.* 4:323–358.

Fuller, R. W., 1978. Structure–activity relationships among the halogenated amphetamines, *Ann. N.Y. Acad. Sci.* 305:147–159.

Fuller, R. W., and Molloy, B. B., 1976. The effect of aliphatic fluorine on amine drugs, in *Biochemistry Involving Carbon–Fluorine Bonds* (R. Filler, ed.), ACS Symposium Series, No. 28, American Chemical Society, Washington, D.C., pp. 77–98.

Ganellin, C. R., 1982. Chemistry and structure–activity relationships of drugs acting at histamine receptors, in *Pharmacology of Histamine Receptors* (C. R. Ganellin and M. E. Parsons, eds.), John Wright and Sons, Bristol, pp. 10–102.

Glennon, R. A., 1987. Central serotonin receptors as targets for drug research, *J. Med. Chem.* 30:1–12.

Goldberg, L. I., Kohli, J. D., Cantacuzene, D., Kirk, K. L., and Creveling, C. R., 1980. Effects of ring fluorination on the cardiovascular actions of dopamine and norepinephrine in the dog, *J. Pharmacol. Exp. Ther.* 213:509–513.

Gottfried-Anacker, J., Preussmann, R., Eisenbrand, G., and Janzowski, C., 1985. Fluoro-substituted *N*-nitrosamines. 8. *N*-Nitrosodibutylamine and ω-fluorinated analogues: *In vivo* metabolism in relation to the induction of urinary bladder cancer in the rat, *Carcinogenesis* 6:1559–1564.

Kaiser, C., and Setler, P. E., 1981. Antipsychotic drugs, in *Burger's Medicinal Chemistry*, Part III (M. E. Wolff, ed.), John Wiley and Sons, New York, pp. 859–980.

Kaiser, C., Colella, D. F., Pavloff, A. M., and Wardell, J. R., Jr., 1974. Adrenergic agents. 2. Synthesis and potential β-adrenergic agonist activity of some ring-chlorinated relatives of isoproterenol, *J. Med. Chem.* 17:1071–1075.

Kinemuchi, H., Arai, Y., Toyoshima, Y., Tadona, T., and Kisara, K., 1987. Studies on 5-fluoro-α-methyltryptamine and *p*-chloro-β-methylphenethylamine: Determination of the MAO A or MAO B selective inhibition *in vitro*, *Jpn. J. Pharmacol.* 46:197–199.

Kirk, K. L., 1976a. Photochemistry of diazonium salts. 4. Synthesis of ring-fluorinated tyramines and dopamines, *J. Org. Chem.* 41:2373–2376.

Kirk, K. L., 1976b. Synthesis of ring-fluorinated serotonins and melatonins, *J. Heterocycl. Chem.* 13:1253–1256.

Kirk, K. L., and Cohen, L. A., 1976. Biochemistry of ring-fluorinated imidazoles, in *Biochemistry Involving Carbon–Fluorine Bonds* (R. Filler, ed.), ACS Symposium Series, No. 28, American Chemical Society, Washington, D.C., pp. 23–36.

Kirk, K. L., and Creveling, C. R., 1984. The chemistry and biology of ring-fluorinated biogenic amines, *Med. Res. Rev.* 4:189–220.

Kirk, K. L., Cantacuzene, D., Nimitkitpaizan, Y., McCulloh, D., Padgett, W. L., Daly, J. W., and Creveling, C. R., 1979. Synthesis and biological properties of 2, 5, and 6-fluoronorepinephrine, *J. Med. Chem.* 22:1493–1497.

Kirk, K. L., Cantacuzene, D., Collins, B., Chen, G. T., and Creveling, C. R., 1982. The synthesis and adrenergic agonist properties of ring-fluorinated isoproterenols, *J. Med. Chem.* 25:680–684.

Kirk, K. L., Olubajo, O., Buchhold, K., Lewandowski, G. A., Gusovsky, F., McCulloh, D., Daly, J. W., and Creveling, C. R., 1986. Synthesis and adrenergic activity of ring-fluorinated phenylephrines, *J. Med. Chem.* 29:1982–1988.

Kirk, K. L., Adejare, A., Calderon, S., Chen, G., Furlano, D. C., and Gusovsky, F., 1988. Molecular basils for adrenergic selectivities of fluorinated biogenic amines, in *Progress in Catecholamine Research, Part A: Basic Aspects and Peripheral Mechanisms* (A. Dahlström, R. H. Belmaker, and M. Sandler, eds.), Alan R. Liss, New York, pp. 393–396.

Knoll, J., 1979. (−)-Deprenyl—the MAO inhibitor without the "cheese effect," *Trends Neurosci.* 2:111–113.

Lee, F. G. H., Dickson, D. E., Suzuki, J., Zirnis, A., and Manian, A. A., 1973. Synthesis of 5,7- and 6,7-disubstituted tryptamines and analogues (1), *J. Heterocycl. Chem.* 10:649–654.

Levy, B., and Ahlquist, R. P., 1961. An analysis of adrenergic blocking activity, *J. Pharmacol. Exp. Ther.* 133:202–210.

Lyles, G. A., Marshall, C. M. S., McDonald, I. A., Bey, P., and Palfreyman, M. G., 1987. Inhibition of rat aorta semicarbazide-sensitive amine oxidase by 2-phenyl-3-haloallyl-amines and related compounds, *Biochem. Pharmacol.* 36:2847–2853.

Martin, J. E., Kirk, K. L., and Klein, D. C., 1980. Effects of 6-hydroxy-, 6-fluoro-, and 4,6-difluoromelatonin on the *in vitro* pituitary response to luteinizing hormone-releasing hormone, *Endocrinology* 106:398–401.

McDonald, I. A., Lacoste, J. M., Bey, P., Wagner, J., Zreika, M., and Palfreyman, M. B.,

1984. (E)-β-(Fluoromethylene)-m-tyrosine: A substrate for aromatic L-amino acid decarboxylase liberating an enzyme-activated irreversible inhibitor of monoamine oxidase, *J. Am. Chem. Soc.* 106:3354–3356.

McDonald, I. A., Lacoste, J. M., Bey, P., Palfreyman, M. G., and Zreika, M., 1985. Enzyme-activated irreversible inhibitors of monoamine oxidases: Phenylallylamine structure-activity relationships, *J. Med. Chem.* 28:186–193.

Miller, D. D., Clark, M. T., Adejare, A., Neidert, K., Hamada, A., Shams, G., Romstedt, K. J., and Feller, D. R., 1988. Fluorotetrahydroisoquinolines as adrenergic and antithrombotic agents, in *Progress in Catecholamine Research, Part A: Basic Aspects and Peripheral Mechanisms* (A. Dahlström, R. H. Belmaker, and M. Sandler, eds.), Alan R. Liss, New York, pp. 403–407.

Mislankar, S. G., Gildersleeve, D. L., Wieland, D. M., Massin, C. C., Mulholland, G. K., and Toorongian, S. A., 1988. [^{18}F]6-Fluorometaraminol: A radiotracer for *in vivo* mapping of adrenergic nerves of the heart, *J. Med. Chem.* 31:362–366.

Mueller, A. C., Kirk, K. L., Hoffer, B. J., and Dunwiddie, T. V., 1983. Noradrenergic responses in rat hippocampus: Electrophysiological actions of direct and indirect-acting sympathomimetics in the *in vitro* slice, *J. Pharmacol. Exp. Ther.* 223:599–605.

Nimit, Y., Cantacuzene, D., Kirk, K. L., Creveling, C. R, and Daly, J. W., 1980. The binding of fluorocatecholamines to adrenergic and dopaminergic receptors in rat brain membranes, *Life Sci.* 27:1577–1585.

Palfreyman, M. G., McDonald, I. A., Fozard, J. R., Mely, Y., Sleight, A. J., Zreika, M., Wagner, J., Bey, P., and Lewis, P. J., 1985. Inhibition of monoamine oxidase selectively in brain monoamine nerves using the bioprecursor (E)-β-fluoromethylene-m-tyrosine (MDL 72394), a substrate for aromatic L-amino acid decarboxylase, *J. Neurochem.* 45:1850–1860.

Palfreyman, M. G., Bey, P., and Sjoerdsma, A., 1987. Enzyme-activated/mechanism-based inhibitors, *Essays Biochem.* 23:28–81.

Pert, C. B., Danks, J. A., Channing, M. A., Eckelman, W. C., Larson, S. M., Bennet, J. M., Burke, T. R., Jr., and Rice, K. C., 1984. 3-[^{18}F]Acetylcyclofoxy: A useful probe for the visualization of opiate receptors in living animals, *FEBS Lett.* 177:281–286.

Powell, C. E., and Slater, I. H., 1958. Blocking of inhibitory adrenergic receptors by a dichloro analogue of isoproterenol, *J. Pharmacol. Exp. Ther.* 122:480–488.

Reppert, S. M., Weaver, D. R., Rivkees, S. A., and Stopa, E. G., 1988. Putative melatonin receptors in a human biological clock, *Science* 242:78–81.

Rudnick, G., Kirk, K. L., Fishkes, H., and Schuldiner, S., 1989. Zwitterionic and anionic forms of a serotonin analogue as transport substrates, *J. Biol. Chem.* 264:14865–14868.

Schiemann, G., and Winkelmüller, W., 1932. Aromatic fluorine compounds XII. Fluorinated amino acids and their derivatives. 3. The first fluorotyrosine and fluorothyronine, also phenethylamines fluorinated on the nucleus, *J. Prakt. Chem.* 135:101–127.

Smith, F. A., 1970. Biological properties of selected fluorine-containing organic compounds, in *Handbook of Experimental Pharmacology*, Vol. XX, *Pharmacology of Fluorides, Part 2* (F. A. Smith, ed.), Springer-Verlag, Heidelberg, pp. 252–408.

Stone, E. A., Platt, J. E., Herrera, A. S., and Kirk, K. L., 1986. Effect of repeated restraint stress, desmethylimipramine, or adrenocorticotropin on the alpha and beta adrenergic components of the cyclic AMP response to norepinephrine in rat brain slices, *J. Pharmacol. Exp. Ther.* 237:702–707.

Thakker, D. R., Kirk, K. L., and Creveling, C. R., 1982. Enzymatic O-methylation of norepinephrine: Studies on the site of methylation by high pressure liquid chromatography, in *Biochemistry of 5-Adenosylmethionine and Related Compounds* (E. Vodin, R. T. Borchardt, and C. R. Creveling, eds.), Macmillan, London, pp. 473–477.

Thakker, D. R., Boehlert, C., Kirk, K. L., Antkowiak, R., and Creveling, C. R., 1986.

Regioselectivity of catechol O-methyltransferase. The effect of pH on the site of O-methylation of fluorinated norepinephrines, *J. Biol. Chem.* 261:178–184.

Thakker, D. R., Boehlert, C., Kirk, K. L., and Creveling, C. R., 1988. Interaction of fluorinated catecholamines with catechol O-methyltransferase, in *Progress in Catecholamine Research, Part A: Basic Aspects and Peripheral Mechanisms* (A. Dahlström, R. H. Belmaker, and M. Sandler, eds.), Alan R. Liss, New York, pp. 397–402.

Triggle, D. J., 1981. Adrenergics: Catecholamines and related agents, in *Burger's Medicinal Chemistry*, Part III (M. E. Wolff, ed.), John Wiley and Sons, New York, pp. 225–283.

Vakkuri, O., Lamsa, E., Rahkamaa, E., Ruotsalainen, R., and Leppaluoto, J., 1984. Iodinated melatonin: Preparation and characterization of the molecular structure by mass and ^{1}H NMR spectroscopy, *Anal. Biochem.* 142:284–289.

Vaughan, M. K., Richardson, B. A., Petterborg, L. J., Vaughan, G. M., and Reiter, R. J., 1986. Reproductive effects of 6-chloromelatonin implants and/or injections in male and female Syrian hamsters (*mesocricetus auratis*), *J. Reprod. Fert.* 78:381–387.

Weaver, D. R., Namboodiri, A. M. A., and Reppert, S. M., 1988. Iodinated melatonin mimics melatonin action and reveals discrete binding sites in fetal brain, *FEBS Lett.* 228:123–127.

Weiner, N., 1985a. Norepinephrine, epinephrine, and the sympathomimetic drugs, in *Goodman and Gilman's The Pharmacological Basis of Therapeutics*, 7th ed. (A. G. Gilman, L. S. Goodman, T. W. Rall, and F. Murad, eds.), Macmillan, New York, pp. 145–180.

Wiener, N., 1985b. Drugs that inhibit adrenergic nerves and block adrenergic receptors, in *Goodman and Gilman's The Pharmacological Basis of Therapeutics*, 7th ed. (A. G. Gilman, L. S. Goodman, T. W. Rall, and F. Murad, eds.), Macmillan, New York, pp. 181–214.

Williams, C. H., 1973. Monoamine oxidase—I. Specificity of some substrates and inhibitors, *Biochem. Pharmacol.* 23:615–628.

Zeynek, 1921. Preparation of chloro- and bromotyrosine and analogues tyramines, *Z. Biol. Chem.* 114:275–285.

Index

353